Mine Automation and New Technologies

Mine Automation and New Technologies

Guest Editors

Roohollah Shirani Faradonbeh
Robert Solomon
Phillip Stothard

MDPI

Basel • Beijing • Wuhan • Barcelona • Belgrade • Novi Sad • Cluj • Manchester

Guest Editors
Roohollah Shirani Faradonbeh
Western Australian School of Mines (WASM)
Curtin University
Kalgoorlie
Australia

Robert Solomon
Western Australian School of Mines (WASM)
Curtin University
Kalgoorlie
Australia

Phillip Stothard
Western Australian School of Mines (WASM)
Curtin University
Kalgoorlie
Australia

Editorial Office
MDPI AG
Grosspeteranlage 5
4052 Basel, Switzerland

This is a reprint of the Special Issue, published open access by the journal *Mining* (ISSN 2673-6489), freely accessible at: https://www.mdpi.com/journal/mining/special_issues/MANT.

For citation purposes, cite each article independently as indicated on the article page online and as indicated below:

Lastname, A.A.; Lastname, B.B. Article Title. *Journal Name* **Year**, *Volume Number*, Page Range.

ISBN 978-3-7258-7194-0 (Hbk)
ISBN 978-3-7258-7195-7 (PDF)
https://doi.org/10.3390/books978-3-7258-7195-7

Cover image courtesy of Roohollah Shirani Faradonbeh

Contents

About the Editors

Roohollah Shirani Faradonbeh

Roohollah Shirani Faradonbeh is a Lecturer (Assistant Professor) at the Western Australian School of Mines (WASM), Curtin University, Australia. He earned his PhD in Mining Engineering from the University of Adelaide in 2021. During his doctoral research, he investigated rockburst hazards and rock failure mechanisms in deep underground mining using experimental and data-driven approaches. At Curtin University, he developed a new course, "Mine Automation and New Technologies (MINE4011)", which covers state-of-the-art technologies for intelligent and sustainable mining, in collaboration with industry partners, thereby enhancing the mining curriculum. He has extensive experience teaching a range of mining engineering courses, including Mine Automation and New Technologies, Rock Excavation, and Rock and Soil Slope Engineering. His current research focuses on developing novel platforms for full mine automation, incorporating technologies such as UAVs, LiDAR, and Virtual Reality (VR), as well as proposing new solutions for the stability assessment of underground structures and rockburst mitigation. For example, in collaboration with Flanders, he is currently developing a system for automated drill-rig path-planning optimisation in open-pit mining operations. Over the past decade, his research has resulted in more than 80 peer-reviewed journal and conference publications, with over 3,800 citations and an h-index of 34. He also actively serves as an Editorial Board Member for several journals, including *Underground Space*, *Geotechnical and Geological Engineering*, *Discover Minerals*, and *Mining*.

Robert Solomon

Dr Rob Solomon received his PhD in Physical Metallurgy from the University of New South Wales in 1984. Over a 40-year career, he has gained extensive experience across multiple industries, with a strong emphasis on mining, as well as significant work in manufacturing, biotechnology, and telecommunications. His professional focus has centred on the application of mathematical modelling techniques to support decision-making and deepen understanding of complex operational environments. He has specialised in discrete-event simulation to analyse, design, and optimise complex stochastic production systems. Between 2000 and 2016, he made major modelling contributions to all key iron ore mining companies in Western Australia during a period of substantial industry expansion. Most recently, Dr Solomon served as Professor of Practice in Data Analysis at the WA School of Mines in Kalgoorlie. He retired in June 2025.

Phillip Stothard

Phillip Stothard is an Adjunct Associate Professor at the Western Australian School of Mines (WASM) at Curtin University's Faculty of Science and Engineering. Originally from an electrical background, Phillip is an Applied Geology Honours graduate of the University of Leicester, has a Master's degree in Mineral Resources Engineering and Management from the University of Nottingham, and a Mining Engineering PhD from the University of Leeds. He is a member of AusIMM. Over his long career, Phillip has developed specialist electrical engineering industrial automation and control projects, and led the research, development, and integration of an advanced virtual reality mine simulation capability into the mining industry. He took mining virtual reality from a concept through to commercialisation via various international industry and academic collaborations, and developed sixteen mining virtual reality simulation modules for industry. Phillip also developed an international research collaboration to promote mine site simulation globally and has worked as an independent technical consultant researching simulation technology options for clients. He has continued to provide specialist consulting services and provided management, technical, and commissioning services on fully autonomous engineering systems. Phillip continues to increase collaborative research capability within the application and development of advanced mine site simulation, visualisation, and automation technologies that will help achieve sustainable mining practices. He has delivered mine automation and new technologies and other mining engineering courses at Curtin University and the University of New South Wales. Phillip is also active on various international committees for simulation and modelling and mining metaverse standards, and is an editor for the *Mining Technology* journal.

Preface

This Reprint presents a focused collection of recent scientific contributions addressing the rapid transformation of the mining industry driven by digitalisation and automation. The subject matter spans intelligent mining systems, autonomous and collaborative operations, data-driven modelling, real-time monitoring, and emerging sensor and computing technologies that enable safer and more efficient resource extraction. The motivation for compiling this Reprint arises from the increasing operational complexity of modern mining environments and the growing demand for technological solutions that enhance safety, productivity, and sustainability. By bringing together methodological advances alongside practical applications and review-based perspectives, this Reprint aims to provide an integrated reference for researchers, engineers, and decision-makers seeking to understand both the current state of mine automation technologies and their future development pathways. The Reprint is intended for academic researchers, mining professionals, technology developers, and graduate students who are engaged in advancing intelligent and automated mining systems.

Roohollah Shirani Faradonbeh, Robert Solomon, and Phillip Stothard
Guest Editors

mining

MDPI

Editorial

Editorial for the Special Issue "Mine Automation and New Technologies"

Roohollah Shirani Faradonbeh *, Phillip Stothard and Robert Solomon

Western Australian School of Mines, Curtin University, Kalgoorlie, WA 6430, Australia; phillip.stothard@curtin.edu.au (P.S.); robert.solomon@curtin.edu.au (R.S.)
* Correspondence: roohollah.shiranifaradonbeh@curtin.edu.au; Tel.: +61-8-9088-6011

1. Introduction

Mining is undergoing a transformation driven by digitalisation and automation, promising improvements in efficiency, sustainability, and safety. Specifically, mine automation plays a key role in removing humans from hazardous environments, thereby reducing accidents and enabling the vision of "zero-entry" mining (i.e., no personnel exposure at the active face in hazardous locations). Automation helps cut emissions and supports greener, more sustainable operations. Meanwhile, the depletion of high-grade mineral resources prompts interest in unconventional mining methods, such as deep-sea mining, urban mining (metal recovery from waste), in situ leaching, and even space mining, to access new reserves as conventional ones dwindle and satisfy the high demand from different industry sectors [1,2]. Therefore, research is essential to assess the long-term stability and efficiency of novel artificial intelligence (AI)-powered technologies in harsh mining environments. Additionally, examining the potential for extracting alternative mineral resources and the challenges associated with them is of great importance.

This Special Issue of *Mining* (MDPI) on "Mine Automation and New Technologies" [3] showcases state-of-the-art research at the intersection of advanced technology and mining engineering. The papers cover a wide range of topics: AI and machine learning for solving complex mining problems, autonomous systems for mine operations, advanced sensing and simulation for monitoring and planning, and explorations of alternative mining approaches. Collectively, these contributions highlight how embracing digital transformation and novel techniques can enhance productivity, safety, and sustainability in the mining sector. The editorial concludes with a summary of cross-cutting insights and future research directions.

2. Machine Learning Accelerating Mining Optimisation

A prominent theme within this issue is the application of machine learning (ML) and AI to model and optimise various mining processes. Several contributions demonstrate how data-driven algorithms may yield more accurate predictions and efficient planning than traditional methods:

Predicting coal abrasiveness: Afrazi et al. (Contribution 1) used tree-based ML models (Random Forest, Gradient Boosting, and XGBoost) to predict the abrasive index of coal from its mineralogical properties. After cleaning the dataset and selecting key features, the best model (XGBoost) achieved a high accuracy ($R^2 \approx 0.92$), substantially outperforming the initial models (R^2~0.63–0.72 before refinement). The analysis identified quartz and ash content as the most influential factors increasing abrasiveness, while calorific value has least impact. This data-driven approach may offer a cost-effective alternative to laboratory tests, allowing industry to predict material wear characteristics in advance and optimise coal blending and maintenance schedules.

Optimising blast vibrations: another paper (Saubi et al., Contribution 2) addressed ground vibration control in open-pit blasting through an AI-based model. Using data from 100 blasts, the researchers trained an artificial neural network (ANN) to capture the complex relationship between eight blast design parameters (burden, spacing, hole depth, charge per delay, etc.) and the resulting vibration. By treating the trained ANN as a "solution surface" in a nine-dimensional space, it was possible to search for the combination of inputs that minimises vibration. The model suggested that, under ideal conditions, peak vibration could be reduced to around 0.1 mm/s—roughly equivalent to the ground motion from a passing vehicle. While such an extreme reduction may be theoretical, the study demonstrates how ML optimisation may guide blast designs to minimise environmental and social impacts.

Process modelling with limited data: a challenge in applying ML to mineral processing is the large data requirement. One innovative contribution (Moraga et al., Contribution 3) showed how to train neural networks with minimal data for a comminution (crushing/grinding) circuit. By carefully selecting only 77 representative input scenarios using a conjoint analysis approach, the authors trained neural nets to predict key process outputs (such as energy and water usage, particle size distribution) based on simulator-generated data. Despite the small training set, the networks achieved excellent predictive performance ($R^2 > 0.99$ on training scenarios) and maintained high accuracy ($R^2 > 0.98$) on new test scenarios within the considered range. The only limitation observed was a loss of accuracy for inputs outside the originally sampled range. This approach reduces the burden of data collection and shows that even complex process dynamics can be learned with limited but well-chosen data. It opens the door to deploying ML models in mineral processing plants where historical data are scarce, thus simplifying and improving process optimisation.

Optimising truck and loader fleets: efficient fleet management in mines is another area enhanced by ML. A case study (Nobahar et al., Contribution 4) from a Kaolin mining operation applied a gradient boosting algorithm to optimise the selection of trucks and loaders needed to meet daily production targets. Five years of operational data (weather, number/type of trucks and loaders, daily tonnage hauled, etc.) were used to train and validate several models, among which gradient boosting performed best (approximately 85% accuracy in prediction). The trained model was then fed over 11,000 hypothetical fleet scenarios to identify the optimal equipment combination for each part of the mine section and for each day. From this, the system recommended the configuration that met the required output with minimal idle time (e.g., specifying ideal truck numbers and loaders for each shift). This data-driven fleet optimisation promises to reduce truck queues and shovel downtime, cutting operational costs. It illustrates how ML can be leveraged as a decision-support tool in mine planning, enabling managers to explore numerous what-if scenarios quickly and choose the most efficient plan without costly trial-and-error in the field.

3. Automation and Human–Machine Collaboration in Mining

Mining equipment automation is another key focus of the Special Issue. Several papers explore the current capabilities and future possibilities of autonomous systems in mining, often stressing the importance of human integration with these technologies:

Human-in-the-loop autonomy: a forward-looking conceptual paper (Ruiz-del-Solar, Contribution 5) advocates for "autonomous collaborative mining" as a new paradigm for mine automation. While completely human-free, fully autonomous mines remain out of reach with today's technology, this concept suggests that safety and efficiency can be

 https://doi.org/10.3390/mining6010011

maximised by having people and automated machines work together rather than separately. The authors note that many "autonomous" machines still require operators, and many tasks remain unautomated. Rather than viewing this as a setback, the authors propose deeper personnel involvement inside the automation loop: for example, remote operators could guide or intervene in autonomous operations when complex judgements are needed, people would make strategic decisions with AI support. This collaborative approach could accelerate progress toward zero-entry mining by leveraging human judgment where automation is weak and using robots where people would be at risk. The paper analyses enabling factors (such as advanced operator interfaces and communication systems) and presents a case study of an autonomous load–haul–dump machine supervised by people to demonstrate the concept. The broader implication is a shift in mindset: future mines may not be 100% automated, but rather symbiotic environments where operators and robotic efficiency complement each other and improve operations.

Autonomous shovels and productivity: a study by Yaghini et al. (Contribution 6) evaluated the impact of automation on an electric rope shovel used in surface mines. Using discrete-event simulation with real operational data, the researchers modelled four scenarios, ranging from a manual operation to fully autonomous operation, with intermediate levels such as operator-assisted movements. The findings were striking; even partial automation (e.g., automating the swing and dump motions) improved performance, and the fully autonomous scenario showed a potential 41% increase in output compared to the manual operation. These gains resulted from more consistent cycle times and reduced idling, demonstrating that automation can significantly boost productivity in load–haul systems. The study suggests that both mining companies and equipment manufacturers stand to benefit: mines achieve higher production and improved safety by removing operators from the pit, while OEMs gain insight into which automated functions deliver the most value for future machine designs. It provides a quantitative case for ongoing investment in automation technology.

UGV navigation underground (review): complementing the above, a comprehensive review by Abdukodirov and Benndorf (Contribution 7) surveyed the state of path-planning algorithms for unmanned ground vehicles (UGVs) in underground mines. Reliable navigation is essential for autonomous trucks, loaders, or inspection robots to operate in the dark, GPS-denied, and cramped conditions of an underground mine. The review outlines the key requirements for effective underground path planning (robust sensors, mapping systems like SLAM, and considerations for the Robot Operating System) and examines a range of algorithmic approaches. Both global path planners (graph-based algorithms such as A* and Dijkstra, and sampling-based methods such as RRT) and local planners (real-time obstacle avoidance using potential fields, ant colony algorithms, or reinforcement learning) are discussed, along with their respective strengths and typical use cases. Notably, the review identifies a gap between simulation and reality: many algorithms have been tested only in research simulations and face challenges such as computational load and sensor noise when applied in actual mines. It concludes with a call for full-scale field trials in real mines and possibly hybrid approaches that combine the best of different algorithms. The insights from this paper guide practitioners in identifying the most promising navigation techniques and underscore the need to move beyond theoretical research to truly enable autonomous underground vehicles.

4. Innovations in Monitoring and Sensing Technologies

Accurate monitoring of mining processes and geotechnical conditions is crucial for efficiency and safety. This Special Issue covers studies which introduce novel approaches

https://doi.org/10.3390/mining6010011

for sensing and measurement, which improve our ability to gather data in challenging mining environments:

Three-dimensional fragmentation measurement via drone: one technical contribution by Toriya et al. (Contribution 8) introduced an improved method for measuring blast fragmentation size distribution using drone-based photogrammetry enhanced with GNSS (satellite positioning). Fragmentation (post-blasting rock size) is traditionally assessed by manual sampling or photographs with reference scales, which can be tedious or inaccurate. The researchers developed a system to automatically generate a scaled 3D model of a post-blast rock pile using only the positional data from the drone's GNSS, without any ground control points (GCPs) or manual scale objects. Field trials at an operating mine showed that this GNSS-aided approach successfully produced accurately scaled rock pile models, enabling reliable particle-size analysis. In surface mines where GPS signals are available, this technique eliminates the need for surveyors to access the muck pile (enhancing safety) and accelerates the survey process. The result is near-real-time feedback on blasting outcomes, which assists engineers in improving blast designs for optimal fragmentation (thereby improving downstream efficiencies). This work exemplifies how combining drone imagery with precise positioning data can automate a task that once required considerable human effort.

Slope stability monitoring (review): ensuring the stability of large open-pit slopes is a vital safety concern, and this Special Issue includes a systematic review of slope monitoring methods and technologies conducted by Le Roux et al. (Contribution 9). The review covers both traditional in situ instruments (extensometers, inclinometers, piezometers, etc.) and modern remote sensing tools (radar interferometry, LiDAR, satellite and drone photogrammetry), as well as emerging digital technologies like IoT sensor networks and AI-based data analysis. By comparing these approaches across attributes such as accuracy, spatial/temporal coverage, complexity, and cost, the review finds that no single monitoring method is universally optimal. Each technology involves trade-offs: for example, radar and satellite methods provide broad coverage but may be limited by weather or update frequency, whereas ground sensors give precise continuous data but only at discrete points. To overcome these limitations, the review suggests future systems must integrate multiple sensing methods, potentially combining remote and in situ data through IoT platforms and using AI to fuse and interpret the information in real time. Such an integrated multi-sensor approach would enhance early-warning capabilities for slope failures. The review serves as a consolidated knowledge base on slope monitoring, guiding mines in designing robust monitoring frameworks and highlighting research directions to develop cost-effective, scalable solutions that improve mine safety.

5. Alternative and Integrated Mining Approaches

Innovative techniques and integrated planning frameworks are also discussed, reflecting the drive to make mining more sustainable and versatile:

Combined conventional and in situ mining: Krassakis et al. (Contribution 10) examined how conventional mining might be combined with an alternative extraction method, specifically, Underground Coal Gasification (UCG), which converts coal to gas in situ without physical mining. The paper detailed the construction of a 3D geological model of the Kardia lignite deposit and used GIS-based analysis to identify zones that met the criteria for UCG (sufficient seam thickness, appropriate depth, etc.). By mapping these zones, the researchers showed where UCG could potentially be implemented alongside or after the open-pit operations. This integrated geospatial approach helps pinpoint suitable areas for unconventional methods and supports planning a combined coal-to-liquid supply chain within a complex deposit. The broader significance is that such modelling may extend mine

https://doi.org/10.3390/mining6010011

life: uneconomical resources that may not be economically viable by traditional methods might economic via techniques such as UCG. Moreover, the study's workflow—building a detailed subsurface model and overlaying alternative method criteria—can save time and cost in evaluating novel mining options, contributing to more sustainable resource utilisation. It underscores the value of advanced geological modelling in decision-making, especially when considering the adoption of emerging mining methods.

6. Broader Perspectives and Future Directions

Two contributions provide a broader perspective on how technology is reshaping mining through extensive literature reviews:

Real-time mining (RTM) review: a comprehensive review by Anvari and Benndorf (Contribution 11) on real-time mining (RTM) examined developments over the last decade that enable mines to adapt operations dynamically based on continuous data. RTM is presented as a response to pressing industry challenges, including lower ore grades, deeper deposits, and higher energy costs, and aims to integrate advanced sensing and analytics to enable immediate decision-making. The review highlights key components of RTM: real-time grade monitoring, dynamic mine planning and scheduling that adjust to new information, and continuous optimisation algorithms that refine production. The paper discusses the role of technologies such as high-resolution online sensors, data assimilation techniques (e.g., Kalman filters), and AI in updating mine models on the fly. Importantly, the authors document how RTM can enhance sustainability and profitability simultaneously by extracting resources more efficiently by reducing waste and energy use, and enabling quicker responses to operational issues, thus minimising downtime. The review proposed a comprehensive framework for adaptive, data-driven mining operations that leverage continuous feedback to support sustainable development and better decision-making. RTM is positioned as a cornerstone of future mining, drawing parallels to the Industry 4.0 revolution in manufacturing and suggests future mines will operate as interconnected, intelligent systems.

Emerging technologies in mining (bibliometric study): another article (Emere et al., Contribution 12) offers a bibliometric analysis of emerging technologies (ETs) in the mining sector, quantitatively tracking research trends from 1986 to 2021. By examining 135 scholarly publications on topics like AI, digital twins, Internet of Things (IoT), and blockchain in mining, the study visualises how interest in these technologies has grown and which areas are most influential. The analysis confirms a significant surge in adoption and research of digital and automation technologies, reflecting the mining industry's drive to improve safety, efficiency, and sustainability through innovation. Key clusters identified include automation and AI in mining operations, smart mining initiatives, and the use of blockchain for supply chain transparency. The study provides stakeholders with a "state-of-the-art" map of technology integration in mining, and it suggests that the sector has reached a tipping point where such technologies are becoming mainstream. The bibliometric review thus not only charts past and present trends but also points to cross-industry learning as a strategy to accelerate mining's technological transformation.

7. Conclusions

The Special Issue papers paint a picture of an evolving mining industry, one that is increasingly data-driven, automated, and technologically innovative. From the detailed case studies to the broad reviews, a common thread is the pursuit of safer and more efficient mining through advanced modelling and technology deployment. Several recurring insights emerge:

- *Integration is key:* no single technology or solution can address all mining challenges. Many contributions highlight the need for integrated approaches, whether combining

https://doi.org/10.3390/mining6010011

multiple monitoring techniques for slope stability or integrating human decision-makers into automated systems to achieve optimal results. The future mine will likely be a synergistic blend of AI algorithms, robotics, sensors, and human expertise working in concert.

- *From reactive to proactive:* tools like predictive ML models and real-time analytics are shifting mining from a reactive mode (responding to problems after they occur) to a proactive, anticipatory mode. For instance, anticipating equipment wear from material properties, or adjusting plans on the fly through real-time mining, can prevent issues and seize opportunities earlier. This predictive capacity will be crucial as mines operate in increasingly complex and challenging conditions.
- *Safety and sustainability:* underlying many of these works is the twin goal of making mining safer for people and less impactful on the environment. Automation and remote operation directly remove people from harm's way, whether by autonomous vehicles or zero-entry concepts. At the same time, optimised processes and better monitoring mean less waste and lower emissions (e.g., efficient blasting reduces vibration and noise, and precise extraction reduces energy use). The Special Issue shows that technology can enable the mining sector to meet production demands while also adhering to stricter safety and sustainability requirements.
- *Bridging the gap to implementation:* a note of pragmatism is visible as well. Several authors point out that despite promising results, challenges remain in moving from research to real-world deployment. Whether it is the need for more on-site trials of autonomous navigation algorithms or ensuring machine learning models remain robust outside their training data, there is work to do to translate these advances into everyday mining practice. Change management and workforce training will also play a role as the mining workforce adapts to new roles (such as managing automated systems or interpreting complex data outputs).

In summary, this Special Issue highlights the potential of combining modern technology with mining engineering. The contributions demonstrate advances, higher prediction accuracy, productivity gains, improved monitoring, and comprehensive frameworks for future operations. Importantly, they suggest the future of mining will be defined by connectivity and collaboration: connecting disparate data sources in real time, connecting people with intelligent machines, and connecting cross-disciplinary innovative ideas to solve mining's challenges. The insights and directions outlined will assist the mining industry in moving towards more sustainable, safer, productive, and cost-effective operations. However, further investigations are required, particularly within alternative/future mining methods, communication systems, mine electrification, and advanced sensor technologies. The authors encourage researchers to submit their latest research findings in these areas to the second edition of this Special Issue [4].

Funding: This research received no external funding.

Conflicts of Interest: The authors declare no conflicts of interest.

List of Contributions:

1. Afrazi, M.; Huat, C.Y.; Onifade, M.; Khandelwal, M.; Shonuga, D.O.; Fattahi, H.; Jahed Armaghani, D. Modeling the Abrasive Index from Mineralogical and Calorific Properties Using Tree-Based Machine Learning: A Case Study on the KwaZulu-Natal Coalfield. *Mining* **2025**, *5*, 48. https://doi.org/10.3390/mining5030048.
2. Saubi, O.; Jamisola, R.S., Jr.; Gaopale, K.; Suglo, R.S.; Matsebe, O. A Solution Surface in Nine-Dimensional Space to Optimise Ground Vibration Effects Through Artificial Intelligence in Open-Pit Mine Blasting. *Mining* **2025**, *5*, 40. https://doi.org/10.3390/mining5030040.

3. Moraga, C.; Astudillo, C.A.; Estay, R.; Maranek, A. Enhancing Comminution Process Modeling in Mineral Processing: A Conjoint Analysis Approach for Implementing Neural Networks with Limited Data. *Mining* **2024**, *4*, 966–982. https://doi.org/10.3390/mining4040054.
4. Nobahar, P.; Pourrahimian, Y.; Mollaei Koshki, F. Optimum Fleet Selection Using Machine Learning Algorithms—Case Study: Zenouz Kaolin Mine. *Mining* **2022**, *2*, 528–541. https://doi.org/10.3390/mining2030028.
5. Ruiz-del-Solar, J. The Road to the Mine of the Future: Autonomous Collaborative Mining. *Mining* **2025**, *5*, 25. https://doi.org/10.3390/mining5020025.
6. Yaghini, A.; Hall, R.; Apel, D. Autonomous and Operator-Assisted Electric Rope Shovel Performance Study. *Mining* **2022**, *2*, 699–711. https://doi.org/10.3390/mining2040038.
7. Abdukodirov, A.; Benndorf, J. Recent Developments in Path Planning for Unmanned Ground Vehicles in Underground Mining Environment. *Mining* **2025**, *5*, 33. https://doi.org/10.3390/mining5020033.
8. Toriya, H.; Tungol, Z.P.L.; Ikeda, H.; Owada, N.; Jang, H.D.; Adachi, T.; Kitahara, I.; Kawamura, Y. Fragmentation Size Distribution Measurement by GNSS-Aided Photogrammetry at Real Mine Site. *Mining* **2022**, *2*, 438–448. https://doi.org/10.3390/mining2030023.
9. Le Roux, R.; Sepehri, M.; Khaksar, S.; Murray, I. Slope Stability Monitoring Methods and Technologies for Open-Pit Mining: A Systematic Review. *Mining* **2025**, *5*, 32. https://doi.org/10.3390/mining5020032.
10. Krassakis, P.; Pyrgaki, K.; Gemeni, V.; Roumpos, C.; Louloudis, G.; Koukouzas, N. GIS-Based Subsurface Analysis and 3D Geological Modeling as a Tool for Combined Conventional Mining and In-Situ Coal Conversion: The Case of Kardia Lignite Mine, Western Greece. *Mining* **2022**, *2*, 297–314. https://doi.org/10.3390/mining2020016.
11. Anvari, K.; Benndorf, J. Real Time Mining—A Review of Developments Within the Last Decade. *Mining* **2025**, *5*, 38. https://doi.org/10.3390/mining5030038.
12. Emere, C.; Oguntona, O.; Ohiomah, I.; Ayorinde, E. Harnessing Emerging Technologies in the Global Mining Sector from a Bibliometric Standpoint. *Mining* **2025**, *5*, 13. https://doi.org/10.3390/mining5010013.

References

1. Shirani Faradonbeh, R.; Shah, M.I.; Bahadori, M.; Jang, H. Toward Sustainable Mining: Exploring Alternative Mineral Resources and Innovative Extraction Techniques. *Mining* **2025**, *5*, 66. [CrossRef]
2. Samaei, M.; Shirani Faradonbeh, R.; Topal, E.; Goodwin, J. Path optimisation for drill rigs: A step toward full automation in mining. In Proceedings of the 2025 Resource Operators Conference (ROC2025), Brisbane, Australia, 6–7 February 2025.
3. Shirani Faradonbeh, R.; Solomon, R.G.; Stothard, P. (Eds.) Mining | Special Issue: Mine Automation and New Technologies. Available online: https://www.mdpi.com/journal/mining/special_issues/MANT (accessed on 18 December 2025).
4. Shirani Faradonbeh, R.; (Ed.) Mining | Special Issue: Mine Automation and New Technologies, 2nd Edition. Available online: https://www.mdpi.com/journal/mining/special_issues/0DQ3H34FE4 (accessed on 18 December 2025).

mining

MDPI

Article

Modeling the Abrasive Index from Mineralogical and Calorific Properties Using Tree-Based Machine Learning: A Case Study on the KwaZulu-Natal Coalfield

Mohammad Afrazi [1], Chia Yu Huat [2], Moshood Onifade [3,*], Manoj Khandelwal [3,*], Deji Olatunji Shonuga [4], Hadi Fattahi [5] and Danial Jahed Armaghani [6]

1 Department of Mechanical Engineering, New Mexico Institute of Mining and Technology, Socorro, NM 87801, USA; mohammad.afrazi@student.nmt.edu
2 Department of Civil Engineering, Faculty of Engineering, University of Malaya, Kuala Lumpur 50603, Malaysia; chiayuhuat@gmail.com
3 Institute of Innovation, Science and Sustainability, Federation University Australia, Ballarat, VIC 3350, Australia
4 Victoria Institute of Technology, Melbourne, VIC 3000, Australia; de58104@student.vit.edu.au
5 Faculty of Earth Sciences Engineering, Arak University of Technology, Arak 3818146763, Iran; h.fattahi@arakut.ac.ir
6 School of Civil and Environmental Engineering, University of Technology Sydney, Ultimo, NSW 2007, Australia; danial.jahedarmaghani@uts.edu.au
* Correspondence: m.onifade@federation.edu.au (M.O.); m.khandelwal@federation.edu.au (M.K.)

Abstract: Accurate prediction of the coal abrasive index (AI) is critical for optimizing coal processing efficiency and minimizing equipment wear in industrial applications. This study explores tree-based machine learning models; Random Forest (RF), Gradient Boosting Trees (GBT), and Extreme Gradient Boosting (XGBoost) to predict AI using selected coal properties. A database of 112 coal samples from the KwaZulu-Natal Coalfield in South Africa was used. Initial predictions using all eight input properties revealed suboptimal testing performance (R^2: 0.63–0.72), attributed to outliers and noisy data. Feature importance analysis identified calorific value, quartz, ash, and Pyrite as dominant predictors, aligning with their physicochemical roles in abrasiveness. After data cleaning and feature selection, XGBoost achieved superior accuracy (R^2 = 0.92), outperforming RF (R^2 = 0.85) and GBT (R^2 = 0.81). The results highlight XGBoost's robustness in modeling non-linear relationships between coal properties and AI. This approach offers a cost-effective alternative to traditional laboratory methods, enabling industries to optimize coal selection, reduce maintenance costs, and enhance operational sustainability through data-driven decision-making. Additionally, quartz and Ash content were identified as the most influential parameters on AI using the Cosine Amplitude technique, while calorific value had the least impact among the selected features.

Keywords: coal processing; abrasive index; Random Forest; XGBoost; cost-effective predictive model

1. Introduction

Coal is a vital natural resource that has been utilized for centuries in various industrial applications, particularly in power generation, steel manufacturing, and cement production [1,2]. Its suitability for these applications depends on several factors, including its composition and physical properties [3,4]. Among the many parameters that influence the efficiency of coal processing, the abrasive index (AI) is one of the most significant [5]. This index provides essential insights into the coal's behavior during milling, influencing both equipment wear and energy consumption during coal grinding [6]. The abrasive index (AI) is defined as the loss of mass (in milligrams) of standard steel blades used in grinding a coal sample under specified test conditions, as per the Wells et al. [7] and Spero [8] methods. Therefore, accurate prediction of AI is crucial for optimizing coal utilization in industrial processes. Coal with high abrasiveness increases maintenance costs due to accelerated wear and tear on machinery. Hence, predicting AI accurately can help industrial operators make better decisions regarding coal selection, minimize equipment downtime, and reduce operational costs [9,10].

AI is influenced by coal's chemical and physical composition, with factors including moisture content (M), Ash content, volatile matter (VM), fixed carbon (FC), Pyrite, and quartz [9,11]. Quartz, being one of the hardest minerals present in coal, significantly contributes to AI by increasing wear on mills, crushers, and pulverizers [12]. Pyrite (FeS_2), another highly abrasive mineral, enhances AI due to its hardness and tendency to accelerate material degradation in coal mills [7,13,14]. Ash content, which comprises various mineral impurities, particularly silica and iron oxides, further increases AI by introducing hard particles that intensify equipment wear [13]. While calorific value does not directly affect AI, it often correlates with coal quality and density, which may influence the presence of harder mineral inclusions and, consequently, abrasiveness [15,16]. Among these, the most common factors contributing to AI are quartz, calorific value, Pyrite, and Ash, as these have the most direct impact on coal abrasiveness [9]. Other properties such as M, VM, and FC also play a role in determining AI. Higher moisture levels can reduce coal abrasiveness by acting as a lubricant, decreasing friction between coal particles and grinding surfaces [17]. However, excessively high moisture can lead to clogging in pulverizers and inefficient milling. Lower VM content is often associated with denser coal structures, potentially increasing AI [9]. FC, which determines coal's density and hardness, can contribute to AI by making coal more resistant to grinding and increasing wear on milling equipment [18].

Given the complexity and variability of these coal properties, there is a growing need for accurate methods to predict AI [19]. One of the traditional common methods to determine AI is using the laboratory testing which is time-consuming and costly. In addition, there are also several existing empirical formulas that can be used to determine the AI [8], however this method is limited to a certain type of data and most of these empirical formulas are also limited to certain behavior of the data such as linear and non-linear relationships. The ability to predict AI is crucial for industries that rely on coal as a fuel or raw material [20]. In power plants, low-abrasiveness coal ensures efficient milling, reducing energy consumption and improving combustion efficiency. Likewise, accurately predicting AI allows for better coal selection, optimizing the milling process and preventing unnecessary wear on machinery. By enabling better resource management, AI prediction also supports cost-effective blending strategies, ensuring that industries meet their desired specifications while minimizing waste and improving sustainability. Given the operational and economic implications of AI, various predictive methods are strongly recommended for further exploration, while traditional empirical models and laboratory analyses remain widely used.

In recent years, the use of machine learning (ML) algorithms for the prediction of various parameters can be seen in various industries such as construction [21], medical, and commercial [22]. ML has been used for the prediction of coal AI by using different coal properties. Some studies have used different ML methods for the prediction of coal AI [9,10] and show reasonably good prediction. However, another algorithm, which is tree-based ML, can be considered as another potential algorithm for the prediction of the AI. The tree-based ML models have the capability to handle large datasets and capture complex relationships between coal properties. Therefore, in this study, tree-based algorithms such as the Random Forest (RF), Gradient Boosting (GBT), and Extreme Gradient Boosting (XGBoost) are explored to assess the capability of the prediction of coal AI. RF enhances predictive stability by aggregating multiple decision trees, whereas GBT and XGBoost iteratively refine predictions by minimizing errors to improve accuracy. The key difference between GBT and XGBoost lies in their optimization strategies, where XGBoost incorporates advanced regularization techniques, parallel processing, and a more efficient handling of missing values, making it computationally faster and more robust than traditional GBT implementations.

These algorithms have shown promising results for different predictions. For example, RF has been used by Yan et al. [23] for the prediction of rock mass classification in tunnel construction and good results were obtained. GBT has also been used to predict the soil compression index (Cc), a key geotechnical parameter that quantifies soil deformation under load [24], and XGBoost was used by He et al. [25] to predict air-overpressure (AOp) resulting from blasting activities in granite quarries and was optimized using Bayesian optimization (BO) and random search (RS) to enhance predictive accuracy and minimize environmental impact. The studies carried out by previous authors have shown the potential of these algorithms; hence, these algorithms are further explored for the prediction of coal AI. By integrating ML with well-characterized coal datasets reported by Onifade et al. [9] based on the KwaZulu-Natal Coalfield, this study aims to enhance AI-driven predictive models for coal quality assessment, ultimately optimizing industrial coal utilization and reducing operational costs. The proposed framework leverages on selected coal properties such as CV (calorific value), Pyrite, Qtz (quartz), Ash, and AI (abrasive index) to refine coal selection, improve milling efficiency, and optimize combustion performance in coal-dependent industries.

In this study, our target is to propose several predictive models, i.e., RF, XGB, and GBT, and provide a comparative study of their prediction capabilities in estimating AI of coal using selected physicochemical properties of coal. Our methodological framework involves comprehensive data preprocessing, feature selection through importance analysis, and rigorous hyperparameter tuning to ensure robust model training. The contributions of this work include the following: establishing a predictive pipeline that integrates domain-driven feature refinement and outlier handling for enhanced accuracy, demonstrating a systematic comparison of tree-based models for AI prediction, and presenting a cost-effective and data-driven alternative model that supports optimized coal selection and operational sustainability.

2. Data Information

2.1. Description of the Study Area

The KwaZulu-Natal Coalfield is in the eastern part of South Africa, within the province of KwaZulu-Natal. It extends across multiple districts, including Utrecht, Newcastle, Vryheid, and Dundee, which have historically been major coal-producing areas. The coalfield lies between latitudes 27° S and 30° S and longitudes 29° E and 32° E, stretching inland from the coastal areas of Richards Bay and Durban. The KwaZulu-Natal is part of the Ecca Group within the Karoo Supergroup, a sedimentary sequence that was deposited during the late Carboniferous to early Jurassic periods (approximately 300–180 million years ago). The coal seams are primarily found in the Vryheid Formation of the Ecca Group, which consists of alternating layers of sandstone, mudstone, shale, and coal. The KwaZulu-Natal Coalfield contains bituminous coal, ranging from high-quality anthracite to lower-rank bituminous coal. The coal seams vary in thickness but are generally 1–3 m thick, with some localized thicker seams. The coal is generally high in Ash content and varies in sulfur content.

2.2. Sample Preparation and Testing

A total of 133 coal data samples were considered in this study before data processing and cleaning. In the same database, the information covers a wide spectrum of physical and chemical characteristics. These samples were prepared and analyzed following ASTM Standards to ensure the accuracy and reliability of the data. Each coal sample was crushed, blended, and subjected to a rotary splitter to generate sub-samples of consistent size fractions for testing according to ASTM [26]. The crushing process ensured that all coal fragments were reduced to a uniform particle size suitable for subsequent analysis. The blending step involved thoroughly mixing the crushed coal to achieve homogeneity and minimize variability within individual samples. A rotary splitter was then employed to divide the homogenized coal into smaller, representative sub-samples, ensuring that each fraction maintained the same compositional characteristics. The proximate analysis was conducted in accordance with the ASTM [27]. Approximately 1 g of coal sample was used for the analyses in determining the inherent moisture, Ash content, and VM present in the coal with FC calculated by difference. The ultimate and total sulfur (TS) analyses of the coal were performed according to ASTM [28] and ISO [29]. The calorific value as the measure of the heat content was determined in accordance with ASTM [30]. The AI was determined according to Wells et al. [7] and Spero [8]. This was accomplished by mechanically grinding 4 kg of coal in a steel pot, thoroughly leveled out over the blades and covered with a steel lid. Having introduced the coal into the steel pot, the stirrer rotated for 12,000 revolutions at 1500 rpm. The AI was determined after the test from the loss of mass in milligrams of the blades from its initial mass and mass after grinding.

2.3. Data Characteristics

The available data used in the initial data preprocessing comprised a total of 133 data samples where CV, Pyrite, M, VM, FC, TS, Qtz, and AI are the available parameters in the data. Onifade et al. [9] considered 89 data samples of the same database where CV, TS, M, VM, FC, Ash, Pyrite, and Hardgrove grindability index were used as inputs and AI was utilized as model output. In this study, a general flowchart of the steps involved in predicting AI is presented in Figure 1. After a series of preprocessing analyses on the data (see Figures 2 and 3 for more information), we found that four of those parameters had more effects on the AI results, which are CV, Ash, Pyrite, and Qtz. These parameters were used in the analysis and modeling of this research as the lower number of input parameters will help increase interest from researchers or industry experts in the proposed model. Therefore, a total of 112 data samples were considered in this study, and their key distribution parameters such as mean, standard deviation, and range are presented in Table 1.

Table 1. Descriptive statistics of the used data in the analysis (Adapted from [9]).

Symbol	N Total	Mean	Std	Min	Max	Unit
CV	112	27.44	2.85	20.10	32.50	MJ/kg
Ash	112	15.85	5.50	7.30	33.30	%
Qtz	112	0.90	0.48	0.13	2.84	%
Pyrite	112	1.75	1.58	0.06	7.25	%
AI	112	132.33	93.55	21	459	%

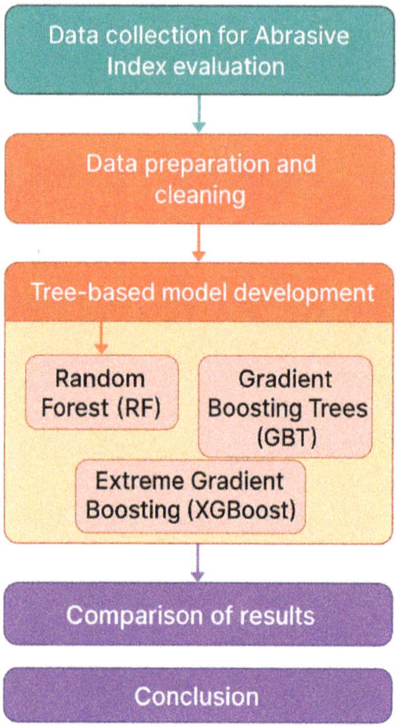

Figure 1. A general flowchart of the study procedure.

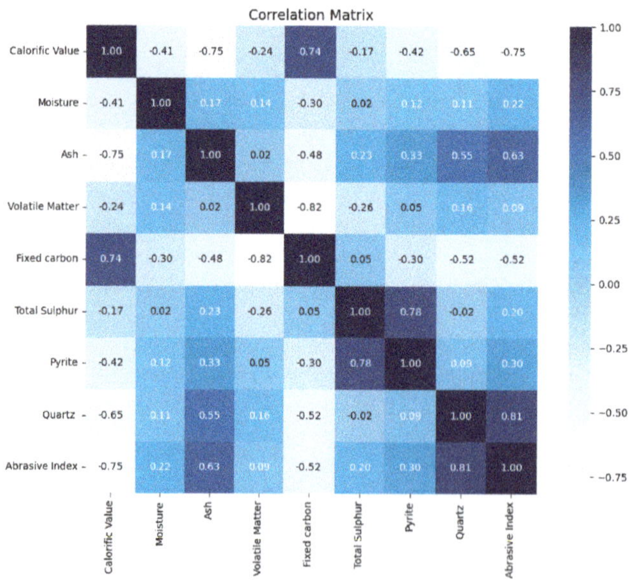

Figure 2. Correlation matrix of the input.

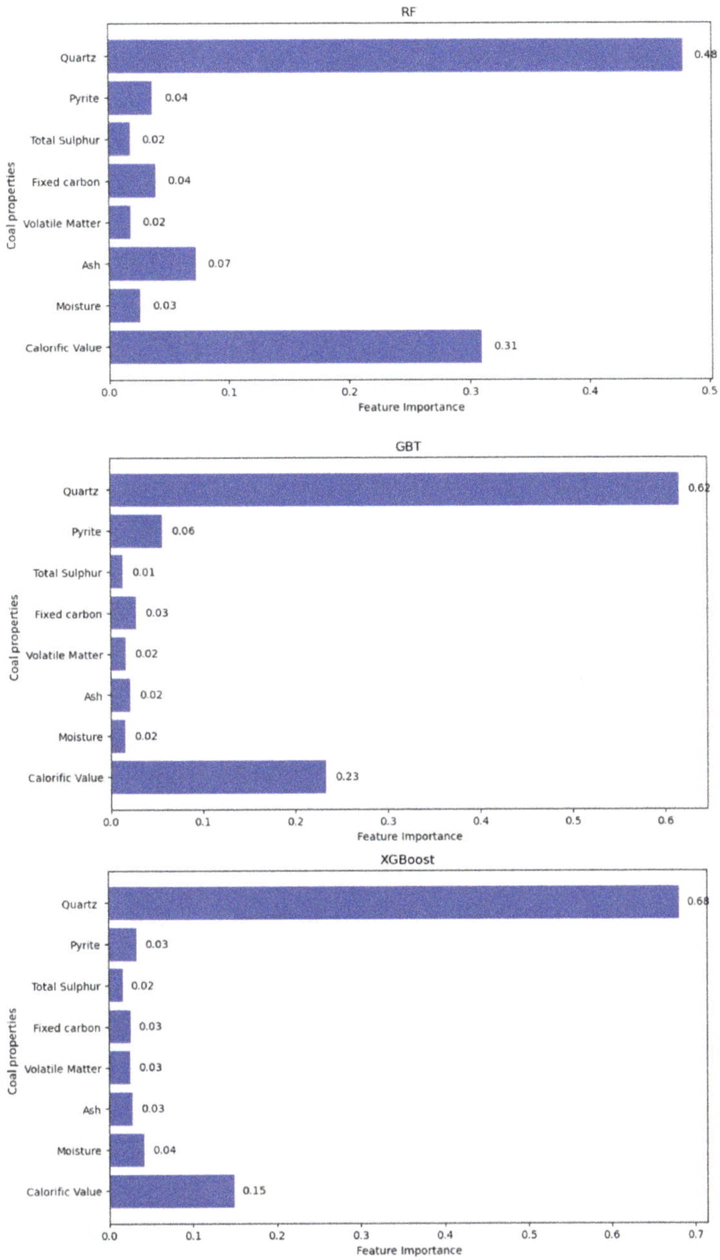

Figure 3. Feature importance (RF, GBT, and XGBoost).

3. Methodology

The present study employs three prominent tree-based ML algorithms: RF, GBT, and XGBoost to develop predictive models for the AI. The selection of these algorithms was based on their documented efficacy in capturing complex and non-linear relationships characteristic of coal property datasets, as well as their capability to process multivariate datasets robustly [31–33]. RF builds an ensemble of decision trees, each trained on a bootstrapped subset of the data with random feature selection at each split. The final prediction is the average (regression) of individual tree outputs. The prediction for a given sample x is as follows:

$$F(x) = \frac{1}{T}\sum_{t=1}^{T} h_t(x) \quad (1)$$

where $h_t(x)$ is the prediction from the tth decision tree, T is the total number of trees, and $F(x)$ is the predicted sample.

GBT builds trees sequentially, where each new tree attempts to correct the residual errors of the ensemble of previously built trees. The model prediction is updated iteratively as follows:

$$F_m(x) = F_{m-1}(x) + \eta\, h_m(x) \quad (2)$$

where $F_m(x)$ is model after m iterations; η is learning rate; and $h_m(x)$ is decision tree trained on residuals.

XGBoost improves upon GBT by introducing regularization, parallel processing, missing value handling, and efficient pruning to prevent overfitting and improve computational performance.

$$L(\phi) = \sum_{i=1}^{n} l\,(y_i,\, \hat{y}_i) + \sum_{k=1}^{K} \Omega\,(f_k) \quad (3)$$

where $L(\phi)$ is the total objective function; n is number of training samples; y_i is actual AI value of coal; $\hat{y}_i$ is predicted AI value from the model; $l\,(y_i,\, \hat{y}_i)$ is the loss function; K is number of trees; f_k is the regression tree; and $\Omega\,(f_k)$ is regularization function that penalizes complexity.

All models were developed using the Python programming language, utilizing widely adopted libraries such as scikit-learn and XGBoost. The development and analysis were carried out in the Jupyter Notebook environment.

A general flowchart of the steps involved in predicting AI is presented in Figure 1.

The experimental dataset consisted of eight key coal properties: CV, M, Ash, VM, FC, total sulfur (TS), pyrite, and Qtz. These input parameters were measured using standardized procedures outlined in Section 2.2, ensuring consistent and accurate laboratory results.

Prior to model training, the data underwent rigorous preprocessing. Initially, data consistency was verified, confirming completeness across all samples. Subsequently, outliers were identified through statistical approaches involving the interquartile range (IQR) and standard deviation criteria. Specifically, data points situated beyond ±3 standard deviations or 1.5 times the IQR from the quartiles were identified as potential outliers. Domain experts evaluated these flagged data points to determine their validity, retaining cases representing genuine geological variations and removing those identified as measurement anomalies or transcription errors. This expert-driven review ensured that removed data points did not unjustly limit the representativeness and diversity inherent within coal properties from the studied coalfield.

Following preprocessing, hyperparameter tuning was meticulously conducted to optimize model performance and avoid overfitting. For each ML algorithm, a systematic hyperparameter optimization approach combining grid search and random search methods was implemented. Cross-validation techniques, specifically five-fold cross-validation, were employed during this tuning phase to reliably estimate generalization performance. Optimal hyperparameters identified through this procedure included the number of trees, maximum tree depth, learning rate, subsample ratio, and regularization parameters specific to each algorithm. These hyperparameters were carefully adjusted based on initial performance feedback to ensure a balance between model complexity and generalization capability, guided by both computational efficiency and domain-specific understanding of coal property interactions.

After hyperparameter optimization, the dataset was partitioned into training and testing subsets using an 80/20 split. The optimized models were then trained on the larger training subset and independently evaluated on the testing subset, ensuring a reliable assessment of each model's predictive capability and generalizability to unseen data. This validation strategy, complemented by cross-validation, provided robust insights into model stability and effectiveness across various data scenarios.

An initial modeling iteration employed all eight coal properties to predict the AI. Subsequently, a feature importance analysis was performed to identify the most influential predictors within each model. Features consistently exhibiting minimal predictive power across all three algorithms were critically reviewed against geological knowledge and literature insights. Following expert evaluation, parameters exhibiting limited contribution to predictive accuracy and lacking geochemical significance were excluded, resulting in a refined set of predictors comprising CV, Qtz, Ash, and Pyrite. This iterative refinement process ensured that the final models relied on scientifically and practically meaningful predictors, enhancing interpretability and reducing model complexity.

A second modeling iteration, using only the refined subset of influential parameters, was then conducted. Performance evaluation criteria included the coefficient of determination (R^2) and correlation coefficient (R), ensuring objective quantification of predictive accuracy. Additionally, cross-validation was employed throughout both model-building stages to rigorously assess the stability and robustness of predictions, mitigating risks associated with potential overfitting. Regularization techniques, subsampling, and depth restrictions were employed as additional measures against overfitting, ensuring that developed models demonstrated both predictive reliability and practical utility.

Finally, results obtained from each modeling stage were interpreted in the context of coal processing implications, emphasizing their potential impact on equipment selection, operational efficiency, and maintenance cost reduction within coal-dependent industries. The comprehensive methodological approach outlined here, encompassing detailed preprocessing, rigorous hyperparameter tuning, systematic feature selection, and careful model evaluation, ensured robust, reproducible, and accurate predictive models capable of effectively predicting coal abrasiveness from physicochemical coal properties.

4. Results and Discussion

In the initial modeling phase, all eight coal properties were used to predict the AI through RF, GBT, and XGBoost. Model performance was assessed using two key statistical measures: the R^2 and the (R). These metrics were calculated as follows:

$$R^2 = 1 - \frac{\sum_{i=1}^{N}(y_i - y_i')^2}{\sum_{i=1}^{N}(y_i - \overline{y_i})^2} \tag{4}$$

$$R = \frac{\sum_{i=1}^{N}(y_i - \overline{y})\left(y_i' - \overline{y'}\right)}{\sqrt{\sum_{i=1}^{N}(y_i - \overline{y})^2}\sqrt{\left(y_i' - \overline{y'}\right)^2}} \tag{5}$$

where y_i is measured AI, y_i' is computed AI, $\overline{y_i}$ is mean measured AI, and n is the number of data points.

The initial evaluation results (Table 2) indicated a notable disparity between the training and testing datasets, with testing dataset R^2 values ranging from 0.63 to 0.72, significantly lower than the corresponding training dataset performance. This discrepancy suggested potential overfitting and data-related issues such as noise or outliers. These results are graphically presented in Figure 2, clearly highlighting the differences between training and testing performance.

Table 2. Coefficient of determination (R^2) for training and testing.

Tree-Based ML	Training Dataset (R^2)	Testing Dataset (R^2)
RF	0.96	0.67
GBT	0.99	0.72
XGBoost	1.00	0.63

To investigate and address the observed discrepancies, feature importance analyses were conducted using the RF, GBT, and XGBoost algorithms. Importance scores are unitless and represent the relative contribution of each feature to the model's predictive performance (normalized to sum to 1). Results of this analysis, depicted in Figure 3, indicated that CV, Qtz, Ash, and Pyrite content were consistently ranked as the most influential variables. Qtz's high impact on AI was attributed to its hardness and consequent abrasive action, while Ash and Pyrite contributed substantially due to their abrasive mineral impurities and intrinsic hardness, respectively. CV, while indirectly linked to abrasiveness, strongly correlated with coal hardness and density.

Based on these insights, a refined modeling iteration was conducted using only these identified critical parameters: CV, Qtz, Ash, and Pyrite. As shown in Figure 4, this refinement markedly improved predictive performance.

XGBoost emerged as the most accurate model, achieving a high testing dataset R^2 of 0.92, while RF and GBT yielded R^2 values of 0.85 and 0.81, respectively. This performance enhancement confirms XGBoost's superior capability in modeling intricate non-linear interactions among influential coal properties.

It can be noticed that the R^2 value of 1.00 on the training dataset from XGBoost suggests a potential risk of overfitting, particularly given the limited size of the dataset. However, the performance on the validation/test set remains consistent with high R^2, which indicates the model has reasonable generalization ability. Future studies with larger datasets or alternative validation methods are recommended to further confirm the model's performance.

The improved predictive accuracy demonstrates considerable advantages over traditional laboratory-based measurements of AI, providing significant operational efficiencies. By enabling rapid and precise AI estimations from readily obtainable coal property data, these predictive models can directly inform and improve coal selection, blending processes, and operational decision-making. This facilitates reduced equipment wear, minimizes downtime, and optimizes maintenance schedules, thus enhancing overall operational cost-effectiveness.

Furthermore, the identification of Qtz, Ash, and Pyrite as major abrasive contributors provides practical guidance for targeted coal beneficiation strategies aimed at reducing abrasive mineral content. This targeted beneficiation strategy can lead to substantial improvements in operational sustainability by reducing both resource consumption and environmental impacts. These findings provide a robust foundation for future research. It is recommended that subsequent studies further validate these predictive models across different coalfields, incorporate additional relevant coal properties, and explore the integration of real-time sensor data to enhance model accuracy and operational applicability.

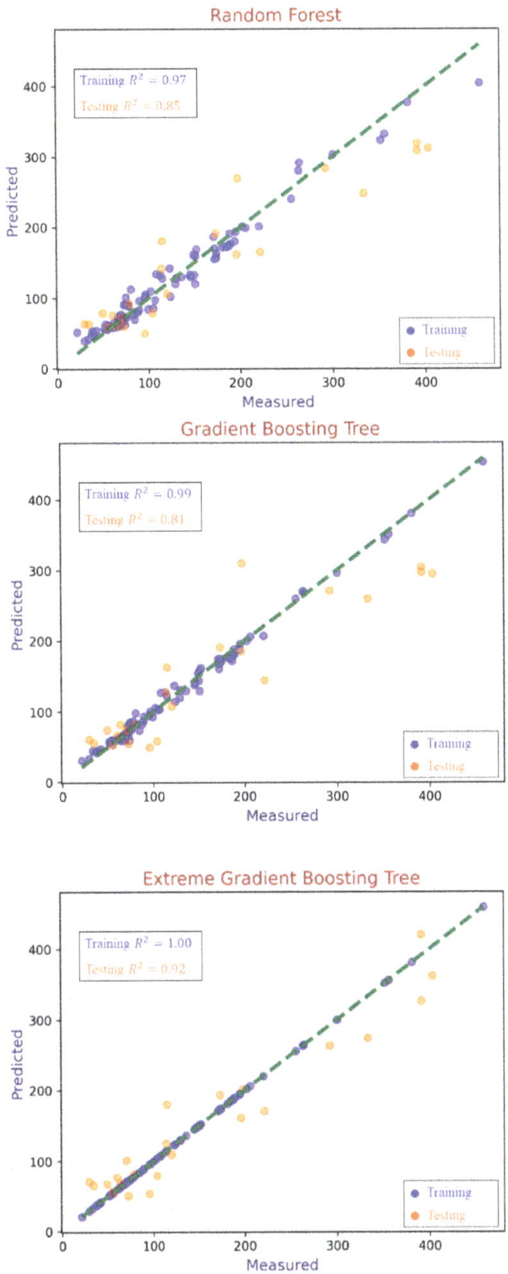

Figure 4. Predictions of the tree-based ML models.

4.1. Taylor Diagram Analysis

To have a better understanding of the behavior of the predictive models and their capacities in predicting AI during training and testing stages, a Taylor diagram was analyzed. This diagram is a compact way to demonstrate predictive model performance in a graphical aspect. In other word, this diagram examines the model capacity based on observed (measured) and predicted values when comparing multiple models or simulations simultaneously. This will give a better understanding of the outcome compared to traditional ways of calculating performance indices. The same diagram was first developed and used in climate science and then was applied in many engineering and science applications.

By using the measured and predicted AI values for the training and testing sets of all predictive models, the Taylor diagrams were generated as shown in Figure 5. According to the comparative performance of the predictive techniques, the XGB model demonstrates superior performance in the training phase among the XGB, GBT, and RF models in estimating AI. This is evident, as indicated by its closer alignment with the reference point. A similar trend is observed in the testing phase; however, the overall accuracy of all three predictive models is lower compared to the training phase. This may be due to the limited amount of data used in this study, potential overfitting, or challenges in generalizing to unseen data. The RF and GBT models were also found to be capable of predicting AI with moderate accuracy compared to XGB in both the training and testing phases. Overall, the ability of these tree-based models to predict AI is evident, and they can be used as substitutes for practical applications in real-world projects.

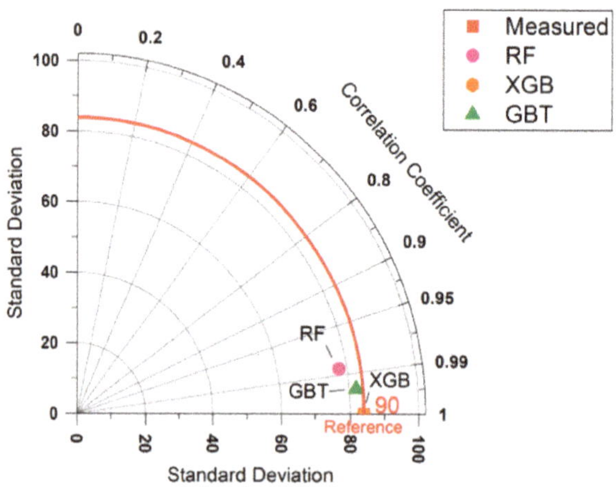

(**A**) Training phase

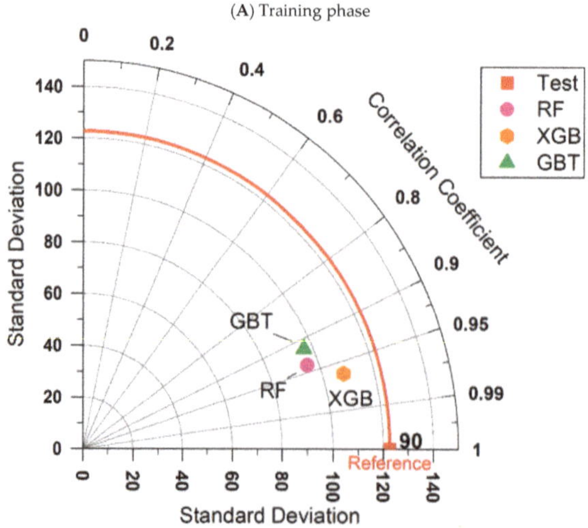

(**B**) Testing phase

Figure 5. Taylor diagrams for all proposed predictive models.

4.2. Model Comparison and Generalization

A fair way of evaluating is to compare the proposed model with similar models published previously. The authors of this study tried to propose a model with a higher level of generalization compared to the original study conducted by Onifade et al. [9]. In this way, this study used 112 data samples from the same site with a wider range of input parameters compared to the study conducted by Onifade et al. [9] with 89 data samples. The range of input parameters used in this study is wider, and this will help to apply this model to a wider range of inputs in future projects.

Another point that needs to be discussed is related to the number of input or effective parameters to propose a new model. The practical application of a new model is to allow other researchers or practitioners a real and easy way of applying the same model. Preparing a larger number of input parameters is more difficult compared to a lower number if someone wishes to use it in the future. Onifade et al. [9] used seven input parameters while this study considered only four parameters as inputs. This again will make our model easier to apply in practice and more attractive for researchers or industry professionals who prefer models with fewer required input parameters. It should be noted that the performance prediction of our model is not as good as Onifade et al. [9]; however, the proposed model provides a balance between simplicity and acceptable accuracy, making it a practical option for cases where data availability is limited or rapid estimation is needed.

4.3. Sensitivity Analysis

Sensitivity analysis is a good way to understand more about the effects of input variables on the output. This will help to identify the most and least effective parameters used in that specific database. Although there are several techniques for analyzing the effects of inputs on output of the system, the Cosine Amplitude technique is one of the simplest and easiest to implement. The Cosine Amplitude technique quantifies the linear association between each input variable and the output (AI) by computing the cosine of the angle between their respective vectors in multi-dimensional space. The value ranges between 0 and 1, where values closer to 1 indicate a stronger association. Since it is a dimensionless metric, no physical unit applies. It can quantify the degree of linear association between input and output parameters. In the area of data analysis, this technique recognizes the greatest influence of each input on the output by measuring the strength and direction of their relationships in a scale-independent manner. The range of cosine values is between 0 and 1, where 1 indicates the greatest influence and 0 indicates the least importance of that specific input parameter. Table 3 shows the effects of input parameters on the system output. The most effective parameter on AI is Qtz with a weight of 0.953, and the least effective parameter on AI is CV with a weight of 0.768.

Table 3. The effects of input features on the system output.

Input Feature	Cosine Amplitude Sensitivity
Ash	0.897
CV	0.768
Pyrite	0.799
Qtz	0.953

Note: Cosine amplitude sensitivity values are unitless.

5. Limitations and Future Studies

Despite the robustness and predictive success of the presented methodology, several limitations need to be acknowledged. Firstly, the coal samples used in this study were sourced exclusively from the KwaZulu-Natal coalfield, potentially constraining the generalizability of these predictive models to other geographic regions with differing geological characteristics. Therefore, the performance of the developed models may vary when applied to other coalfields with distinct geochemical and geological profiles. Secondly, although careful measures were employed during data preprocessing and outlier detection, the manual evaluation by domain experts introduced an element of subjectivity. This approach, while ensuring data integrity, may limit reproducibility across different research contexts, particularly where expert judgment criteria vary.

Another limitation pertains to hyperparameter tuning. While extensive grid and random search methods were implemented, advanced techniques such as Bayesian optimization could potentially further enhance model performance by efficiently exploring the hyperparameter space and potentially discovering more optimal configurations. Such advanced optimization strategies might uncover configurations capable of even greater predictive performance.

Additionally, the current study focused primarily on coal physicochemical properties without considering other potentially influential factors, such as hardness, microscopic coal textures, and particle size distribution, all of which might provide further predictive insights and model enhancement.

In light of these limitations, future research directions are suggested to enhance both the depth and breadth of this research field. It is recommended that subsequent studies validate these predictive models using coal samples from diverse geographic regions, thus assessing generalizability and broadening the practical applicability of the models. Moreover, future research could leverage automated outlier detection methods to enhance reproducibility and minimize subjective biases introduced by manual assessments.

Advanced hyperparameter tuning methods, such as Bayesian optimization and genetic algorithms, are encouraged for future exploration to further refine predictive capabilities. Incorporating additional mechanical and microscopic properties of coal into predictive models could offer enhanced insights into the complex relationships governing coal abrasiveness, potentially leading to improved predictive accuracy and practical applicability.

Lastly, the integration of real-time monitoring technologies and sensor-based data collection within coal processing plants could facilitate the development of dynamic, real-time predictive systems, thereby enabling proactive maintenance strategies, real-time coal blending optimization, and enhanced operational decision-making. Such advancements would significantly contribute to operational efficiency, sustainability, and economic viability in the coal processing industry.

6. Conclusions

This study successfully demonstrates the potential of tree-based ML algorithms in predicting the coal AI. Through a comprehensive analysis of RF, GBT, and XGBoost, we identified CV, Qtz, Ash, and Pyrite as the most influential factors in coal AI prediction. The results indicate that XGBoost provides the highest accuracy (R^2 = 0.92), outperforming RF (R^2 = 0.89) and GBT (R^2 = 0.85). This highlights XGBoost's capability in capturing complex, non-linear relationships between coal properties and abrasiveness, making it a valuable tool for predictive modeling in coal processing industries. The main findings are summarized as follows:

Testing dataset performance was lower than the training dataset at the first stage of the analysis, indicating noisy or outlier data that necessitated cleaning and filtering.

CV, Qtz, Ash, and Pyrite are the most influential properties for the prediction of coal AI in the original database; however, quartz was the most effective one among these four parameters.

Beyond the predictive accuracy, these ML-based approaches offer a cost-effective and efficient alternative to traditional laboratory methods, which are often time-consuming and resource-intensive. By leveraging data-driven models, industries can optimize coal selection, minimize equipment wear, reduce maintenance costs, and enhance operational sustainability. With further refinements and real-world validation, this approach has the potential to become an industry standard, streamlining decision-making and improving efficiency in coal-related operations.

Author Contributions: Conceptualization, All authors; methodology, M.A., C.Y.H. and D.J.A.; software, C.Y.H. and D.J.A.; validation, D.J.A., M.O. and M.K.; formal analysis, C.Y.H.; D.J.A., H.F. and M.K.; investigation, M.O. and M.K.; resources, H.F. and D.J.A.; data curation, M.O., M.K. and D.O.S.; writing All authors; writing-review and editing, All authors; visualization, M.A., C.Y.H., M.O. and D.J.A.; supervision, M.O., M.K. and D.J.A. All authors have read and agreed to the published version of the manuscript.

Funding: This research received no external funding.

Data Availability Statement: The raw data supporting the conclusions of this article will be made available by the authors on request.

Conflicts of Interest: The authors declare that they have no known competing financial interests or personal relationships that could have appeared to influence the work reported in this paper.

References

1. Finkelman, R.B.; Wolfe, A.; Hendryx, M.S. The future environmental and health impacts of coal. *Energy Geosci.* **2021**, *2*, 99–112. [CrossRef]
2. Jiang, L.; Xue, D.; Wei, Z.; Chen, Z.; Mirzayev, M.; Chen, Y.; Chen, S. Coal decarbonization: A state-of-the-art review of enhanced hydrogen production in underground coal gasification. *Energy Rev.* **2022**, *1*, 100004. [CrossRef]
3. Alam Munshi, T.; Jahan, L.N.; Howladar, M.F.; Hashan, M. Prediction of gross calorific value from coal analysis using decision tree-based bagging and boosting techniques. *Heliyon* **2024**, *10*, e23395. [CrossRef] [PubMed]
4. Rehman, A.; Ma, H.; Ozturk, I.; Alvarado, R.; Oláh, J.; Liu, R.; Qiang, W. The enigma of environmental sustainability and carbonization: Assessing the connection between coal and oil rents, natural resources, and environmental quality. *Gondwana Res.* **2024**, *128*, 1–13. [CrossRef]
5. Babu, K.A.; Lawrence, A.; Sivashanmugam, P. Grindability studies on blended coals of high-ash Indian coals with low-ash imported coals. *Int. J. Coal Prep. Util.* **2018**, *38*, 433–442. [CrossRef]
6. Idris, A.; Man, Z.; Bustam, A.; Rabat, N.E.; Uddin, F.; Mannan, H.A. Grindability and abrasive behavior of coal blends: Analysis and prediction. *Int. J. Coal Prep. Util.* **2022**, *42*, 1143–1169. [CrossRef]
7. Wells, J.; Wigley, F.; Foster, D.; Livingston, W.; Gibb, W.; Williamson, J. The nature of mineral matter in a coal and the effects on erosive and abrasive behaviour. *Fuel Process. Technol.* **2005**, *86*, 535–550. [CrossRef]
8. Spero, C. Assessment and prediction of coal abrasiveness. *Fuel* **1990**, *69*, 1168–1176. [CrossRef]
9. Onifade, M.; Lawal, A.I.; Bada, S.O.; Khandelwal, M. Predictive modelling for coal abrasive index: Unveiling influential factors through Shallow and Deep Neural Networks. *Fuel* **2024**, *374*, 132319. [CrossRef]
10. Chen, Y.; Khandelwal, M.; Onifade, M.; Zhou, J.; Lawal, A.I.; Bada, S.O.; Genc, B. Predicting the hardgrove grindability index using interpretable decision tree-based machine learning models. *Fuel* **2025**, *384*, 133953. [CrossRef]
11. Bandopadhyay, A. A study on the abundance of quartz in thermal coals of India and its relation to abrasion index: Development of predictive model for abrasion. *Int. J. Coal Geol.* **2010**, *84*, 63–69. [CrossRef]
12. Alekhnovich, A.N.; Artemieva, N.V.; Bogomolov, V.V. Definition and Assessment of Coal Abrasivity. *Power Technol. Eng.* **2021**, *55*, 96–102. [CrossRef]
13. Wells, J.; Wigley, F.; Foster, D.; Gibb, W.; Williamson, J. The relationship between excluded mineral matter and the abrasion index of a coal. *Fuel* **2004**, *83*, 359–364. [CrossRef]
14. Nahvi, S.; Shipway, P.; McCartney, D. Effects of particle crushing in abrasion testing of steels with ash from biomass-fired powerplants. *Wear* **2009**, *267*, 34–42. [CrossRef]

15. Hower, J.C.; Finkelman, R.B.; Eble, C.F.; Arnold, B.J. Understanding coal quality and the critical importance of comprehensive coal analyses. *Int. J. Coal Geol.* **2022**, *263*, 104120. [CrossRef]
16. Vilakazi, L.; Madyira, D. Estimation of gross calorific value of coal: A literature review. *Int. J. Coal Prep. Util.* **2024**, *45*, 390–404. [CrossRef]
17. Grzegorzek, W.; Adamecki, D.; Głuszek, G.; Lutyński, A.; Kowol, D. Technique to Investigate Pulverizing and Abrasive Performance of Coals in Mineral Processing Systems. *Energies* **2021**, *14*, 7300. [CrossRef]
18. Peisheng, L.; Youhui, X.; Dunxi, Y.; Xuexin, S. Prediction of grindability with multivariable regression and neural network in Chinese coal. *Fuel* **2005**, *84*, 2384–2388. [CrossRef]
19. Tshiongo, N.; Mulaba-Bafubiandi, A. South African coal and its abrasiveness index determination: An account of challenges. In Proceedings of the Southern African Universities Engineering Conference (SAUPEC), Johannesburg, South Africa, 24–26 January 2013; pp. 288–293.
20. Höök, M.; Aleklett, K. A review on coal-to-liquid fuels and its coal consumption. *Int. J. Energy Res.* **2010**, *34*, 848–864. [CrossRef]
21. Deepa, G.; Niranjana, A.; Balu, A. A hybrid machine learning approach for early cost estimation of pile foundations. *J. Eng. Des. Technol.* **2025**, *23*, 306–322. [CrossRef]
22. Abdelfattah, E.; Joshi, S. Comparison of Machine Learning Classification and Clustering Algorithms for TV Commercials Detection. *IEEE Access* **2023**, *11*, 116741–116751. [CrossRef]
23. Yang, B.; Armaghani, D.J.; Fattahi, H.; Afrazi, M.; Koopialipoor, M.; Asteris, P.G.; Khandelwal, M. Optimized Random Forest Models for Rock Mass Classification in Tunnel Construction. *Geosciences* **2025**, *15*, 47. [CrossRef]
24. Chia, Y.H.; Armaghani, D.J.; Lai, S.H. Predicting Soil Compression Index Using Random Forest and Gradient Boosting Tree. In Proceedings of the Chinese Institute of Engineers (CIE), the Hong Kong Institute of Engineers (HKIE), and the Institution of Engineers Malaysia (IEM) Tripartite Seminar, Taipei, Taiwan, 1 November 2023; Volume 1.
25. He, B.; Armaghani, D.J.; Lai, S.H.; Mohamad, E.T. Application of an expert extreme gradient boosting model to predict blast-induced air-overpressure in quarry mines. In *Applications of Artificial Intelligence in Mining, Geotechnical and Geoengineering*; Elsevier: Amsterdam, The Netherlands, 2024; pp. 269–289. [CrossRef]
26. *ASTM D-2013*; Standard Practice for Preparing Coal Samples for Analysis. ASTM: West Conshohocken, PA, USA, 2013.
27. *ASTM D5142*; Standard Test Methods for Proximate Analysis of the Analysis Sample of Coal and Coke by Instrumental Procedures. ASTM: West Conshohocken, PA, USA, 1998.
28. *ASTM D5373-14*; Standard Test Methods for Determination of Carbon, Hydrogen, and Nitrogen in Analysis Samples of Coal and Carbon in Analysis Samples of Coal and Coke. ASTM: West Conshohocken, PA, USA, 2014.
29. *ISO 19579: 2006*; Solid Mineral Fuels Determination of Sulphur by IR spectrometry. ISO: Geneva, Switzerland, 2006.
30. *ASTM D5865-04*; Standard Test Method for Gross Calorific Value of Coal and Coke. ASTM: West Conshohocken, PA, USA, 2004.
31. Breiman, L. Random forests. *Mach. Learn.* **2001**, *45*, 5–32. [CrossRef]
32. Friedman, J.H. Greedy function approximation: A gradient boosting machine. *Ann. Stat.* **2001**, *29*, 1189–1232. [CrossRef]
33. Chen, T.; Guestrin, C. August. Xgboost: A scalable tree boosting system. In Proceedings of the 22nd ACM SIGKDD International Conference on Knowledge Discovery and Data Mining, New York, NY, USA, 13–17 August 2016; pp. 785–794. [CrossRef]

mining

MDPI

Article

A Solution Surface in Nine-Dimensional Space to Optimise Ground Vibration Effects Through Artificial Intelligence in Open-Pit Mine Blasting

Onalethata Saubi [1,*], Rodrigo S. Jamisola, Jr. [2], Kesalopa Gaopale [1], Raymond S. Suglo [1] and Oduetse Matsebe [2]

[1] Department of Mining Engineering, Botswana International University of Science and Technology, Palapye Private Bag 16, Botswana; gaopalek@biust.ac.bw (K.G.); suglor@biust.ac.bw (R.S.S.)
[2] Department of Mechanical, Energy and Industrial Engineering, Botswana International University of Science and Technology, Palapye Private Bag 16, Botswana; jamisolar@biust.ac.bw (R.S.J.J.); matsebeo@biust.ac.bw (O.M.)
* Correspondence: so13000604@biust.ac.bw

Abstract: In this study, we model a solution surface, with each point having nine components using artificial intelligence (AI), to optimise the effects of ground vibration during blasting operations in an open-pit diamond mine. This model has eight input parameters that can be adjusted by blasting engineers to arrive at a desired output value of ground vibration. It is built using the best performing artificial neural network architecture that best fits the blasting data from 100 blasting events provided by the Debswana diamond mine. Other AI algorithms used to compare the model's performance were the k-nearest neighbour, support vector machine, and random forest—together with more traditional statistical approaches, i.e., multivariate and regression analyses. The input parameters were burden, spacing, stemming length, hole depth, hole diameter, distance from the blast face to the monitoring point, maximum charge per delay, and powder factor. The optimised model allows for variations in the input values, given the constraints, such that the output ground vibration will be within the minimum acceptable value. Through unconstrained optimisation, the minimum value of ground vibration is around 0.1 mm/s, which is within the vibration range caused by a passing vehicle.

Keywords: optimisation; prediction; ground vibration; artificial neural network; blast-induced

1. Introduction

Blasting is commonly used in mining, quarrying, and construction to break apart hard rocks and enable easier excavation and removal. Only 20 to 30 percent of the explosive energy during a blast is used for rock breaking, with the remainder contributing to negative environmental impacts, such as ground vibration, airblast, and flyrock [1,2]. The vibrations induced by blasting affect neighbouring residents and may damage structures [3]. Consequently, predicting blast-induced ground vibrations becomes imperative to manage and mitigate the undesirable effects of blasting. Ground vibrations originating from a blast are quantified in terms of peak particle velocity (PPV), measured in mm/s, at a designated ground position. Influencing factors for ground vibrations are categorised into controllable and uncontrollable parameters. The former pertains to human-made elements, such as explosive properties and blast design parameters, while the latter involves natural occurrences, such as geological conditions and rock-mass properties [4,5].

We present a method to establish the relationship of ground vibration versus input parameters as shown below

$$GV = f(S, B, T, L, D, MC, Pf, Di) \quad (1)$$

where GV = ground vibration, S = spacing, B = burden, T = stemming, L = hole depth, D = hole diameter, MC = maximum charge per delay, Pf = powder factor, and Di = distance from the blast point to the monitoring point.

Equation (1) above has been the subject of many studies aiming to define the function $f()$, that is, to establish the effects of input parameters versus the output parameter in a blasting event in the mines. Most mining companies in the world use different software in estimating the function $f()$, namely, Blast Logic, Mining Excellence, O-Pit Blast, Surpac, etc. These can help them in their blast design efforts, that is, provide values of the input parameters and allow them to estimate the outputs of blasting, e.g., rock fragmentation, ground vibration, airblast, and flyrock. Most of these software use empirical methods in defining the function $f()$.

Our solution is to use machine learning to model the function $f()$, and we are not the only ones who attempted to use this approach. What is unique about our approach is that we provide a solution surface, which is the problem's solution space, to further define the function $f()$. In previous studies, researchers have only defined the function $f()$ as a black box that is capable of predicting the output parameter given the input parameters, without providing a solution space. Our solution space enables us to perform prediction as well as two other things as follows:

1. Perform optimisation computations to find the minimum ground vibration using random initial points.
2. Perform blast design by setting the desired value of ground vibration and searching in the solution space for the values of the corresponding input parameters. Alternatively, we can assign the value of ground vibration and some input parameters as constraints, and we can search in the solution space for the values of the remaining input parameters.

To the best of our knowledge, no other previous studies could accomplish this.

1.1. Empirical Methods

Numerous empirical methodologies for predicting ground vibration due to blasting have been developed by researchers [6–8]. These techniques use data obtained from different blasting operations, in certain ground conditions which cannot be generalised for different rock types. There are two major shortfalls related to empirical methods. First, they have a limited number of inputs, normally only two—the maximum charge per delay and distance from the blast point to the monitoring point. Second, they have very low accuracy. Artificial intelligence methods provide generalisability, especially in terms of the number of inputs to the model. However, in terms of accuracy, the results may vary depending on the algorithm used. Moreover, machine learning methods generally provide much higher accuracy compared to empirical methods. From our study, accuracy comparisons are stated as follows. The United States Bureau of Mines (USBM) empirical model has an accuracy of 12.6%, the accuracy of our artificial intelligence model ranges from 54.6% to 93.6%, and the statistical method has an accuracy of 66%.

Adding more input parameters in modelling ground vibration using empirical methods can be very difficult to achieve with high accuracy because of the complex, non-linear relationship between the input and output parameters. This could be the reason why empirical methods typically account for only two parameters and fail to incorporate the full range of factors affecting ground vibration. New approaches to estimating ground vibrations from blasting include artificial intelligence systems that consider many parameters influencing the outcomes of blasting [9–12].

1.2. Machine Learning Methods

Lawal et al. [13] utilised an ordinary artificial neural network (ANN), particle swarm optimisation (PSO), and dragonfly-optimised ANN to forecast how tunnel blasting may affect the environment. The PSO-ANN model, with an architecture of 4-15-15-1, yielded the best performance, exhibiting a coefficient of determination (R^2) of 1, a mean square error (MSE) of 0.125, and a variance accounted for (VAF) of 98.29. In another study, Lawal et al. [14] used blast data from five granite quarries to predict ground vibration by employing gene expression programming (GEP), a sine cosine algorithm-optimised artificial neural network (SCA-ANN), and adaptive neuro-fuzzy inference system (ANFIS) models, using a total of 100 datasets for model development. The obtained R^2 values for the SCA-ANN, GEP, and ANFIS models were 0.999, 0.989, and 0.997, respectively.

Zhang et al. [15] applied five machine learning classifiers for predicting ground vibration, namely, the chi-squared automatic interaction detection (CHAID), ANN, support vector machine (SVM), random forest (RF), and classification regression tree (CART), finding RF demonstrates substantially superior performance over the other regression models investigated. Qiu et al. [16] applied an extreme gradient boosting (XGBoost) model to predict blast-induced ground vibration using 150 data sets with thirteen input parameters. The grey wolf optimisation (GWO), whale optimisation algorithm (WOA), and Bayesian optimisation algorithm (BO) were used to fine-tune the hyper-parameters of the XGBoost model. Their investigation revealed that the suggested WOA-XGBoost model performed better in comparison to other machine learning (ML) models.

Nguyen et al. [17] utilised a novel hybrid model combining the hierarchical k-means clustering algorithm with an ANN (HKM-ANN). The model was compared to fuzzy c-means clustering support vector regression (FCM-SVR), finding that the HKM-ANN model performed better than the FCM-SVR model in forecasting ground vibration. Yang et al. [18] employed ANFIS optimised by PSO and a genetic algorithm (GA) to forecast ground vibration resulting from blasting, utilising data from two Iranian quarry mines. Their model demonstrated an accuracy of 0.979, a root mean square error ($RMSE$) of 0.240, and a mean absolute error (MAE) of 0.199 in predicting ground vibration. Other researchers have employed hybrid artificial intelligence methodologies, predominantly ANNs, to model blast-induced ground vibrations and other blast-induced impacts and rock engineering problems [19–28].

This study uses an ANN model as a solution surface in a nine-dimensional space to minimise the effects of ground vibration during blasting operations in an open-pit diamond mine. This solution surface becomes the solution space such that each point on the surface is defined by nine components, i.e., eight input parameters and one output parameter ground vibration. An ANN model is designed for predicting ground vibrations during open-pit mine blasting, using data derived from the Debswana Orapa Diamond Mine in Botswana. The ANN model was chosen as the best-performing model among three other machine learning methods. The model has eight input parameters, which include both blast design parameters and explosive properties. The model's performance is compared against other machine learning algorithms and multivariate regression analysis. Sensitivity analysis is also conducted to determine the relative influence of the input parameters on the output.

We explicitly list our contributions in this paper as follows:

- We present data from Debswana Diamond Company recorded from 100 blast events.

- We develop machine learning models, each with five different architectures, eight input parameters, and one output of ground vibration. The four models are compared against a statistical method.
- We optimise the architecture of the best performing machine learning model using the Monte Carlo method.
- A solution space is created from the optimised machine learning model. Other machine learning models did not yield a solution space, as their results only predicted the output parameter ground vibration. Our solution space is capable of inverse solution, i.e., we can search for the solution space, given the input parameters and the expected output, to help blast design engineers in adjusting the input parameters to arrive at an expected output of ground vibration.
- We optimise ground vibration using the gradient descent method from the created solution space.
- Sensitivity analysis is performed using a statistical method and the results are confirmed from the created solution space.

The machine learning model allows us to predict the output ground vibration from the given eight input parameters by locating a unique point in the solution space. Because of this model, one can also predict not just the output parameter, but also an input parameter when seven other input parameters are given together with their corresponding output ground vibration. In this way, this model can be very useful to blast design engineers by setting a target output ground vibration and searching for the corresponding input parameters in the solution space. This is a kind of inverse solution approach and can be very useful to blast design engineers.

1.3. Mine Case Study

The Orapa Diamond Mine in Botswana was discovered in 1967 by De Beers. The kimberlite igneous rocks span an area of 1.18 square kilometres, with two merging diatremes, each with preserved crater sedimentary facies known as the southern and northern lobes. These lobes overlay Archaean granite and belong to the Karoo Supergroup, incorporating Stormberg-aged basalt lava and tracing back approximately 93 million years. The anticipated yield of the Orapa mine is around 153 million carats from a total of 205 million tonnes of mineral reserves. Figure 1 depicts the current layout of the Orapa Diamond Mine.

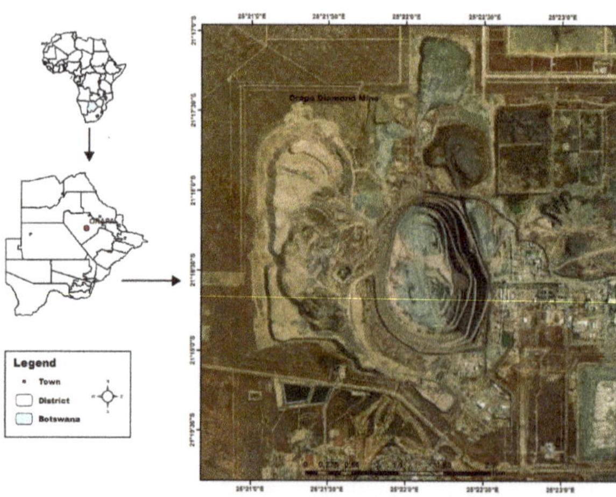

Figure 1. Layout of the Orapa Diamond Mine.

The mine utilises a staggered blasting pattern. The predominant explosives used are S135B emulsion and pentolite boosters.The mine has a bench height of 15 m and makes use of blast holes with diametres of 127, 165, and 250 millimetres. Each blast comprises between 15 and 25 rows, with each row containing between 40 and 60 holes. The loading of the fragmented materials is carried out using shovels and hydraulic excavators, while dump trucks are used for hauling.

2. Materials and Methods

This section explores the materials and methodologies employed in this study, focusing on the datasets collected from the mining operation, the machine learning techniques applied, and the optimisation processes implemented.

2.1. Materials

A dataset with 100 blasting events was collected from the blasting operations of the Debswana Orapa Diamond Mine. The dataset consists of eight inputs, including the blast design parameters and explosive properties, and the ground vibration as the output, as shown in Table 1.

Table 1. Description of the input and output parameters for the models.

Parameter	Type	Unit	Symbol	Min	Max
Burden	Input	m	B	4	6
Spacing	Input	m	S	5	8
Stemming length	Input	m	T	4	6
Hole depth	Input	m	L	12	15
Hole diameter	Input	mm	D	165	250
Distance from blast to monitoring point	Input	m	DI	438	1500
Maximum charge per delay	Input	kg	MC	216	552.6
Powder factor	Input	kg/m^3	Pf	0.3	1.17
Ground vibration	Output	mm/s	GV	0.163	6.5

Ground vibration data in this study were recorded using seismographs placed at varying distances from the blast sources. These devices measured vibrations in three orthogonal components—radial, transverse, and vertical. The radial component corresponds to the direction of propagation of the vibration wave from the blast source, while the transverse and vertical components provide a complete three-dimensional representation of the vibration effects. This approach ensures that the dataset captures the full spectrum of vibrational impacts resulting from blasting activities. "Starting from 1984 to the present, Orapa had an average of around two blast events per month. The data used in the study span the period from 2018 to 2024, covering all blasting conditions and potential biases in this six-year period. There were a total of 144 blast events. However, some events have incomplete information and need to be discarded. This results into 100 events with complete information that are included in our dataset. In terms of the completeness and representativeness of the data collected, we present the range of values for each of the following parameters included in the dataset, as shown on Table 1: (a) distance from blast point to monitoring stations (ranging from 438 m to 1500 m), (b) hole depths (12 m to 15 m), and (c) hole diameter (127 mm, 165 mm, and 250 mm). These parameters are consistent with the range of values the Orapa mine used throughout its blast operations in the stated six-year period. Thus, this ensures the relevance and representativeness of the dataset used within the operational context".

2.2. Methods

Data preprocessing involved cleaning and normalising the data to prepare it for analysis; the data was divided into training (80%), validation (10%), and testing (10%) sets. The machine learning models employed are the random forest, support vector machine, k-nearest neighbour (k-NN), and artificial neural network models. Additionally, the statistical model multivariate regression analysis (MVRA) was utilised. These models were built using Python 3.9.16 Specifically, scikit-learn libraries were used to model the SVM, k-NN, and RF models, and Keras library was used in Tensor Flow for modelling using an ANN. The gradient descent method, coupled with the Monte Carlo method, was used for the optimisation of ground vibration based on the best-performing ANN model. The cosine amplitude approach was applied for sensitivity analysis to determine the relative impact of input parameters on the output. The metrics used for evaluating the performance of all predictive models are the $RMSE$ and R^2. The $RMSE$ is calculated using the following formula:

$$RMSE = \sqrt{\frac{1}{N}\sum_{i=1}^{N}(y - y')^2} \quad (2)$$

The (R^2) is determined by the following:

$$R^2 = \left[\frac{\sum_{i=1}^{N}(y-\bar{y})(y'-\bar{y}')}{\sum_{i=1}^{N}(y-\bar{y})^2\sum_{i=1}^{N}(y'-\bar{y}')^2}\right]^2 \quad (3)$$

In these equations, y and y' represent the measured and forecasted values, respectively, while $\bar{y}$ and $\bar{y}'$ denote the mean values.

For composite score calculation, R^2 is normalised. We note that higher values of R^2 indicate better performance. For the $RMSE$, lower values indicate better performance. To normalise the $RMSE$ values, we used the following:

$$\text{Normalized RMSE} = 1 - \frac{\text{RMSE}}{\text{Max RMSE}} \quad (4)$$

Once both metrics are normalised, we compute a composite score by adding the normalised R^2 and normalised $RMSE$.

The Monte Carlo technique was employed to randomly navigate the solution space and find the best ANN configuration. This optimal ANN structure comprises 34 neurons in the first hidden layer, 15 in the second layer, and 50 in the third hidden layer. Upon establishing the optimal ANN model, we then use the gradient descent approach to find the minimum ground vibration and its corresponding input parameters. Gradient descent is a commonly employed iterative optimisation technique with applications spanning various computational fields [29]. The mathematical equation of the gradient descent method is shown by (5).

$$p_i^{\text{new}} = p_i - \eta\left(\frac{\text{nnet}(p_1,\ldots,p_i,\ldots,p_N) - \text{nnet}(p_1,\ldots,p_i^{\text{old}},\ldots,p_N)}{p_i - p_i^{\text{old}}}\right) \quad (5)$$

In this equation, p_i^{new} represents the updated parameter, while p_i is the current parameter value being updated. The term p_i^{old} denotes the parameter value from the previous iteration. The learning rate is represented by η. The function $\text{nnet}(\cdot)$ represents the neural network model.

3. Results and Discussion

Table 2 shows the performance indices of the models. The ANN model with an 8-10-1 architecture showed the best performance, with the lowest $RMSE$ (0.286), the highest accuracy R^2 (0.941), and the highest composite score (1.864), establishing it as the best model. The order of performance of the models according to their composite score is ANN (1.864), RF (1.768), *k*-NN (1.454), SVM (1.445), and MVRA with the lowest composite score of 0.664.

Table 2. Calculated performance indices for all the algorithms.

Method	R^2	RMSE	Composite Score
k-nearest neighbor			
n-neighbors = 10	0.728	1.030	1.454
n-neighbors = 20	0.643	1.180	1.329
n-neighbors = 30	0.528	1.670	1.084
n-neighbors = 40	0.550	1.884	1.049
n-neighbors = 50	0.582	2.002	1.050
Support vector machine			
sigma = 1	0.675	1.130	1.374
sigma = 3	0.696	1.145	1.391
sigma = 5	0.720	1.035	1.445
sigma = 7	0.596	1.215	1.275
sigma = 9	0.590	1.245	1.260
Random forest			
n-estimators = 5	0.904	0.510	1.768
n-estimators = 15	0.892	0.545	1.748
n-estimators = 25	0.865	0.605	1.705
n-estimators = 35	0.856	0.680	1.676
n-estimators = 45	0.848	0.760	1.646
Artificial neural network			
model 1 (10 neurons)	0.941	0.286	1.864
model 2 (20 neurons)	0.912	0.315	1.829
model 3 (30 neurons)	0.898	0.456	1.778
model 4 (40 neurons)	0.890	0.410	1.800
model 5 (50 neurons)	0.862	0.472	1.737
Multivariate regression analysis	0.664	3.760	0.664

Figure 2 shows the variation in the $RMSE$ for different models with their hyperparameters. Figure 2a illustrates the variation in the $RMSE$ with the number of neighbours for the *k*-NN model. As the number of neighbours decreases, the $RMSE$ also decreases, with the optimum performance found at $k = 10$ (lowest $RMSE$). For the RF model, increasing the number of trees results in a higher $RMSE$. The optimum model was found at 10 trees (lowest $RMSE$), as shown in Figure 2b. In Figure 2c, the $RMSE$ decreases as the number of neurons decreases, leading to an optimum ANN model with 10 neurons (lowest $RMSE$). For the SVM model, smaller sigma values lead to an increased $RMSE$, which is pattern consistent with larger sigma values as well. Medium sigma values result in a lower $RMSE$, with the optimum model being the SVM polynomial kernel, with $\sigma = 5$.

Figure 3a–e depicts the correlation between the predicted and measured ground vibrations. The solid line represents a 1:1 slope, and it can be observed that the data points for the ANN model cluster closely around this line, indicating superior prediction capability compared to other models. Furthermore, a comparison of the predicted and measured ground vibration for each algorithm is shown in Figure 4. The graph illustrates that the predictions generated by the ANN closely align with the measured values, indicating the ANN's proficiency in predicting these parameters compared to other models.

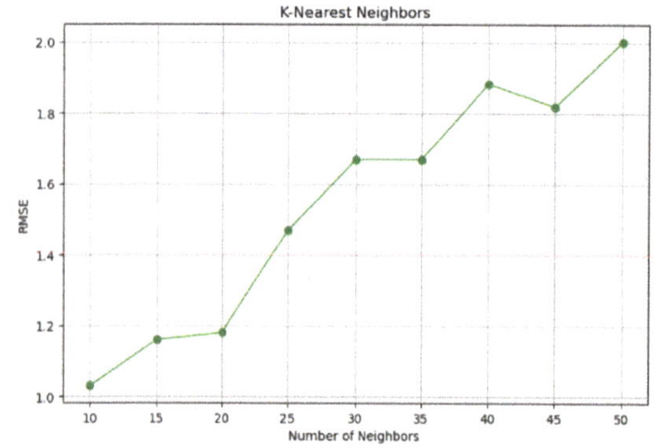

(**a**) *RMSE* for KNN model with different k.

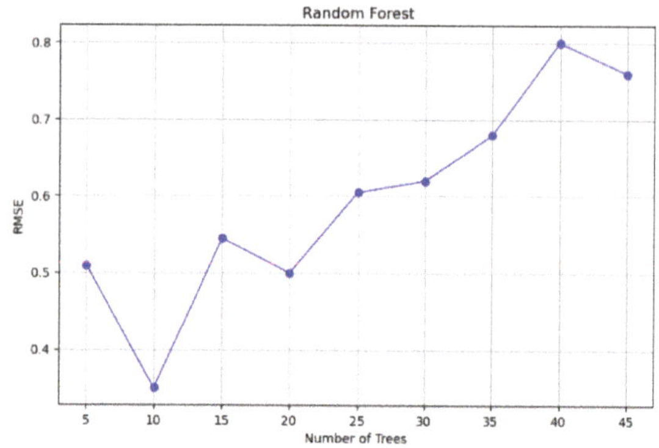

(**b**) *RMSE* for RF model with different trees.

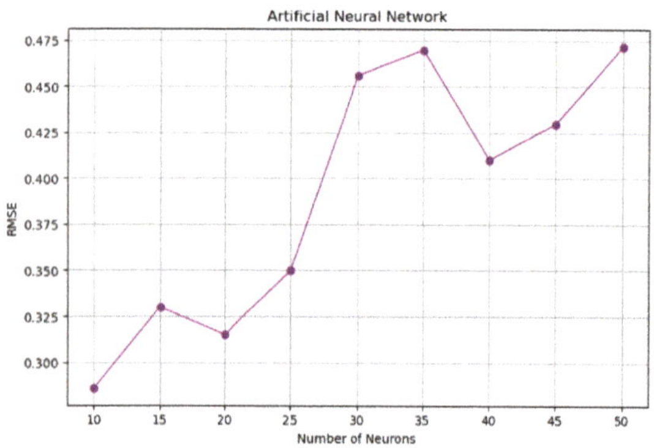

(**c**) *RMSE* for ANN model with different neurons.

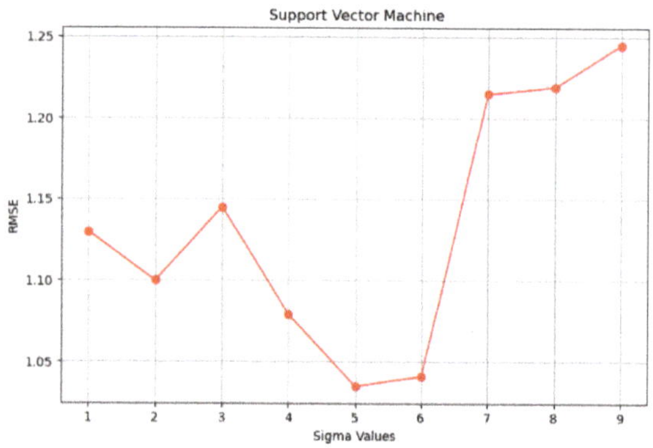

(**d**) *RMSE* for SVM model with different sigma values.

Figure 2. *RMSE* values for the following different models: (**a**) KNN, (**b**) RF, (**c**) ANN, and (**d**) SVM.

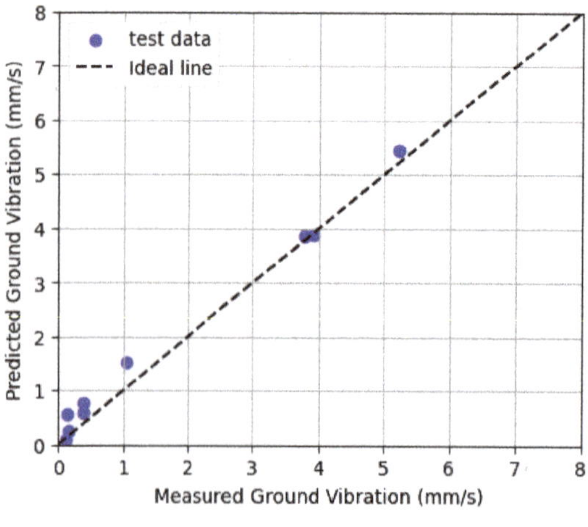

(**a**)

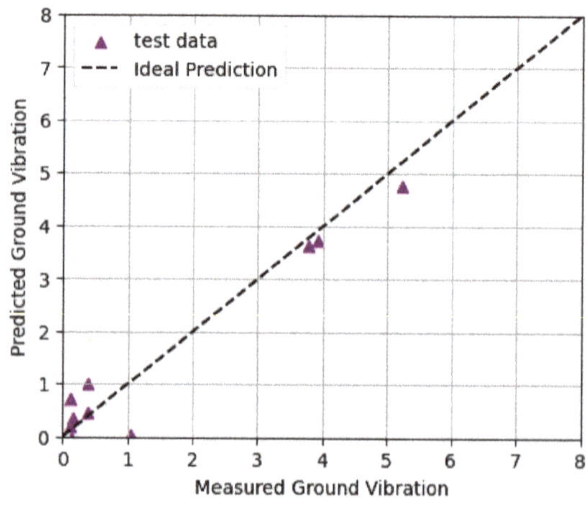

(**b**)

Figure 3. *Cont.*

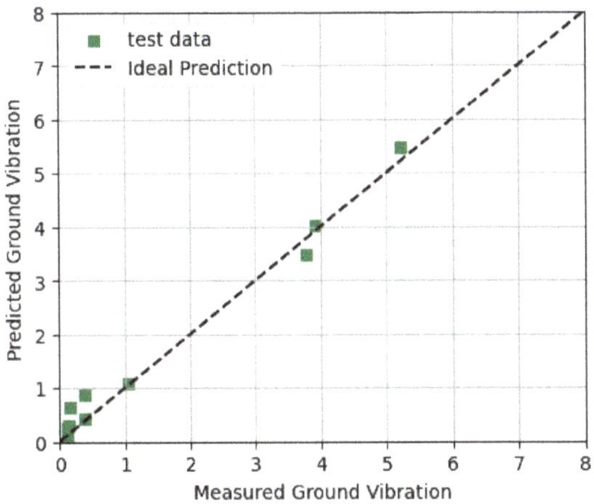

(c)

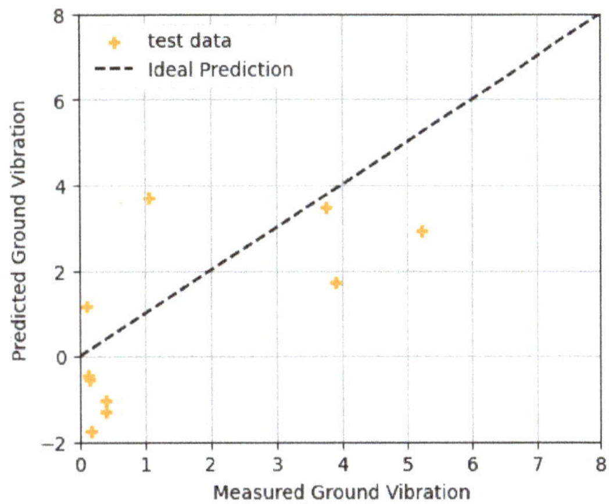

(d)

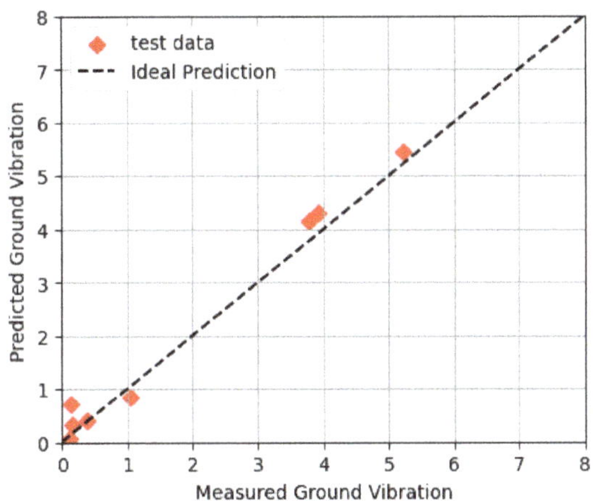

(e)

Figure 3. Scatter plots for all the models. (**a**) Comparison between measured and predicted GV for ANN model 1. (**b**) Comparison between measured and predicted GV for the SVM model. (**c**) Comparison between measured and predicted GV for the k-NN model. (**d**) Comparison between measured and predicted GV for the MVRA model. (**e**) Comparison between measured and predicted GV for the RF model.

The ANN's ability to capture more complex, non-linear, and high-dimensional relationships arises from its multi-layer architecture with several neurons per layer, enabling it to approximate intricate functions within high-dimensional solution spaces. This is compared to RF and SVM, which also handle non-linearity but rely on ensemble averaging or kernel tricks, respectively. During training, the ANN continuously adjusts multiple weights, which can reach up to thousands or even millions, depending on the complexity of the solution. This is where the computational power of ANN lies; the number of adjustable weights is parallel to the degrees of freedom of the designed solution. Our chosen ANN architecture (8-10-1) demonstrated the best generalisation on unseen data, achieving an R^2 of 0.941, due to its capacity to form flexible decision boundaries and learn hierarchical feature representations that align closely with the blast-induced ground vibration patterns. The RF, while also capable of handling non-linearities through ensemble learning, is slightly less effective than ANN in capturing very complex interactions. The *k*-NN, which relies on the proximity of data points, performs adequately but struggles with high-dimensional data. The SVM, though powerful in high-dimensional spaces and capable of non-linear modelling using kernel functions, often requires the meticulous tuning of hyper-parameters to perform optimally. Lastly, the MVRA, being a traditional statistical approach, is limited by its linear assumption and cannot adequately capture the non-linear relationships present in the data, resulting in the poorest performance among the models.

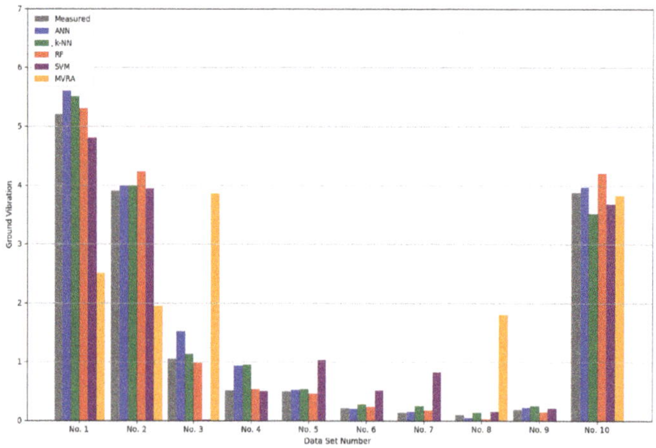

Figure 4. Comparison between measured and predicted GV for all algorithms.

The best-performing model in Table 3 is the PSO-ANN by Lawal et al. [13], which achieved an R^2 of 1.00 using 56 datasets with four inputs, namely, charge per delay (Q), number of holes (N), DI, and rock mass rating (RMR). This model's superior performance can be attributed to the optimisation capabilities of the PSO technique combined with the ANN, which likely captures non-linearities more effectively than other models. In contrast, the lowest-performing model is the ANN by Dindarloo [30], with an R^2 of 0.81 using only 15 datasets and five inputs (D, L, N, B, and S). Its limited dataset size and potentially less complex model architecture might have restricted the model's ability to generalise and capture variability in the data, leading to lower predictive accuracy. In the middle range, the HKM-ANN by Nguyen et al. [17] achieved an R^2 of 0.98 using 185 datasets with six inputs (S, PF, B, DI, and maximum charge per delay (MC)). This model combines hierarchical k-means clustering with an ANN, enabling it to handle larger and more diverse datasets effectively. Its performance is close to the best-performing models, indicating that hybrid techniques can significantly enhance predictive capabilities.

Comparing these with our results, our ANN model achieved an R^2 of 0.941 with 100 datasets and eight inputs. It performs well and aligns closely with other high-performing models, such as the ANN by Armaghani et al. [31], which achieved an R^2 of 0.97 with 109 blasting events. This suggests that our model may attain comparable accuracy with minor adjustments, such as including more input parameters and increasing the dataset size. However, hybrid models that use advanced techniques and larger datasets, including PSO-ANN (R^2 = 1.0, Lawal et al. [13]) and HKM-ANN (R^2 = 0.98, Nguyen et al. [17]), perform slightly better than our model. These results imply that the accuracy and generalisability of our model can be improved by investigating hybrid models and adding more inputs, allowing it to capture more complex interactions and improve generalisability, as seen in the best-performing models.

Table 3 highlights the significant impact of dataset size, the number of input parameters, and the type of model used on the performance of the predictive models. Models using hybrid techniques and larger datasets tend to perform better, as they can capture complex patterns and interactions within the data more effectively.

Table 3. Performance indices from other studies.

Best Model	Other Models	Inputs	Dataset	R^2	Reference
ANFIS		L, B, S, T, Q, DI	25	0.89	[25]
RF	CART, CHAID,	BS, DI, T	102	0.94	[15]
	ANN, SVM, RF	MC, PF, L			
HKM-ANN	ANN, SVR, FCM-ANN	S, Pf, B	185	0.98	[17]
	HKM-SVR, FCM-SVR	DI, MC			
ANFIS-GA	ANFIS, ANFIS-PSO	B, S, T	86	0.98	[18]
	USBM, Indian Standard	Pf, MC, DI			
GEP		BS, L, T, Pf, Q, DI	102	0.88	[22]
SVM	USBM, MLR, PSO power, PSO linear	DI, MC	80	0.96	[32]
WOA-XGBoost	GWO-XGBoost, BO-XGBoost,	D, L, B, S, MC	150	0.97	[16]
	CatBoost, RF, GBR	CL, DI, BI			
		E, PR, Pv, VOD, DOE			
ANFIS	ANN	DI, MC	109	0.97	[31]
PSO-ANN	ANN, DA-ANN	Q, Nh, DI, RMR	56	1.00	[13]
ANN		MC, DI, TC	20	0.93	[33]
ANN	MVRA, Indian Standard	L, MC	174	0.99	[34]
	Langefors-Kihlstrom, USBM	B, S			
	General Predictor, Ambraseys-Hendron	DI, BI, E, PR, Pv			
GRNN	USBM, CMRI, Indian Standard	MC, DI	14	0.99	[35]
	Langefors-Kihlstrom, Ambraseys-Hendron				
ANN	USBM, Indian Standard, MVRA	MC, DI, BS, L, T, D, Pf	180	0.99	[36]
	Langefors-Kihlstrom, Ambraseys-Hendron				
FIS	USBM, Indian Standard	B, S	120	0.94	[37]
	Langefors-Kihlstrom, Ambraseys-Hendron	T, N			
	CMRI, Ghosh-Daemen 1, Ghosh-Daemen 2	MC, DI			
	MVRA, General Predictor				
ANN	GEP	D, L, N ,B, S	15	0.81	[30]
		RDI, HDI, T, Q			
SVM	USBM, Indian Standard, MVRA	DI, MC	174	0.96	[38]
	Langefors-Kihlstrom, Ambraseys-Hendron				
	Ghosh-Daemen, CMRI, General Predictor				
FS-RF	FS-BN, Langefors-Kihlstrom	MC, DI, B, D, TC, S	102	0.90	[5]
	Ghosh-Daemen, Roy, Indian Standard	L, T, Sd, N, Pf, Q			
ANN-KNN	ANN, USBM	MC, DI	75	0.88	[23]
RVR-GWO	BA-GWO	MC, BS, D, T	95	0.84	[28]

USBM, United States Bureau of Mines; MLR, Multi-Linear Regression; GBR, Gradient Boosting Regression; GRNN, Generalised Regression Neural Network; GEP, Gene Expression Programming; CMRI, Central Mining Research Institute; FIS, Fuzzy Inference System; FS, Feature Selection; RVR, Relevance Vector Regression; GWO, Grey Wolf Optimisation; BA, Bat-inspired Algorithm; BN, Bayesian Network; BS, Burden to Spacing Ratio; Q, Charge per Delay; Sd, Subdrill; TC, Total Charge; DI, Distance; N, Number of holes; MC, Maximum Charge per Delay; BI, Blastability Index; E, Young's Modulus; PR, Poisson's Ratio; VOD, Velocity of Detonation; DOE, Density of Explosives; Pv, P-wave Velocity.

3.1. Sensitivity Analysis

Sensitivity analysis was used to determine the relative influence of the input parameters on the output using the cosine amplitude method [39]. The strength of the relation between the dataset x_i and xj is given by the following:

$$r_{ij} = \frac{\sum_{k=1}^{m} X_{ik} X_{jk}}{\sqrt{\sum_{k=1}^{m} X_{ik}^2 \sum_{k=1}^{m} X_{jk}^2}} \tag{6}$$

where x_i and xj are the inputs and outputs, respectively, and m is the dataset number.

In Figure 5, it can be inferred that spacing is the most influential input parameter on ground vibration. It is noted that the hole diameter is the least influential parameter on ground vibration. This observation is consistent with findings from other studies, which have also demonstrated that spacing exerts a greater influence on ground vibration than hole diameter [5,16,36]. The high sensitivity of spacing suggests that even minor adjustments can lead to significant changes in vibration levels, providing a powerful lever for engineers to control blast outcomes. On the other hand, the relatively low sensitivity of hole diameter indicates that it can be considered a secondary factor, allowing engineers to focus their optimisation efforts on more influential parameters.

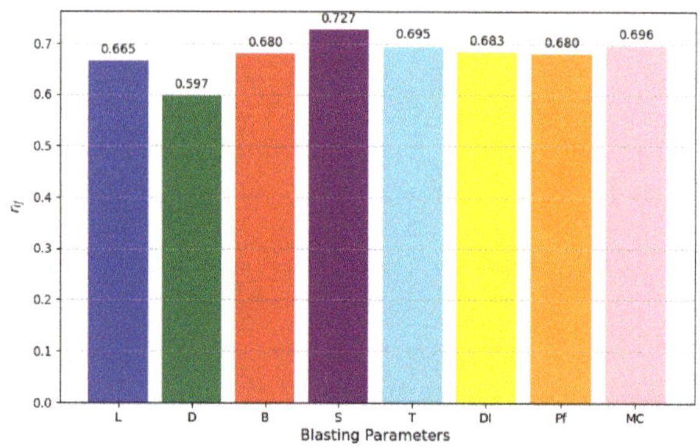

Figure 5. Sensitivity analysis for ground vibration.

The second most influential parameter is maximum charge per delay, followed by stemming, burden, and powder factor. The first four most influential parameters have the greatest effect compared to the rest of the parameters, and therefore, they must be carefully considered in the blast design when ground vibration is a desired output to be optimised. Furthermore, within this group itself, the order of parameter influence must be considered to effectively achieve the desired ground vibration optimisation. The same method of sensitivity consideration towards the influence of ground vibration output can be applied to the four least influential parameters. The sensitivity analysis approach in this section is validated in the sensitivity analysis of the solution space in Section 3.2 "Analysis of the Optimisation Results".

3.2. Analysis of the Optimisation Results

Figure 6a through Figure 7a display the results of the optimisation procedure, starting with seven initial points (depicted as red dots) chosen at random and converging towards the optimised value (illustrated as a blue dot), which is the lowest point in the surface. The optimised blast design parameters are as follows: powder factor (0.58), maximum charge per delay (454.09), burden (8.22), spacing (8.30), stemming (7.16), distance from blast point to monitoring point (2399), hole depth (18.25), and diameter (183). Therefore, the lowest point on the solution surface corresponds to the minimum value of ground vibration predicted by the model (approximately 0.1 mm/s). Every point in the solution surface has nine components, i.e., eight components of the eight input parameters plus the output component of the ground vibration.

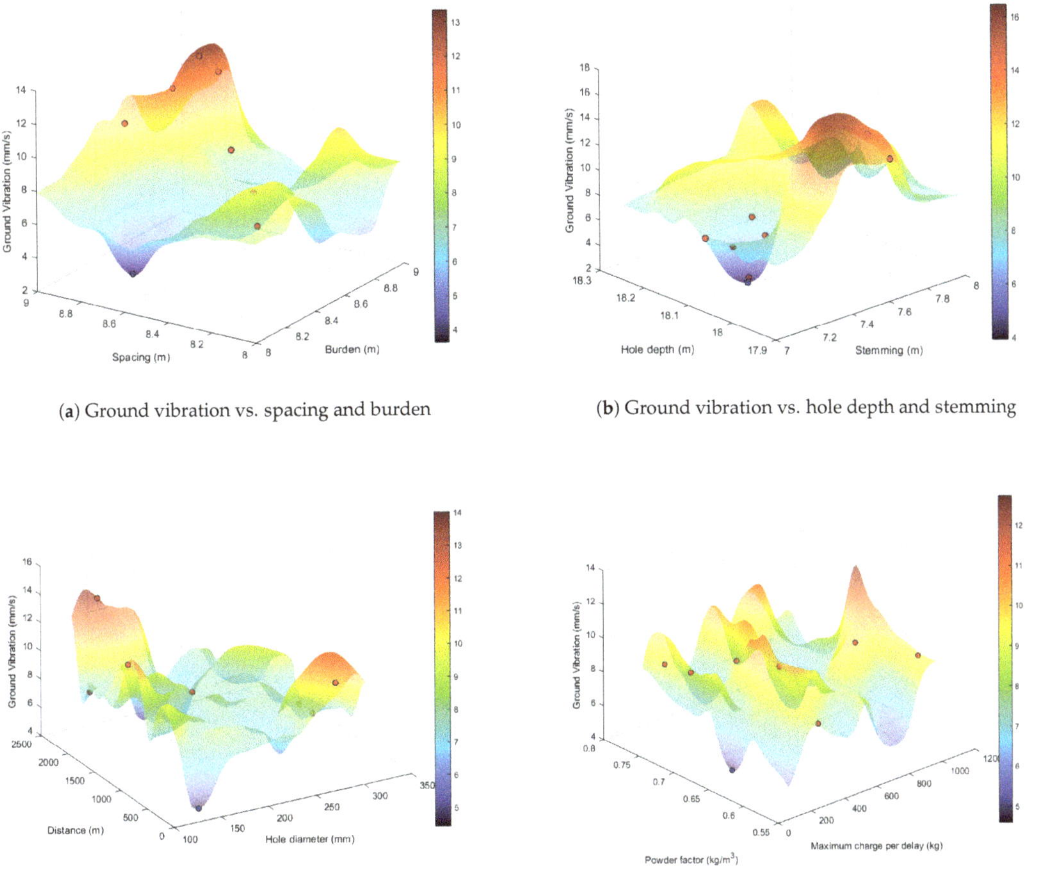

(**a**) Ground vibration vs. spacing and burden

(**b**) Ground vibration vs. hole depth and stemming

(**c**) Ground vibration vs. distance and hole diameter

(**d**) Ground vibration vs. powder factor, charge per delay

Figure 6. Solution space for ground vibration as the output parameter with eight input parameters, taken two at a time. The red dots are random initial values that always converge to the optimal point, while the blue dots represent the values for every minimisation computation.

The solution surface generated from the optimal ANN model (with 34 neurons in the first hidden layer, 15 in the second layer, and 50 in the third hidden layer) using gradient descent was used to predict ground vibration. The R^2 and $RMSE$ values are 0.96 and 0.21, respectively, showing high accuracy and low error. Table 4 shows the values for the measured and predicted ground vibration from the optimal ANN model.

From the sensitivity analysis, spacing is the most influential parameter. Because of this, we compare spacing with all other seven input parameters kept constant to observe the variations in the values of the output parameter in a two-dimensional plot. We assign spacing to be the x-axis, while ground vibration is assigned as the y-axis.

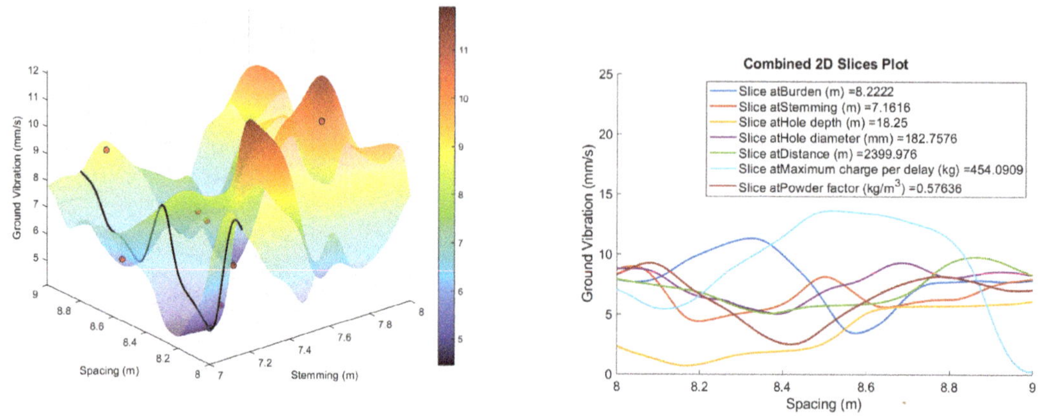

(**a**) Ground vibration vs. spacing and stemming (**b**) 2D plot of ground vibration

Figure 7. Ground vibration versus spacing, the most influential input parameter, presented in a 2D plot. The values of the other seven input parameters at the optimal point are kept constant, while spacing is plotted in its entire range. Each curve plotted in subfigure (**b**) is one of the seven input parameters. Subfigure (**a**) shows a 3D plot where stemming, the second most influential parameter, is kept constant in the traced curve that includes the blue dot which is the minimum point. The red dots are random initial points converging to the blue dot at every optimisation computation.

Table 4. Measured and predicted ground vibration values using the optimal ANN model.

Measured Ground Vibration	Predicted Ground Vibration
0.312	0.480
1.990	1.880
3.195	3.077
3.551	3.576
0.145	0.180
0.388	0.361
0.511	0.491
3.915	4.076
5.226	5.063

Figure 7a shows that as spacing decreases with stemming length fixed at 7.16, ground vibration initially fluctuates and finally decreases towards an optimal spacing value for minimal vibration, after which it increases again. The solution space also confirms the results of the sensitivity analysis. As can be seen in Figure 7b, generated by varying spacing, fixing all other parameters, and taking a slice through the lowest, optimised point in the solution space, the maximum charge per delay, stemming, burden, and powder factor are the most influential input parameters after spacing, and they have the most fluctuations in Figure 7b. This indicates high sensitivity, as changes in these parameters significantly impact ground vibration. Hole diameter is the least influential parameter, as showing the least variations suggests lower sensitivity.

The significance of these results lies in their practical application for blasting engineers seeking to minimise the environmental and structural impacts of ground vibrations during blasting operations. By providing a set of optimised parameters, the model enables engineers to design blasts that adhere to safety regulations and reduce the risk of damage to nearby structures and communities. Reducing ground vibrations not only mitigates the risk of structural damage but also enhances the efficiency of the blasting process. Minimising ground vibration may lead to a greater part of the total energy in blasting to be concentrated more towards its main goal, that is, rock fragmentation.

4. Conclusions

This study successfully demonstrates the application of machine learning techniques to predict and optimise ground vibrations induced by blasting operations at the Debswana Orapa diamond open-pit mine. Among the evaluated models—ANN, k-NN, SVM, RF, and the MVRA—the ANN model emerged as the most effective, evidenced by its highest predictive accuracy, with an R^2 value of 0.941 and a minimal $RMSE$ of 0.286. Using the ANN algorithm, we optimise the architecture of the ANN model using the Monte Carlo method, resulting in an architecture consisting of three hidden layers with 34, 15, and 50 neurons, respectively. We then optimise the resulting ANN model using the gradient descent method to find the minimum ground vibration, which is 0.1 mm/s. The solution space is therefore capable of predicting ground vibration given the input parameters, as well as setting the desired output value of ground vibration and looking for the corresponding values of the input parameters. Thus, the solution space demonstrates its efficacy in minimising, predicting, and designing ground vibrations. Sensitivity analysis confirmed the results provided by the solution space. Although the findings of this study are specific to the Orapa Diamond Mine, the methodologies employed are designed to be scalable and adaptable to other mining operations. To model other mining sites, we will use the results of this study as the baseline model. Then, we will incorporate site-specific data to correct the existing model and iteratively validate the results until the error becomes minimal. When this is achieved, the newly modified model would apply to the specific mining site.

Author Contributions: Conceptualization, K.G., O.S. and R.S.J.J.; methodology, O.S. and R.S.J.J.; software, O.S.; validation, O.S. and R.S.J.J.; formal analysis, O.S. and R.S.J.J.; writing—original draft preparation, O.S.; writing—review and editing, O.S., R.S.S., R.S.J.J. and O.M.; project; supervision, R.S.S., R.S.J.J. and O.M.; project administration, R.S.J.J.; funding acquisition, R.S.J.J. All authors have read and agreed to the published version of the manuscript.

Funding: This research was funded by Debswana Diamond Company with grant number P00064. The APC was funded by Debswana.

Data Availability Statement: The datasets analysed during this study are not available to the public because of a non-disclosure agreement signed with Debswana Mining Company (Gaborone, Botswana).

Acknowledgments: The authors would like to thank Orapa Diamond Mine for supplying the data for this research. Special appreciation is extended to Thuso Mogotsi and Chite Joe Matenga for their valuable contributions to this work.

Conflicts of Interest: The authors declare no conflicts of interest.

References

1. Cheng, G.; Huang, S. Analysis of ground vibration caused by open pit production blast. In *Explosive and Blasting Technique*; A.A. Balkema: Rotterdam, The Netherlands, 2000; pp. 63–70.
2. Ak, H.; Konuk, A. The effect of discontinuity frequency on ground vibrations produced from bench blasting: A case study. *Soil Dyn. Earthq. Eng.* **2008**, *28*, 686–694. [CrossRef]
3. Nateghi, R.; Kiany, M.; Gholipouri, O. Control negative effects of blasting waves on concrete of the structures by analyzing of parameters of ground vibration. *Tunn. Undergr. Space Technol.* **2009**, *24*, 608–616. [CrossRef]
4. Dumakor-Dupey, N.K.; Arya, S.; Jha, A. Advances in blast-induced impact prediction—A review of machine learning applications. *Minerals* **2021**, *11*, 601. [CrossRef]
5. Zhou, J.; Asteris, P.G.; Armaghani, D.J.; Pham, B.T. Prediction of ground vibration induced by blasting operations through the use of the Bayesian Network and random forest models. *Soil Dyn. Earthq. Eng.* **2020**, *139*, 106390. [CrossRef]
6. Ambraseys, N.; Hendron, A. *Dynamic Behavior of Rock Masses, Rock Mechanics in Engineering Practice*; John Wiley: London, UK, 1968.

7. Hustrulid, W.A. *Blasting Principles for Open Pit Mining, Volume 1: General Design Concepts*; A.A. Balkema: Rotterdam, The Netherlands, 1999.
8. Langefors, U.; Kihlström, B. *The Modern Technique of Rock Blasting*; John Wiley & Sons: Hoboken, NJ, USA, 1963.
9. Lawal, A.I.; Kwon, S. Application of artificial intelligence to rock mechanics: An overview. *J. Rock Mech. Geotech. Eng.* **2021**, *13*, 248–266. [CrossRef]
10. Mitchell, T.M. Does machine learning really work? *AI Mag.* **1997**, *18*, 11.
11. Samuel, A.L. Some studies in machine learning using the game of checkers. *IBM J. Res. Dev.* **2000**, *44*, 206–226. [CrossRef]
12. Hu, J.; Niu, H.; Carrasco, J.; Lennox, B.; Arvin, F. Voronoi-based multi-robot autonomous exploration in unknown environments via deep reinforcement learning. *IEEE Trans. Veh. Technol.* **2020**, *69*, 14413–14423. [CrossRef]
13. Lawal, A.I.; Kwon, S.; Kim, G.Y. Prediction of an environmental impact of tunnel blasting using ordinary artificial neural network, particle swarm and Dragonfly optimized artificial neural networks. *Appl. Acoust.* **2021**, *181*, 108122. [CrossRef]
14. Lawal, A.I.; Kwon, S.; Hammed, O.S.; Idris, M.A. Blast-induced ground vibration prediction in granite quarries: An application of gene expression programming, ANFIS, and sine cosine algorithm optimized ANN. *Int. J. Min. Sci. Technol.* **2021**, *31*, 265–277. [CrossRef]
15. Zhang, H.; Zhou, J.; Jahed Armaghani, D.; Tahir, M.; Pham, B.T.; Huynh, V.V. A combination of feature selection and random forest techniques to solve a problem related to blast-induced ground vibration. *Appl. Sci.* **2020**, *10*, 869. [CrossRef]
16. Qiu, Y.; Zhou, J.; Khandelwal, M.; Yang, H.; Yang, P.; Li, C. Performance evaluation of hybrid WOA-XGBoost, GWO-XGBoost and BO-XGBoost models to predict blast-induced ground vibration. *Eng. Comput.* **2021**, *38*, 4145–4162. [CrossRef]
17. Nguyen, H.; Drebenstedt, C.; Bui, X.N.; Bui, D.T. Prediction of blast-induced ground vibration in an open-pit mine by a novel hybrid model based on clustering and artificial neural network. *Nat. Resour. Res.* **2020**, *29*, 691–709. [CrossRef]
18. Yang, H.; Hasanipanah, M.; Tahir, M.; Bui, D.T. Intelligent prediction of blasting-induced ground vibration using ANFIS optimized by GA and PSO. *Nat. Resour. Res.* **2020**, *29*, 739–750. [CrossRef]
19. Lawal, A.I.; Olajuyi, S.I.; Kwon, S.; Onifade, M. A comparative application of the Buckingham π (pi) theorem, white-box ANN, gene expression programming, and multilinear regression approaches for blast-induced ground vibration prediction. *Arab. J. Geosci.* **2021**, *14*, 1073. [CrossRef]
20. Toraño, J.; Ramírez-Oyanguren, P.; Rodríguez, R.; Diego, I. Analysis of the environmental effects of ground vibrations produced by blasting in quarries. *Int. J. Min. Reclam. Environ.* **2006**, *20*, 249–266. [CrossRef]
21. Kaklis, K.; Saubi, O.; Jamisola, R.; Agioutantis, Z. Machine learning prediction of the load evolution in three-point bending tests of marble. *Min. Metall. Explor.* **2022**, *39*, 2037–2045. [CrossRef]
22. Faradonbeh, R.S.; Monjezi, M.; Armaghani, D.J. Genetic programing and non-linear multiple regression techniques to predict backbreak in blasting operation. *Eng. Comput.* **2016**, *32*, 123–133. [CrossRef]
23. Amiri, M.; Amnieh, H.B.; Hasanipanah, M.; Khanli, L.M. A new combination of artificial neural network and K-nearest neighbors models to predict blast-induced ground vibration and air-overpressure. *Eng. Comput.* **2016**, *32*, 631–644. [CrossRef]
24. Saubi, O.; Gaopale, K.; Jamisola, R.S.; Suglo, R.S.; Matsebe, O. Enhancing Blast Design Efficiency for Rock Fragmentation with Gradient Descent and Artificial Neural Networks: An Optimization Study. In Proceedings of the 2023 4th International Conference on Computers and Artificial Intelligence Technology (CAIT), Macau, China, 13–15 December 2023; pp. 1–5.
25. Xue, X.; Yang, X.; Li, P. Evaluation of ground vibration due to blasting using fuzzy logic. *Geotech. Geol. Eng.* **2017**, *35*, 1231–1237. [CrossRef]
26. Lawal, A.I.; Kwon, S.; Kim, G.Y. Prediction of the blast-induced ground vibration in tunnel blasting using ANN, moth-flame optimized ANN, and gene expression programming. *Acta Geophys.* **2021**, *69*, 161–174. [CrossRef]
27. Zhou, J.; Zhang, Y.; Qiu, Y. State-of-the-art review of machine learning and optimization algorithms applications in environmental effects of blasting. *Artif. Intell. Rev.* **2024**, *57*, 5. [CrossRef]
28. Fattahi, H.; Hasanipanah, M. Prediction of blast-induced ground vibration in a mine using relevance vector regression optimized by metaheuristic algorithms. *Nat. Resour. Res.* **2021**, *30*, 1849–1863. [CrossRef]
29. Fitch, F.B. Warren S. McCulloch and Walter Pitts. A logical calculus of the ideas immanent in nervous activity. Bulletin of mathematical biophysics, vol. 5 (1943), pp. 115–133. *J. Symb. Log.* **1944**, *9*, 49–50. [CrossRef]
30. Dindarloo, S.R. Prediction of blast-induced ground vibrations via genetic programming. *Int. J. Min. Sci. Technol.* **2015**, *25*, 1011–1015. [CrossRef]
31. Armaghani, D.J.; Hajihassani, M.; Monjezi, M.; Mohamad, E.T.; Marto, A.; Moghaddam, M.R. Application of two intelligent systems in predicting environmental impacts of quarry blasting. *Arab. J. Geosci.* **2015**, *8*, 9647–9665. [CrossRef]
32. Hasanipanah, M.; Naderi, R.; Kashir, J.; Noorani, S.A.; Zeynali Aaq Qaleh, A. Prediction of blast-produced ground vibration using particle swarm optimization. *Eng. Comput.* **2017**, *33*, 173–179. [CrossRef]
33. Monjezi, M.; Hasanipanah, M.; Khandelwal, M. Evaluation and prediction of blast-induced ground vibration at Shur River Dam, Iran, by artificial neural network. *Neural Comput. Appl.* **2013**, *22*, 1637–1643. [CrossRef]
34. Khandelwal, M.; Singh, T. Prediction of blast-induced ground vibration using artificial neural network. *Int. J. Rock Mech. Min. Sci.* **2009**, *46*, 1214–1222. [CrossRef]

35. Ragam, P.; Nimaje, D. Assessment of blast-induced ground vibration using different predictor approaches-a comparison. *Chem. Eng. Trans.* **2018**, *66*, 487–492.
36. Tiile, R.N. *Artificial Neural Network Approach to Predict Blast-Induced Ground Vibration, Airblast and Rock Fragmentation*; Scholars' Mine, Missouri S&T Library and Learning Resources; Missouri University of Science and Technology: Rolla, MO, USA, 2016.
37. Ghasemi, E.; Ataei, M.; Hashemolhosseini, H. Development of a fuzzy model for predicting ground vibration caused by rock blasting in surface mining. *J. Vib. Control* **2013**, *19*, 755–770. [CrossRef]
38. Khandelwal, M. Evaluation and prediction of blast-induced ground vibration using support vector machine. *Int. J. Rock Mech. Min. Sci.* **2010**, *47*, 509–516. [CrossRef]
39. Yang, Y.; Zhang, Q. A hierarchical analysis for rock engineering using artificial neural networks. *Rock Mech. Rock Eng.* **1997**, *30*, 207–222. [CrossRef]

mining

MDPI

Article

Enhancing Comminution Process Modeling in Mineral Processing: A Conjoint Analysis Approach for Implementing Neural Networks with Limited Data

Carlos Moraga [1,*], CésarA. Astudillo [2], Rodrigo Estay [3] and Alicia Maranek [1]

[1] Department of Mining Engineering, Faculty of Engineering, Universidad de Talca, Curicó 3340000, Chile; amaranek18@alumnos.utalca.cl
[2] Department of Computer Science, Faculty of Engineering, Universidad de Talca, Curicó 3340000, Chile; castudillo@utalca.cl
[3] Mining Engineering Department, Faculty of Engineering, University of Santiago of Chile (USACH), Santiago 9170022, Chile; rodrigo.estay.h@usach.cl
* Correspondence: carmoraga@utalca.cl

Abstract: Mineral processing is a crucial stage in the mining process, involving comminution and concentration stages. Comminution is modeled using various ore variables and operational parameters, representing a complex system. An alternative to simplifying the complexity of these stages is adopting machine learning (ML) techniques; however, ML often requires a substantial amount of data for effective training and validation. The conjoint analysis methodology was used to develop a procedure for discretizing input variables and reducing the data needed for training neural networks, requiring only 77 different scenarios. Using the results from a comminution plant simulator built in Matlab Simulink, neural networks were trained to predict the key output parameters, such as the water consumption, energy consumption, operational parameters, and particle size generated by the plant. The predictive capability of the neural networks was excellent, achieving $R^2 > 0.99$ in all cases. The networks were tested with a new set of scenarios to assess their response to values not categorized in the discretization process, achieving $R^2 > 0.98$. However, the prediction capability was lost for out-of-range input variables. This approach is attractive for developing easy-to-implement ML tools capable of representing complex systems without needing large amounts of input data, thereby simplifying the modeling process in mineral processing.

Keywords: mineral processing; comminution; machine learning; neural networks; conjoint analysis; simulation; water; energy

1. Introduction

Mineral processing involves numerous variables and parameters, making the modeling of these stages highly complex. Alternatively, machine learning (ML) offers a way to simplify the modeling and simulation processes; however, a large amount of data is typically required to train and validate these tools. By using categorization techniques, it may be possible to reduce the amount of data needed to build effective ML tools. In this work, the traditional conjoint analysis method was adapted to discretize variables that influence comminution processes by categorizing them into different levels and generating an orthogonal design to obtain a series of scenarios. These scenarios were then used to simulate the operation of a comminution plant, characterizing its response, and the results were used to train artificial neural networks that can represent the plant's response in a simplified manner. Finally, the possibility of adapting such techniques to simplify the characterization and analysis of complex systems, such as mineral-processing stages, is discussed.

1.1. Mineral Processing

Mineral processing is the process of separating commercially valuable minerals. It involves several steps designed to physically or chemically separate the minerals from the waste rock or gangue [1]. The main stages in mineral processing include a particle size reduction to liberate the mineral and concentration processes like flotation to produce a final product enriched with the mineral of interest. Achieving an adequate final concentrate requires rigorous control of the particle size reduction or comminution stages [2]. Comminution processes consist of crushing and grinding stages, where size classifiers, such as screens and hydrocyclones, are used to control the particle size sent to subsequent stages [3]. Currently, the most commonly used technologies in crushing processes include jaw crushers, gyratory crushers, and cone crushers, among others. Similarly, there are various technologies and equipment for the grinding stages, such as conventional ball mills, SAG mills, and high-pressure grinding rolls, among others [4]. These pieces of equipment are selected based on the type of ore being processed and the requirements of the designed mineral-processing plant.

Typically, crushing is performed in the absence of water, while grinding requires the use of water to make the process more efficient, which, among other advantages, helps reduce the energy consumed by the process [5–7]. This is extremely important, considering that the comminution stages are highly energy-intensive [8,9]. Water and energy are critical inputs for carrying out the comminution stages, and therefore, the proper and rational use of these resources is essential for the sustainable development of mining operations [10,11]. In this regard, it is crucial to have tools that accurately estimate the consumption of these resources based on the treated ore and the operational conditions of the comminution stages.

Currently, there are validated and widely accepted models that represent comminution and particle size classification operations. Comminution models are primarily based on population balances, while classification models are empirical [12–15]. These models serve as the foundation for simulation tools implemented in various software platforms [16–19]. In a previous work [20], mineral-processing models were implemented in Matlab's Simulink, resulting in a robust simulator capable of capturing the full behavior of a plant, generating predictions of the particle size distributions, the residence time distributions, the concentrate and tailing composition, a blending estimation, and the water and energy consumption, among other outputs. Additionally, using this simulator, it was possible to develop operational strategies to reduce the water consumption by integrating meteorological information into the mining planning process [21]. The versatility of this tool allows for the accurate and reliable representation of any mineral-processing plant configuration, making it a valid instrument for representing a real-world operation.

1.2. Machine Learning Applications in the Mining Industry

Machine learning (ML) is increasingly being used in mineral processing to enhance the predictability, optimize the performance, and strengthen maintenance practices [22]. This approach enables the creation of material 'fingerprints', which help forecast the processing behavior and identify the root causes of variability [23]. In flotation processes, ML techniques are now central to monitoring and optimizing operations, with deep learning networks gaining traction for froth image analyses [24]. Additionally, ML methods are valuable in processing remote sensing data for mineral exploration, facilitating geological mapping and the development of mineral prospectivity maps [25]. Nonetheless, significant challenges remain, such as data scarcity, characterizing rare events, and modeling complex processes [22]. Overcoming these barriers calls for innovative learning approaches, the incorporation of phenomenological models, the development of reliable sensors, and the training of skilled personnel [22]. Despite these hurdles, ML continues to show substantial potential for enhancing the efficiency in mineral processing and success in exploration.

Machine learning and neural networks have emerged as powerful tools for addressing various challenges in mining, particularly those related to predictive modeling, process optimization, and managing the vast amounts of data generated in mining operations. In essence, ML refers to a subset of artificial intelligence (AI) that allows computers to learn from data and make predictions or decisions without being explicitly programmed. Neural networks, a key ML technique, are inspired by the human brain's structure and excel at recognizing patterns in complex, non-linear data. These capabilities make ML and neural networks particularly suited for the mining industry, where processes such as ore grade estimations, equipment failure predictions, and energy consumption optimization require the analysis of large, intricate datasets [26].

In a study conducted by Avalos et al. [27], an approach based on long short-term memory (LSTM) networks was proposed to predict the operational relative hardness in semi-autogenous grinding (SAG) mills using real-time operational data. The study emphasized the ability of LSTM to capture temporal relationships by leveraging variables such as the energy consumption and feed tonnage. The model achieved an accuracy of over 93% in predicting the operational hardness. This model significantly enhances the efficiency of comminution process control, which is crucial for mining operations that depend on optimizing energy consumption and operational efficiency [27].

Posteriorly, Avalos and Ortiz [28] introduced a novel technique for simulating complex geostatistical patterns using recursive convolutional neural networks (RCNNs). Their method leverages CNNs' ability to extract features from training images and applies them recursively to simulate two-dimensional and three-dimensional geological structures. This approach demonstrates a superior accuracy and flexibility when compared to traditional multiple-point statistics (MPS) methods, particularly in handling spatial complexity. Their study also emphasized the potential of deep learning to improve geological simulations, offering a promising direction for future applications in geosciences.

In another study, Lee et al. [29] explored the use of artificial neural networks (ANNs) to predict the particle size distribution (PSD) of grinding products. Their approach employed a kinetic model to generate training data, which improved the model's accuracy. After fine-tuning the model with experimental data, the coefficient of determination reached 0.99, indicating highly precise predictions of the particle size distribution. This method offers an efficient and accurate alternative to traditional grinding techniques, which are typically more time-consuming and costly [29].

The study presented by Kwanghui [30] applied an ensemble ML approach to predict the efficiency of multi-stage froth flotation for complex lead–zinc ores. The authors used both neural networks and random forest models to analyze flotation data, achieving a high accuracy in predicting the grade and yield across different flotation stages. The model successfully outperforms traditional methods, offering significant insights for optimizing the flotation processes in mining operations.

Amoako et al. [31] developed a hybrid ML model combining artificial neural networks and support vector regression (SVR) to enhance the prediction accuracy of rock fragmentation in blasting operations. By leveraging variables such as the explosive properties, geometric parameters, and rock characteristics, the model outperformed traditional methods like the Kuz–Ram model. This hybrid approach achieved a higher predictive accuracy with a significant improvement in the R^2 value and a notable reduction in the mean square error (MSE), demonstrating its potential in optimizing the blasting design efficiency.

The study by Al-Bakri and Sazid [32] explored the application of artificial neural networks in predicting and optimizing blast-induced impacts, such as flyrock, ground vibrations, and back-break, in mining operations. The authors highlighted the inefficiency of traditional blasting, where only 30% of the explosive energy is used for rock fragmentation, while the rest leads to environmental and safety concerns. ANN models show superior predictive capabilities, optimizing blast operations by reducing undesirable effects and improving the overall efficiency, despite the need for more research on multi-impact predictions.

In another topic related to mining operations and safety, Vallejos and McKinnon [33] used neural networks and logistic regression to classify microseismic events as blasting operations based on the seismic source parameters. Using training data, they achieved an accuracy of over 95% in classifying seismic records as blasts. On the same topic, Vallejos and Estay [34] employed spatio-temporal clustering methodologies to identify and monitor areas with a higher microseismic risk and identify seismic sequences.

On a broader perspective, Jung and Choi [35] examined ML applications in mining, focusing on mineral exploration, exploitation, and reclamation. By analyzing 109 papers from 2011–2020, their study found that ML applications have increased significantly since 2018, with support vector machines and deep learning models being the most utilized. The key evaluation metrics include the root mean square error and the coefficient of determination. Their review highlights the potential of ML for improving the operational efficiency in the mining industry.

A more recent systematic review [26] provides a comprehensive overview of how deep learning (DL) techniques are applied across different stages of the mining process, including exploration, extraction, and reclamation. It emphasizes that, despite the slower adoption of deep learning in mining compared to other industries, recent advancements have shown great potential. The review categorizes existing studies based on DL architectures, such as convolutional neural networks (CNNs) and recurrent neural networks (RNNs), highlighting their use in geotechnical mapping, fault detection, and ore classification. The authors identified significant gaps in the availability of large, publicly accessible datasets and the need for more diverse sensor data in mining. Recommendations for future research include an improved generalization of models, adversarial testing, and the broader integration of DL in mining automation.

1.3. Conjoint Analysis

A conjoint analysis is a research tool used to assess the market potential, predict the market share, and forecast the sales of new or improved products and services [36–38]. The process involves estimating utilities for varying levels of product features and simulating marketplace preferences for established, improved, and/or new products. A conjoint analysis is widely used in marketing research for new product development, pricing, branding, market simulations, and market segmentation [36,38,39]. It has evolved from traditional rating- or ranking-based methods to a choice-based conjoint analysis, making it a versatile tool for studying discrete choice behavior [36].

There are different types of conjoint analysis models, including the following:

Traditional conjoint analysis: Widely used for measuring consumer preferences for product or service features, but limited by integrating a small number of attributes [40].

Choice-based conjoint analysis: A general experimental method to study individual's discrete choice behavior, not limited to classical applications in marketing, and can be applied to study research questions from related disciplines [36].

Hierarchical information integration: A variation of a conjoint analysis that resolves the limitations arising in a traditional conjoint analysis and establishes ways to proceed in complex situations with many attributes [41].

However, a conjoint analysis has its limitations, such as complexity, as a traditional conjoint analysis is limited to handling a small number of attributes, while complex decision-making processes often involve between 10 and 20 relevant criteria [40]. Additionally, experiments with a conjoint analysis can face issues with information overload, which may impact the validity of the results [41]. Furthermore, the predictive power of conjoint applications can be compromised by response biases and measurement errors, potentially affecting the accuracy of the models [42].

A conjoint analysis can be represented using an orthogonal design, which generates an experimental plan that captures the possible combinations of categorized variables. An orthogonal design refers to a type of experimental design where the factors being studied are independent and uncorrelated, which is crucial in various fields such as statistics, coding theory, and computer experiments [43–45]. The orthogonal design ensures that the factors being studied are independent and uncorrelated. Various methods and algorithms have been proposed for constructing orthogonal designs, which are included as part of commercial software.

This research addresses a critical limitation in the application of ML to mineral processing: the substantial data requirements that often render ML methods impractical for small and medium-sized operations. By leveraging a conjoint analysis to discretize and categorize input variables, this study presents a novel approach that requires a minimal amount of data for effective neural network training. This methodology allows for the accurate modeling of complex comminution processes while overcoming the traditional data and resource barriers associated with ML applications in mineral processing.

2. Materials and Methods

2.1. Models Used in Simulation

The methodology implemented to obtain results was based on the use of a comminution plant simulator. The plant consisted of two crushing stages with screening classification and a conventional grinding stage with hydrocyclone classification, as shown in Figure 1. The first crushing stage operated in an open circuit with a gyratory crusher (feed opening of 1000 mm), while the second stage utilized a cone gyratory crusher (feed opening of 350 mm) paired with a screen for size classification at 51 mm. Lastly, the setup included a ball mill (measuring 3.5 × 4.5 m) followed by a size classification stage consisting of five hydrocyclones, each with a diameter of 76 cm.

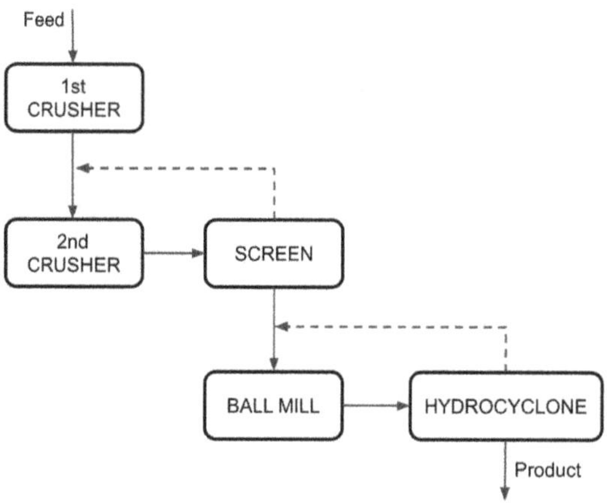

Figure 1. Diagram of the comminution plant used for the simulation process.

It was assumed that the feed to the plant followed a Rosin–Rammler distribution, where the complete size distribution can be constructed using the following expression [46]:

$$Y = 1 - exp\left(-\frac{x}{x_0}\right)^n \tag{1}$$

where Y is the cumulative fraction of material by weight less than size x; n is the uniformity constant; and x_0 is the characteristic particle size. In this way, the parameters n and x_0 become input variables for the simulation process.

The energy consumed in the comminution processes was calculated using the Bond equation, which relates the hardness of the ore to the size reduction capacity of the equipment through the following expression [47]:

$$P_C = 10 \cdot G_S \cdot W_i \cdot \left(\frac{1}{\sqrt{P_{80}}} - \frac{1}{\sqrt{F_{80}}}\right) \tag{2}$$

where P_C is the power consumption expressed in [kW]; G_S is the fresh material flow in [t/h]; W_i is the work index in [kWh/t]; and F_{80} and P_{80} are the feed and product characteristic sizes in [μm].

The models implemented in the simulation are presented in Table 1. These models are based on population balances for comminution processes and empirical models for size classification. The detailed formulation of the models is described in a previous work [20]. These models are widely accepted and used to simulate operations in mineral processing, being part of the commercially available simulators currently in use. The simulation was carried out in Matlab Simulink R2024a due to its advantage of easily connecting stages, making it possible to implement complex systems with material recirculation.

Table 1. Models used in the comminution plant simulation.

Model/Stage	Crushing	Ball Grinding	References
Fragmentation	Population Balance	Population Balance	[12,13,48–50]
Classification	Karra	Plitt	[14,15]

Once the simulator was built, it was possible to manipulate the input variables and parameters to capture the plant's response. The input variables and the operational results of interest provided by the simulator are indicated in Table 2. The simulator generated a series of other operational results that allowed the plant's performance to be characterized; however, in this work, we used those presented in Table 2 for simplicity to exemplify the adjustment of the methodology.

Table 2. Input variables for the simulation and output variables provided by the simulation.

Input Variables	Characteristic particle size X_0	Uniformity constant	Work index	Feed flow	Ore density	1st crusher setting	2nd crusher setting	Mill speed	Solid concentration by weight	Hydrocyclone feed pressure
Output Variables	1st crusher P_{80} size	1st crusher energy consumption	2nd crusher P_{80} size	2nd crusher circulating load	2nd crusher energy consumption	Mill P_{80} size	Mill energy consumption	Circuit P_{80} size	Mill circulating load	Makeup water

2.2. Conjoint Analysis Implementation

It was necessary to explore the response space for the various output variables provided by this tool to implement neural network tuning based on the mineral-processing simulator. Due to the considerable number of identified input variables, characterizing the simulator's response can be very resource-intensive. The conjoint analysis technique was used to reduce the number of scenarios needed to characterize the simulator's response.

The selection of input parameters for the neural network model was grounded in their critical influence on the mineral-processing and comminution stages within the mining industry. The key parameters, including the feed granulometry, the slope of the granulometric curve, the mineral hardness, the feed flow rate, the mineral density, the primary and secondary crusher discharge settings, the mill rotation speed, the solids concentration, and the feed pressure, were chosen due to their direct impact on system performance. These variables were identified through a combination of industry relevance and simulated process scenarios, generating a comprehensive dataset of representative scenarios. This approach ensured that the model captured a broad range of operational conditions, enhancing its predictive accuracy and applicability across diverse processing settings.

The operational range of the input variables and operational parameters was defined based on typical data reported in the literature [47,51–55]. Then, a discretization of these variables was defined, generating 5 categories for each of them. This was a crucial step because the variables are continuous, and the conjoint analysis method requires discrete categories. The ranges and discretization of the variables are presented in Table 3.

Table 3. Categorization of operational ranges for input variables in the comminution plant simulation.

Variables	Very Low	Low	Medium	High	Very High
Particle size distribution X_0 [mm]	750	925	1100	1275	1450
Particle size distribution n	0.8	0.9	1.0	1.1	1.2
Work index [kWh/t]	9	11	13	15	17
Feed flow [t/h]	1000	1100	1200	1300	1400
Ore density [t/m^3]	2.0	2.25	2.5	2.75	3
1st crusher setting [cm]	5	10	15	20	25
2nd crusher setting [cm]	1	2	3	4	5
Percent of critical mill speed [%]	70	72.5	75	77.5	80
Solids concentration by weight [%]	57	59	61	63	65
Hydrocyclone feed pressure [kPa]	50	87.5	125	162.5	200

Since 10 variables were defined with 5 categories each, the possible combinations amounted to more than 9.7 million. Using the orthogonal design tool in the SPSS software (version IBM SPSS 27.0) [56], a reduced number of scenarios was generated, ensuring the representation of each category of the input variables and capturing their variation within the defined ranges. In this way, a sample of 77 scenarios from the total possible combinations was established, provided as Supplementary Data to this work in Table S1.

2.3. *Network Architecture and Training Justification*

The neural network structure used in this study consisted of a single hidden layer with 10 neurons, along with an output layer. This architecture was selected to optimize the computational efficiency and enable faster model training and evaluation—important factors for industrial applications, especially in the mining sector. Prior research shows that a single hidden layer is often sufficient to capture complex, non-linear relationships in predictive tasks within similar fields. Al-Bakri et al. [32] reviewed the recent studies on blast-induced rock fragmentation using ANNs, noting that many solutions successfully employ a single hidden layer. For example, Dimitraki et al. [57] utilized 5 neurons in the hidden layer and achieved a high predictive accuracy (R^2 = 0.88), while Rosales-Huamani et al. [58] used 13 neurons, reaching an accuracy of R^2 = 0.87. These findings underscore that single-layer networks can effectively capture complex data patterns with minimal depth, thereby reducing the risk of overfitting and computational costs.

To determine the most effective training approach, three algorithms were tested: Levenberg–Marquardt, Bayesian regularization, and scaled conjugate gradient. The dataset was split into 70% for training, 15% for validation, and 15% for testing. Bayesian regularization, however, inherently optimized without requiring a separate validation set, making it particularly suitable for achieving strong generalization while preventing overfitting. Among the tested methods, Bayesian regularization showed the lowest average error and a superior generalization performance, proving to be robust across diverse operational scenarios, while the other methods exhibited higher cross-validation error rates, especially under extreme operational conditions.

In conclusion, the combination of a single hidden layer and Bayesian regularization training offered an optimal balance between the predictive accuracy and the computational efficiency. This architecture and training strategy ensured that the model is highly applicable in real-world, time-sensitive mining operations, providing reliable, efficient predictions in a challenging industrial context.

3. Results

In this study, continuous variables were discretized into five levels (very low, low, medium, high, and very high) to simplify the input space and optimize computational resources while preserving essential distinctions between operational states. This choice allowed the model to maintain its predictive accuracy by capturing the most relevant aspects of each variable's influence on the process outcomes without overwhelming the model with excessive detail. Selecting an appropriate number of discretization levels is crucial: too few levels creates the risk of oversimplifying the input data and omitting important nuances, while overly granular levels could lead to increased computational demands without substantial gains in accuracy.

For practitioners, we recommend aligning the granularity of discretization with the significance and variability of each variable. Variables that have a substantial impact on process outcomes may benefit from finer discretization, whereas others can be grouped into broader categories. This approach helps ensure a balance between the computational efficiency and the model fidelity.

3.1. Neural Network Fitting

The neural network training process is exemplified in Figure 2, where the fit of the training, validation, test, and overall fit for a neural network predicting the water consumption is shown. It can be observed that the neural network fit well with the data generated by the simulation, an effect that was replicated for each of the remaining defined output variables.

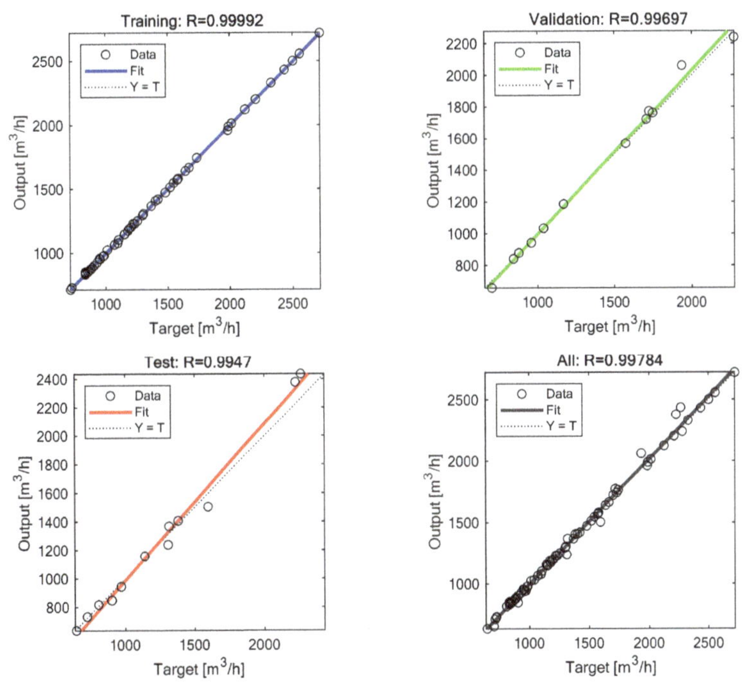

Figure 2. Training performance of the neural network generated for makeup water characterization.

To evaluate the fit quality of the neural networks, the behavior of six output variables representing the operational efficiency of the proposed comminution circuit was analyzed. These variables included the makeup water, energy consumption, circulating load, and characteristic particle sizes (P_{80}) resulting from each comminution stage. Figure 3 presents the neural network predictions contrasted with the values obtained from the simulations for the output variables. The fit measured by the coefficient of determination (R^2) provided very high values for each of the variables, demonstrating a high predictive capability of the neural networks. These results allow for the precise estimation of the particle size evolution generated in each of the comminution stages, as well as the consumption of inputs such as water and energy. Additionally, the circulating load generated in the grinding circuit was reported as an operational control parameter, verifying that the neural network can characterize this behavior.

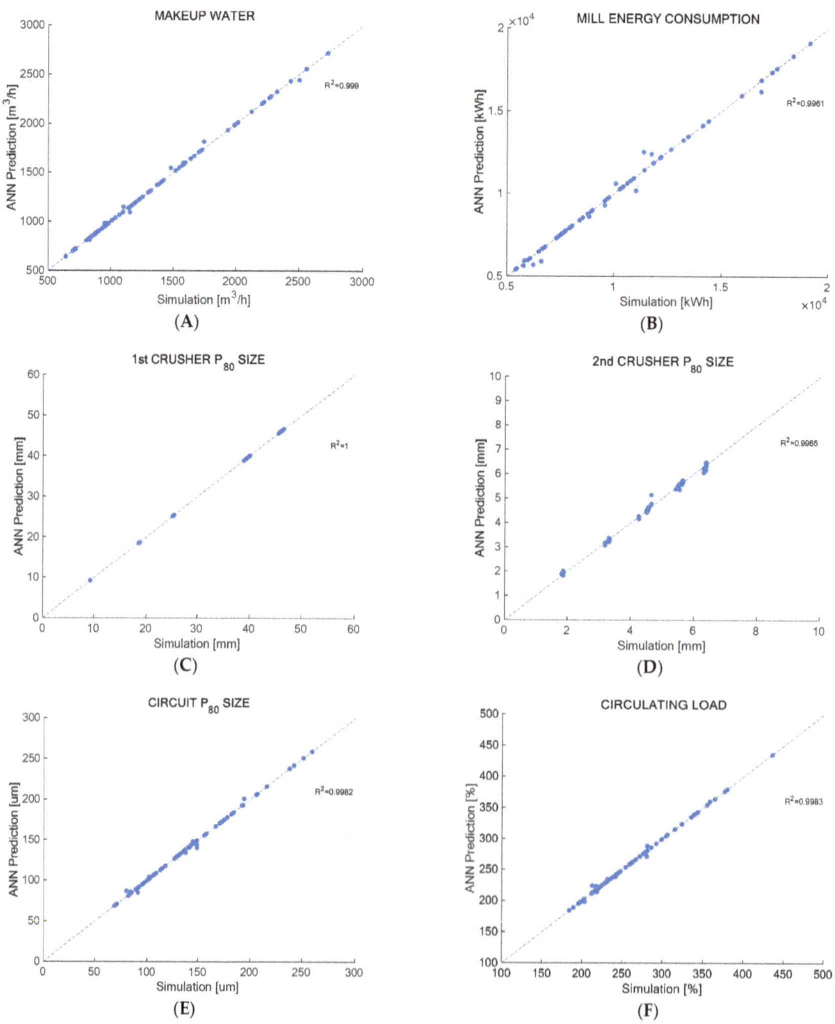

Figure 3. Neural network predictions vs. simulation responses for output variables: **(A)** makeup water; **(B)** mill energy consumption; **(C)** 1st crusher P_{80} size; **(D)** 2nd crusher P_{80} size; **(E)** circuit's P_{80} size; and **(F)** circulating load.

3.2. Predictions for Uncategorized Input Variable Values

To verify the effectiveness of the created neural networks, a new random set of scenarios was generated, where the input variables took values that were not categorized as part of the conjoint analysis. This was implemented because, in practice, the input variables are continuous and can take any value within the defined validity ranges. The new scenarios are provided as Supplementary Data to this work in Table S2. This additional test set allowed us to examine the model's ability to generalize beyond the training data, focusing on both the accuracy and robustness. The neural networks, trained using the Bayesian regularization method, achieved a high precision, with determination coefficients (R^2) consistently exceeding 0.99 and a minimal mean squared error (MSE) across all the key output variables. These results underscore the model's effectiveness in replicating the system's behavior under new operational scenarios, confirming its reliability and strong predictive capacity within the predefined operational boundaries.

The simulator and neural networks were executed using these values, and the results are presented in Figure 4. To simplify the analysis, the particle size behavior was exemplified through the P_{80} generated by the entire circuit, understanding that the results for the crushing stages are analogous. The results show that the neural networks correctly predicted the system's behavior represented by the simulator, even when tested with values that were not categorized for training. This demonstrates that it is possible to successfully train neural networks to capture the behavior of a comminution plant by using a variable discretization process that reduces the amount of input information and decreases the number of scenarios needed for their construction.

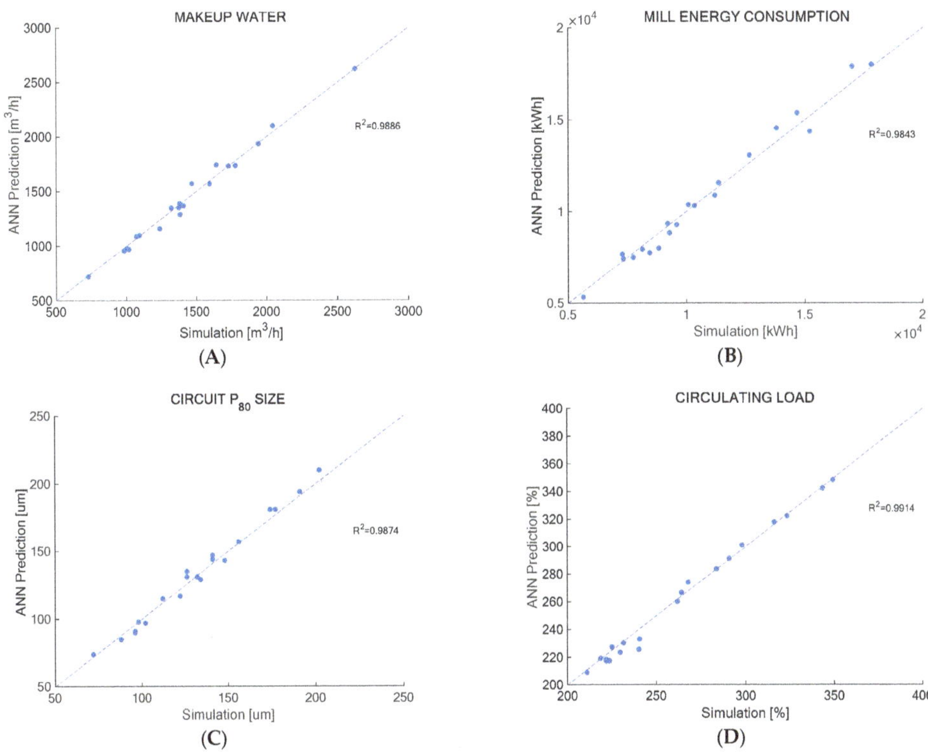

Figure 4. Neural network predictions vs. simulation responses for uncategorized input variable values: (**A**) makeup water; (**B**) mill energy consumption; (**C**) circuit's P_{80} size; and (**D**) circulating load.

3.3. Sensitivity Analysis for Out-of-Range Input Variable Values

The aim was to explore whether the neural networks can capture the comminution plant's behavior when the input variable values are outside the established training ranges. To achieve this, a sensitivity analysis of the neural networks was conducted using variables with values below the lower limit and above the upper limit of the established range. The sensitivity analysis was performed based on the work index and the ore density variables to exemplify the neural networks' response. The input values used for the sensitivity analysis are presented in Table 4. The sensitivity analysis results for the work index and the ore density are presented in Figures 5 and 6, respectively.

Table 4. Input variable values of work index and ore density for sensitivity analysis.

Low Work Index Values [kWh/t]	High Work Index Values [kWh/t]	Low Ore Density Values [t/m^3]	High Ore Density Values [t/m^3]
4	18	1.1	3.5
5	19	1.2	4
6	20	1.4	4.5
7	21	1.6	5
8	22	1.8	5.5

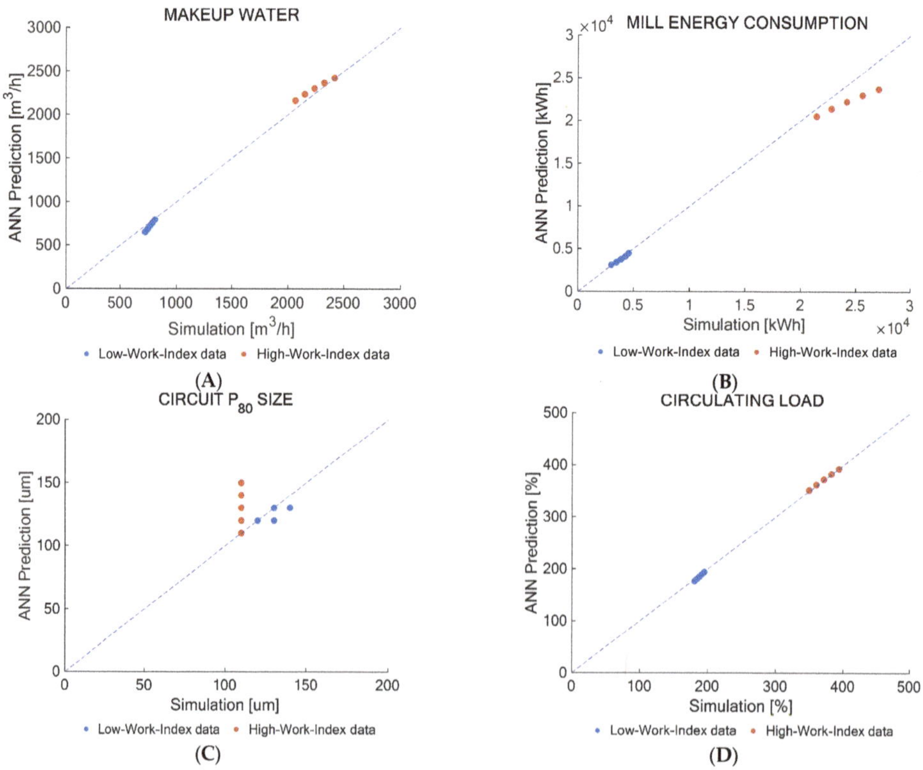

Figure 5. Results of the sensitivity analysis when varying the work index: (**A**) makeup water; (**B**) mill energy consumption; (**C**) circuit's P_{80} size; and (**D**) circulating load.

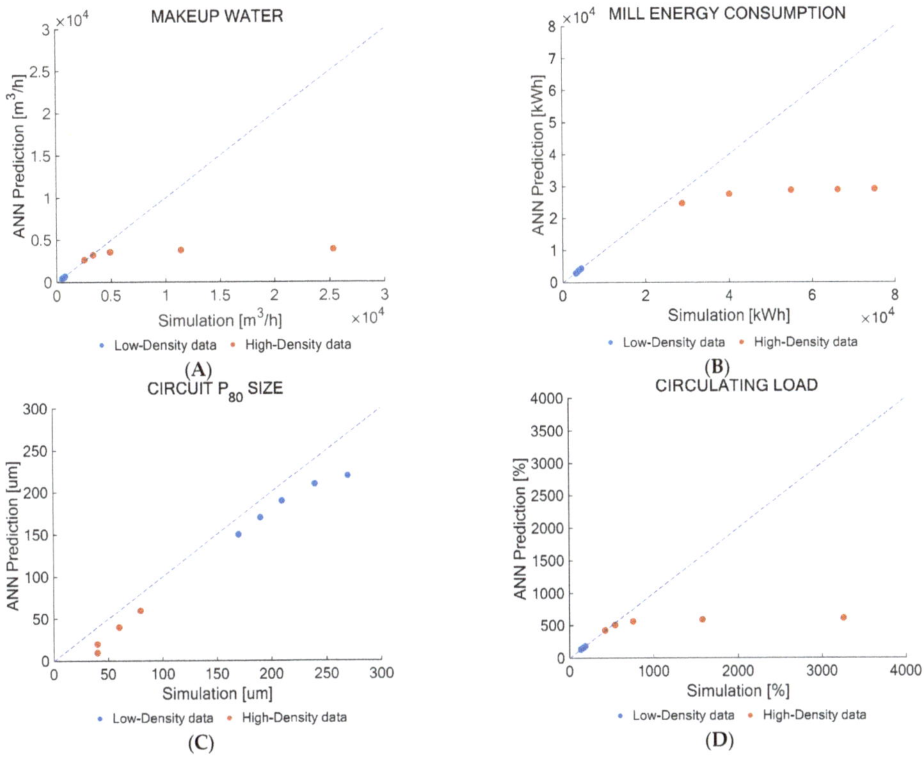

Figure 6. Results of the sensitivity analysis when varying the ore density: (**A**) makeup water; (**B**) mill energy consumption; (**C**) circuit's P_{80} size; and (**D**) circulating load.

The makeup water and circulating load predictions still aligned with the simulation values. However, in the case of the mill energy, the neural network failed to predict it accurately when tested with high work index values, underestimating the results obtained by the simulation. Additionally, the prediction of the particle size generated by the circuit was poor and not accurately represented by the neural network, showing no discernible trend toward fitting.

For the ore density variation, the neural networks preserved their predictive capability only for the output variables of the makeup water, mill energy, and circulating load when tested with low-density values. In cases of very high densities, the neural network's fit was completely lost. For the particle size, although a fitting trend was observed, the neural network underestimated the values obtained by the simulator, leading to the conclusion that an adequate fit was not achieved.

In general terms, it was observed that, when testing the neural networks with input values outside the established training ranges, the predictive capability was deficient and did not reflect the behavior of the comminution plant. Therefore, it was established that the validity of the neural networks is limited to the training ranges, and it is not advisable to extend their use to conditions beyond these limits.

4. Discussion

This study used neural networks as a first approach to address the modeling problem in the context of comminution processes. Given the complexity and novelty of this problem, the research team chose to explore the neural network approach to understand the system's behavior and its operational parameters. As this understanding progresses, future research could incorporate other machine learning techniques, such as random forest and support vector machines, to assess their accuracy and efficiency in comparison with this initial approach. It is noteworthy that this is the first work to address the specific modeling problem of comminution processes using a simulator and detailed parameters through machine learning methods, as no prior published results exist on this application.

These results show that it is possible to generate predictive neural networks for the behavior of a complex comminution plant. This process can be accomplished by adapting the conjoint analysis technique and discretizing and categorizing the input variables to generate an orthogonal experimental design. This approach reduces the number of scenarios needed to train the neural networks, allowing for a good fit between the input data (simulations) and the predictions made by the neural networks. This outcome is significant because it demonstrates that, with a small sample of the plant's response, it is possible to reliably model its behavior.

This methodology allows results to be obtained on the consumption of inputs such as the makeup water and the energy used in size-reduction processes. This information could be used to estimate the expected consumption for processing a given ore precisely and promptly by evaluating the neural networks with input values associated with a planned feed. Additionally, it is possible to predict the particle sizes generated at each stage of the plant, enabling the tracking of particle size evolution as the ore progresses through the plant, thereby providing information that can be used to improve or optimize operations.

In this regard, when applied to a real mineral-processing plant, it is estimated that generating or collecting an exhaustive amount of data on the plant's response is not necessary. Instead, the key is to adequately select data that represents the plant's variability. Thus, it is suggested that the quality of the data is more important than its quantity. This observation is significant for the evaluation of small to medium-sized plants, where financial resources may be a limiting factor for process characterization and the implementation of ML tools. Therefore, using the conjoint analysis method could be an efficient and easy-to-implement alternative for diagnostic, control, and operational improvement purposes.

As observed in the results, the generalization of the neural network outcomes was only valid within the range of the values established for the input variables. Based on this, it is recommended to create a robust geometallurgical characterization of all the variables associated with the ore to be processed, covering the entire range of values encountered in operation. This information can be incorporated into the block model representing the exploited deposit and used as temporal data to feed a processing plant. In this way, developing neural networks that are specifically valid for each mining deposit would be possible.

In this study, additional tests were conducted with input values outside the trained operational range to assess the neural network's generalization capabilities under extreme conditions. While the network demonstrated a high predictive accuracy within the trained range, its performance declined when tested with out-of-range values, leading to an underestimation of the mill energy and less precise particle size predictions. This outcome aligns with typical neural network behavior, as models trained within specific data ranges often lack reliability when extrapolated beyond those boundaries. Therefore, while the neural network model is highly accurate and dependable within the predefined operational parameters, its applicability under extreme conditions is limited, emphasizing the need to respect the defined training range for optimal performance.

This type of study based on a conjoint analysis can be easily extended to other stages of the mining process, such as extraction operations, processing, concentration, tailings disposal, etc. The tool used proved to be versatile and user-friendly for characterizing complex systems through variable discretization. Thus, it opens the possibility of representing complex mining systems through the interaction between stages of the production process, intending to provide a robust representation using ML tools.

Additionally, it is important to consider that simulation tools based on mineral-processing modeling often consume significant computational resources due to the large number of variables and parameters involved in the models. As a more comprehensive representation of the plant is required, more stages than those addressed in this work may be included, further complicating the simulation system and consuming more resources. In this context, ML offers an alternative for generating simple models that consume fewer computational resources, which could be significant for conducting quick and reliable analyses for control or process optimization purposes.

The proposed methodology is expected to yield neural network models capable of accurately predicting critical outputs, such as the water and energy consumption, the particle size distribution, and the circulating load, achieving a high coefficient of determination ($R^2 > 0.98$). By utilizing a conjoint analysis to reduce the data volume required, this study anticipates that this approach will enable the effective modeling of comminution processes, even with limited datasets. This data-efficient technique is particularly advantageous for small to medium-sized operations, where resources for extensive data collection may be limited. Moreover, while the neural networks are expected to perform well within their defined input ranges, this study acknowledges that the predictive capability may decrease when applied to scenarios beyond these training boundaries, highlighting the importance of adhering to the established range for reliable outcomes.

5. Conclusions

The conjoint analysis method was successfully adapted to discretize and categorize the variables involved in a complex comminution system, which includes crushing, grinding, and size classification stages. Using this technique, an orthogonal experimental design was developed and used to obtain the system's response, referencing a comminution plant simulator. With these data, predictive neural networks were trained for various output variables, including the most relevant ones: the makeup water, energy consumption, particle size, and circulating load.

The neural networks achieved a high level of fit, accurately predicting the values reported by the simulator. Scenarios with input variable values that were not categorized during training were tested to assess the predictive capability of the neural networks. The results successfully predicted the behavior of the comminution plant. Additionally, a sensitivity analysis using values outside the training range confirmed that the neural networks are only valid within the specified range of input variables.

The methodology used proved effective in representing complex processes. Based on this, a conjoint analysis can be projected as a tool to characterize any mining and metallurgical processes with minimal input data or limited information volume. In this context, it is considered feasible for use in small to medium-sized operations, which often have limited financial resources for conducting an exhaustive characterization of the mineral to be processed.

In future work, it is planned to continue exploring this technique to adapt it to other stages of the mining process, integrate them, and generate interactions between these stages. This approach could provide an integrated view of the entire mining value chain through a simple, predictive tool that is less demanding in terms of resource consumption.

Supplementary Materials: The following supporting information can be downloaded at: https://www.mdpi.com/article/10.3390/mining4040054/s1, Table S1: Set of scenarios for neural network training defined by the conjoint analysis technique; Table S2: Set of scenarios for uncategorized input variable values.

Author Contributions: Conceptualization, C.M. and C.A.A.; methodology, C.M. and A.M.; validation, A.M., C.M. and C.A.A.; formal analysis, C.M., C.A.A. and R.E.; investigation, A.M. and C.M.; data curation, C.M. and A.M.; writing—original draft preparation, C.M. and C.A.A.; writing—review and editing, C.A.A. and R.E. All authors have read and agreed to the published version of the manuscript.

Funding: This research received no external funding.

Data Availability Statement: The raw data supporting the conclusions of this article will be made available by the authors upon request.

Conflicts of Interest: The authors declare no conflicts of interest.

References

1. Pereira, L.; Schach, E.; Tolosana-Delgado, R.; Frenzel, M. All About Particles: Modelling Ore Behaviour in Mineral Processing. *Elements* **2023**, *19*, 359–364. [CrossRef]
2. Pokrajcic, Z. Designing crushing and grinding circuits for improved energy efficiency. In Proceedings of the TMS Annual Meeting, Seattle, WA, USA, 14–18 February 2010; p. 933.
3. Concha, A.F.; Bouso, A.J.L. Classification Fundamentals. In *Fluid Mechanics Fundamentals of Hydrocyclones and Its Applications in the Mining Industry*; Springer: Berlin, Germany, 2021; Volume 126. [CrossRef]
4. Erickson, M.T. Innovations in comminution equipment: Crushers, high pressure grinding rolls, semi-autogenous grinding, ball mills, and regrind mills. In *Mineral Processing and Extractive Metallurgy: 100 Years of Innovation*; Society for Mining, Metallurgy, and Exploration, Inc.: Englewood, CO, USA, 2014; pp. 65–75.
5. Bu, X.; Chen, Y.; Ma, G.; Sun, Y.; Ni, C.; Xie, G. Wet and dry grinding of coal in a laboratory-scale ball mill: Particle-size distributions. *Powder Technol.* **2020**, *359*, 305–313. [CrossRef]
6. Kotake, N.; Kawaguchi, T.; Koizumi, H.; Kanda, Y. A fundamental study of dry and wet grinding in bending tests on glass—Effect of repeated impact on fracture probability. *Miner Eng.* **2004**, *17*, 1281–1285. [CrossRef]
7. Chelgani, S.C.; Parian, M.; Parapari, P.S.; Ghorbani, Y.; Rosenkranz, J. A comparative study on the effects of dry and wet grinding on mineral flotation separation—A review. *J. Mater. Res. Technol.* **2019**, *8*, 5004–5011. [CrossRef]
8. Abelhaffez, G.S. Rationalization of energy consumption in the grinding of some ores by using additives. *Mater. Test.* **2017**, *59*, 395–401. [CrossRef]
9. Hacksteiner, M.; Peherstorfer, H.; Bleicher, F. Energy efficiency of state-of-the-art grinding processes. *Procedia Manuf.* **2018**, *21*, 717–724. [CrossRef]
10. Daniel, M.J. Energy Use in Comminution in a Global Context. In *Sustainability in the Mineral and Energy Sectors*; CRC Press: Boca Raton, FL, USA, 2016. [CrossRef]
11. Taghipour, A.; Stefanakis, A. Water Footprint of the business Sector and Circular Economy. In *Water in Circular Economy*; Springer: Berlin, Germany, 2023. [CrossRef]
12. Napier-Munn, T.J.; Morrell, S.; Morrison, R.D.; Kojovic, T. *Mineral Comminution Circuits, Their Operation and Optimisation*; Julius Kruttschnitt Mineral Research Centre: Indooroopilly, QLD, Australia, 1999; p. 413.
13. Whiten, W.J. A matrix theory of comminution machines. *Chem. Eng. Sci.* **1974**, *29*, 589–599. [CrossRef]
14. Karra, V.K. Development of a model for predicting the screening performance of a vibrating screen. *CIM Bull.* **1979**, *72*, 167–171.
15. Nageswararao, K.; Wiseman, D.M.; Napier-Munn, T.J. Two empirical hydrocyclone models revisited. *Miner. Eng.* **2004**, *17*, 671–687. [CrossRef]
16. Brochot, S.; Wiegel, R.L.; Ersayin, S.; Touze, S. Modeling and simulation of comminution circuits with USIM PAC. In *Advances in Comminution*; Kawatra, S.K., Ed.; SME Inc.: Littleton, CO, USA, 2006; pp. 495–511.
17. Silva, A.C.; Silva, E.M.S.; Silva, J.A.d.O. Ball Mill Simulation with Moly-Cop Tools. In Proceedings of the 2015—Sustainable Industrial Processing Summit, Antalaya, Turkey, 4–9 October 2015; Flogen Star Outreach: Montreal, QC, Canada, 2015; pp. 215–224.
18. Aleksandrova, T.N.; Chanturiya, A.V. Ore preparation process selection for ferruginous quartzites based on simulation modeling. *Obogashchenie Rud* **2023**, *2023*, 3–9. [CrossRef]
19. Aaltonen, A.; Izart, C.; Lyyra, M.; Lang, A.; Saari, E.; Dahl, O. Simulating the Impact of Ore and Water Quality on Flotation Recovery during the Life of a Mine. *Minerals* **2023**, *13*, 1230. [CrossRef]
20. Moraga, C.; Kracht, W.; Ortiz, J.M. Process simulation to determine blending and residence time distribution in mineral processing plants. *Miner. Eng.* **2022**, *187*, 107807. [CrossRef]
21. Moraga, C.; Kracht, W.; Ortiz, J.M. Water consumption assessment in mineral processing integrating weather information and geometallurgical modeling. *Miner. Eng.* **2023**, *201*, 108162. [CrossRef]
22. Estay, H.; Lois-Morales, P.; Montes-Atenas, G.; Ruiz del Solar, J. On the Challenges of Applying Machine Learning in Mineral Processing and Extractive Metallurgy. *Minerals* **2023**, *13*, 788. [CrossRef]
23. Van Duijvenbode, J.R.; Buxton, M.W.N.; Shishvan, M.S. Performance Improvements during Mineral Processing Using Material Fingerprints Derived from Machine Learning—A Conceptual Framework. *Minerals* **2020**, *10*, 366. [CrossRef]
24. Szmigiel, A.; Apel, D.B.; Skrzypkowski, K.; Wojtecki, L.; Pu, Y. Advancements in Machine Learning for Optimal Performance in Flotation Processes: A Review. *Minerals* **2024**, *14*, 331. [CrossRef]
25. Shirmard, H.; Farahbakhsh, E.; Müller, R.D.; Chandra, R. A review of machine learning in processing remote sensing data for mineral exploration. *Remote Sens. Environ.* **2022**, *268*, 112750. [CrossRef]
26. Azhari, F.; Sennersten, C.C.; Lindley, C.A.; Sellers, E. Deep learning implementations in mining applications: A compact critical review. *Artif. Intell. Rev.* **2023**, *56*, 14367–14402. [CrossRef]
27. Avalos, S.; Kracht, W.; Ortiz, J.M. An LSTM Approach for SAG Mill Operational Relative-Hardness Prediction. *Minerals* **2020**, *10*, 734. [CrossRef]

28. Avalos, S.; Ortiz, J.M. Recursive convolutional neural networks in a multiple-point statistics framework. *Comput. Geosci.* **2020**, *141*, 104522. [CrossRef]
29. Lee, D.; Je, J.; Kwon, J. Prediction of particle size distribution of grinding products using artificial neural network approach. *Miner. Eng.* **2024**, *216*, 108831. [CrossRef]
30. Jo, K.; Je, J.; Lee, D.; Cho, H.; Kim, K.; You, K. Prediction of multi-stage froth flotation efficiency of complex lead–zinc sulfide ore using an integrated ensemble neural network–random forest model. *Miner. Eng.* **2024**, *210*, 108669. [CrossRef]
31. Amoako, R.; Jha, A.; Zhong, S. Rock Fragmentation Prediction Using an Artificial Neural Network and Support Vector Regression Hybrid Approach. *Mining* **2022**, *2*, 233–247. [CrossRef]
32. Al-Bakri, A.Y.; Sazid, M. Application of Artificial Neural Network (ANN) for Prediction and Optimization of Blast-Induced Impacts. *Mining* **2021**, *1*, 315–334. [CrossRef]
33. Vallejos, J.A.; McKinnon, S.D. Logistic regression and neural network classification of seismic records. *Int. J. Rock Mech. Min. Sci.* **2013**, *62*, 86–95. [CrossRef]
34. Vallejos, J.A.; Estay, R.A. Seismic Parameters of Mining-Induced Aftershock Sequences for Re-entry Protocol Development. *Pure Appl. Geophys.* **2018**, *175*, 793–811. [CrossRef]
35. Jung, D.; Choi, Y. Systematic Review of Machine Learning Applications in Mining: Exploration, Exploitation, and Reclamation. *Minerals* **2021**, *11*, 148. [CrossRef]
36. Eggers, F.; Sattler, H.; Teichert, T.; Völckner, F. Choice-Based Conjoint Analysis. In *Handbook of Market Research*; Springer: Berlin, Germany, 2021. [CrossRef]
37. Whitlark, D.B.; Smith, S.M. Sales forecasting with conjoint analysis by addressing its key assumptions with sequential game theory and macro-flow modeling. In *Conjoint Measurement*; Springer: Berlin, Germany, 2007. [CrossRef]
38. Mukherjee, S.; Malakar, A.; Roy, A. Conjoint Analysis: A Heuristic Decision-Making Model of Product Selection. In *Advances on Mathematical Modeling and Optimization with Its Applications*; CRC Press: Boca Raton, FL, USA, 2024. [CrossRef]
39. Camilleri, L.; Azzopardi, L.M. Market segmentation through conjoint analysis using latent class models. In Proceedings of the ESM 2011—2011 European Simulation and Modelling Conference: Modelling and Simulation, Guimaraes, Portugal, 24–26 October 2011; pp. 112–116.
40. Herbst, U.; Voeth, M. An Empirical Comparison of Computer-Based Conjoint Variants in Multi-Attributive Purchase Decisions. In Proceedings of the 2008 Academy of Marketing Science (AMS) Annual Conference, Vancouver, BC, Canada, 29–31 May 2008; Springer: Berlin, Germany, 2015. [CrossRef]
41. Ramírez-Hurtado, J.M. Measuring preferences: From conjoint analysis to integrated conjoint experiments. *Rev. Metodos Cuantitativos Econ. Empresa* **2010**, *9*, 28–43.
42. Mishra, D.P.; Min, J.; Deniz Dalman, M. An empirical assessment of stimulus presentation mode bias in conjoint analysis. *Innov. Mark.* **2011**, *7*, 108–121.
43. Chu, X.R.; Chen, L.; Gao, S. Optimal Orthogonal Designs. 2015. Available online: https://api.semanticscholar.org/CorpusID:183841511 (accessed on 26 August 2024).
44. Karunanayaka, R.C.; Tang, B. On the existence and constructions of orthogonal designs. *Aust. N. Z. J. Stat.* **2018**, *60*, 471–480. [CrossRef]
45. Georgiou, S.; Koukouvinos, C.; Seberry, J. Generalized orthogonal designs. *Ars Comb.* **2004**, *71*, 33–47.
46. Vesilind, P.A. The Rosin–Rammler particle size distribution. *Resour. Recovery Conserv.* **1980**, *5*, 275–277. [CrossRef]
47. Wills, B.A.; Finch, J.A. Chapter 5—Comminution. In *Wills' Mineral Processing Technology*, 8th ed.; Wills, B.A., Finch, J.A., Eds.; Butterworth–Heinemann: Boston, MA, USA, 2016; pp. 109–122. [CrossRef]
48. King, R.P. *Modeling and Simulation of Mineral Processing Systems*; Elsevier: Amsterdam, The Netherlands, 2001.
49. Herbst, J.A.; Fuerstenau, D.W. Scale-up procedure for continuous grinding mill design using population balance models. *Int. J. Miner. Process.* **1980**, *7*, 1–31. [CrossRef]
50. Austin, L.G.; Luckie, P.T. The estimation of non-normalized breakage distribution parameters from batch grinding tests. *Powder Technol.* **1972**, *5*, 267–271. [CrossRef]
51. Wills, B.A.; Finch, J.A. Chapter 8—Industrial Screening. In *Wills' Mineral Processing Technology*, 8th ed.; Wills, B.A., Finch, J.A., Eds.; Butterworth–Heinemann: Boston, MA, USA, 2016; pp. 181–197. [CrossRef]
52. Wills, B.A.; Finch, J.A. Chapter 7—Grinding Mills. In *Wills' Mineral Processing Technology*, 8th ed.; Wills, B.A., Finch, J.A., Eds.; Butterworth–Heinemann: Boston, MA, USA, 2016; pp. 147–179. [CrossRef]
53. Wills, B.A.; Finch, J.A. Chapter 6—Crushers. In *Wills' Mineral Processing Technology*, 8th ed.; Wills, B.A., Finch, J.A., Eds.; Butterworth–Heinemann: Boston, MA, USA, 2016; pp. 123–146. [CrossRef]
54. Wills, B.A.; Finch, J.A. Chapter 9—Classification. In *Wills' Mineral Processing Technology*, 8th ed.; Wills, B.A., Finch, J.A., Eds.; Butterworth–Heinemann: Boston, MA, USA, 2016; pp. 199–221. [CrossRef]
55. Wills, B.A.; Finch, J.A. Chapter 17—Modeling and Characterization. In *Wills' Mineral Processing Technology*, 8th ed.; Wills, B.A., Finch, J.A., Eds.; Butterworth–Heinemann: Boston, MA, USA, 2016; pp. 449–462. [CrossRef]
56. IMB Documentation. Generating an Orthogonal Design. Available online: https://www.ibm.com/docs/en/spss-statistics/saas?topic=preference-generating-orthogonal-design (accessed on 26 August 2024).

57. Dimitraki, L.; Christaras, B.; Marinos, V.; Vlahavas, I.; Arampelos, N. Predicting the average size of blasted rocks in aggregate quarries using artificial neural networks. *Bull. Eng. Geol. Environ.* **2019**, *78*, 2717–2729. [CrossRef]
58. Rosales-Huamani, J.A.; Perez-Alvarado, R.S.; Rojas-Villanueva, U.; Castillo-Sequera, J.L. Design of a Predictive Model of Rock Breakage by Blasting Using Artificial Neural Networks. *Symmetry* **2020**, *12*, 1405. [CrossRef]

mining

MDPI

Article

Optimum Fleet Selection Using Machine Learning Algorithms—Case Study: Zenouz Kaolin Mine

Pouya Nobahar [1,*], Yashar Pourrahimian [2] and Fereidoun Mollaei Koshki [1]

1 Iran China Clay industries Co., Marand 5418916571, Iran
2 School of Mining and Petroleum Engineering, University of Alberta, Edmonton, AB T6G 1H9, Canada
* Correspondence: pouya.nobahar@gmail.com

Abstract: This paper presents the machine learning (ML) method, a novel approach that could be a profitable idea to optimize fleet management and achieve a sufficient output to reduce operational costs, by diminishing trucks' queuing time and excavators' idle time, based on the best selection of the fleet. The performance of this method was studied at the Zenouz kaolin mine to optimize the type of loader and the number of trucks used to supply the processing plant's ore demands. Accordingly, five years' data, such as dates, weather conditions, number of trucks, routes, loader types, and daily hauled ore, were collected, adapted, and processed to train the following five practical algorithms: linear regression, decision tree, K-nearest neighbour, random forest, and gradient boosting algorithm. By comparing the results of the algorithms, the gradient boosting decision tree algorithm was determined to be the best fit and predicted test data values with 85% accuracy. Subsequently, 11,322 data were imported into the machine as various scenarios and daily hauled minerals as output results were predicted for each working zone individually. Finally, the data which had the minimum variation from the selected required scheduled value, and its related data concerning loader type and the number of demanded trucks, were indicated for each day of the working year.

Keywords: machine learning; mining optimization; fleet management; mine planning

1. Introduction

In recent years, interest in artificial intelligence (AI) has dramatically increased among researchers and practitioners from all fields, with successful real-world applications in consumer products, like digital assistants or content recommendation, as well as in manufacturing environments, like autonomous machinery and robotics. One aspect of AI that has attracted the most interest and enthusiasm in recent years is machine learning (ML), which is the foundation for many successful real-world AI applications. ML techniques are a group of algorithms that can find intricate patterns in data and use them to forecast future events in a variety of industries. To illustrate, in 2021, Zhao et al. [1] presented and proposed a novel method for ground fissure identification and exploration by infrared remote sensing onboard an unmanned aerial vehicle (UAV). Using this method, a region of interest (ROI) that includes ground fissures directly above the middle of a long wall face, No. 12401 in the Shangwan coal mine was identified. In the same year, Zheng et al. [2] proposed a Multi-class Oil Palm Detection approach (MOPAD) to reap both accurate detection of oil palm trees and accurate monitoring of their growing status. Based on a faster region-based convolutional neural network (faster R-CNN), MOPAD combines a refined pyramid feature (RPF) module and a hybrid class-balanced loss module to achieve a satisfying observation of the growing status for individual oil palms. In addition, Puniach et al. [3] proposed a workflow for automatically determining the field of horizontal displacements caused by underground mining using ultra-high resolution orthomosaics. The study included a comparison of the effectiveness of image registration algorithms for matching of multi-temporal orthomosaics.

On the other hand, fleet management and scheduling are the most significant components of operations in the mining cycle. So, hauling costs, accounting for 60% of operating costs, play a crucial role in mining economics and can influence production costs and final product price [4]. So, this paper aimed to make a correct prediction and selection of the fleet using the ML method and real data from the work environment. In open-pit mining, the complexity of operations, coupled with an uncertain and dynamic environment, limits the certainty of predictions. Consequently, to achieve production targets and decrease operational costs, the best accuracy in predictions with a minimum of opportunity lost in fleet management should be reflected by considering all the factors, no matter however small, which are related to each other. Accordingly, for many years, various methods have been performed and accomplished by many scientists and industrial companies to optimize fleet management by analyzing multiple situations. Lizotte and Bonates [5] proposed a method to minimize shovel idle time, maximizing immediate truck use and allotting trucks to shovels to meet specific production purposes. However, in this study, all situations were considered stable. Hashemi and Sattarvand [6] presented a dispatching simulation model in ARENA simulation software with the objective function of minimizing truck waiting times for trucks having a developed hauling cycle and obtained a 7.8% improvement by applying a flexible assignment of the trucks for the loaders, compared to the fixed assignment system. Temeng and Otuonye [7] used the goal-programming-based dispatching model to maximize production rate and maintain ore quality compared to linear programming. Rodrigo et al. [8] performed a novel system productivity simulation and optimization modeling framework. In their model, equipment availability was a variable in the expected productivity function of the system. The framework was used for allocating trucks by route. according to their operating performances, in a truck–shovel system of an open-pit mine so as to maximize the overall productivity of the fleet. In these three studies, only productivity was considered the main goal and idle time was neglected. In 2010, Topal and Ramazan [9] presented a mixed-integer programming model (MIP). Their model provides substantial cost savings for equipment scheduling by optimizing truck usage. However, this study focused on decreasing the maintenance cost of mining. Gu et al. [10] presented a dynamic management system of ore blending in an open-pit mine based on GIS/GPS/GPRS using technologies from space, wireless location, wireless communication, and computers to control ore quality and ensure the stability of ore grade. They just focused on ore grade control instead of fleet management. Cox et al. [11] used a genetic algorithm to develop cyclic automata for dispatching trucks in mines, but, they should have focused on the real-time evolution of schedules and generalized the problem to include blending constraints. Ahangaran et al. [12] discussed the changing trend of programming and dispatching control algorithms and automation conditions. Finally, a real-time dispatching model, compatible with the requirement of trucks with different capacities, was developed using flow-networks techniques and integer programming (IP). Additionally, the use of innovative methods in recent years has improved the performance of the transport systems in mines. This model was presented for blending purposes, too. Upadhyay and Askari-Nasab [13] presented a framework using a discrete event simulation model (DES) of mine operations, which interacts with a goal programming (GP)-based mine operational optimization tool to develop an uncertainty-based short-term schedule. This framework allows the planner to make proactive decisions to achieve the mine's operational and long-term objectives. Baek and Choi [14] proposed a deep neural network (DNN)-based method for predicting ore production by truck-haulage systems in open-pit mines, which assisted comprehension of truck-haulage-system characteristics along with discrete haulage-operation sequences and supported the prediction of ore production through training of DNN-based deep learning models without the need to develop additional algorithms. This method needs to determine the optimum period for collecting training data. Moradi-Afrapoli et al. [15] presented a new mixed-integer linear programming model (MILP) to solve the truck dispatching problem in surface mines. They showed that the fuzzy linear programming (FLP) model improved the ore production and truck wait time in the queues by more than 15%. However, further study by considering mixed truck sizes was needed. In 2021, Mohtasham et al. [16] presented a multi-objective optimization model based upon a mixed-integer linear goal programming (MILGP) model, which determines the optimal production plan of the shovels and allocation plan of the trucks and shovels in order to maximize production, to meet desired head grade and tonnage at the ore destinations, and to minimize fuel consumption of trucks. Yeganejoo et al. [17] developed, implemented, and validated an integrated simulation and optimization tool capable of predicting truck fleet productivity and determining optimal fleet size based on historical data collected from the active mine. Mohtasham et al. [18] proposed new strategies based on mixed-integer non-linear programming (MINLP) models for the equipment sizing (ES) problem to verify the overall efficiency of the fleet. The developed models estimate the optimal size of trucks concerning the match factor value with two different strategies. The first strategy deals with each loader type, and the second strategy is applied simultaneously with all types of loaders. These studies did not integrate sources of uncertainty, including uncertainties related to crushers' capacity, truck cycle times, shovel's output, equipment failures, and ore quality into their proposed models. Upadhyay et al. [19] presented a simulation-based fleet productivity estimation and fleet size determination algorithm developed to be used in open-pit mines to estimate fleet productivity and predict the required fleet size to meet production schedules in the presence of technical uncertainties. Their results showed that the developed simulation-based algorithm could predict fleet productivity with more than 20% higher accuracy and had lower dependency on haulage distances.

The mentioned studies have individual problems, including disregarding past expertise in mining operations, limited flexibility for change in the production process, and ignoring actual working situations in mines.

This paper uses machine ML, a novel approach known as a subfield of AI methods, which can be a beneficial approach to best fit environmental conditions and work situations to optimize fleet management and attain an adequate output. While fleet management is related to several factors and procedures, ML methods consider work situations, like routes, types of machinery, time, and weather conditions. Furthermore, these methods also help planners to make reliable and accurate predictions. Considering that this method uses historical data, one can be assured that the results of the method can be updated. On the other hand, with the progression of time and the adding of more data, the accuracy of the algorithms increases. Also, in this method, in contrast to other initiatives, the planning can be updated by considering various situations, even the difference in plant demand in a short time that does not require any costs for the mining department. In addition, this method can be used to predict machinery in the short and long terms.

2. Machine Learning (ML)

ML has become one of the most critical topics within development organizations looking for innovative ideas to leverage data assets to help the business gain a new level of understanding. ML is a form of AI that enables a system to learn from data, rather than through explicit programming. Resurging interest in ML is due to the growing volumes and varieties of available data, and the fact that ML's computational processing is cheaper and more powerful, and provides affordable data storage. Machines that learn can more quickly highlight or find patterns in data that human beings would have otherwise missed. Consequently, ML techniques can be used to enhance humans' abilities to solve problems and make informed inferences on a wide range of problems. ML techniques are divided into three sections, each of which has individual performance: supervised learning, unsupervised learning, and reinforcement learning. ML uses various algorithms that iteratively learn from data to improve and describe data, and predict outcomes. As the algorithms ingest training data, it is possible to produce more precise models based on that data. An ML model is the output generated when a machine learns by a learning algorithm with data. Then, when the predictive model is provided with data, it predicts based on the data that trained the model [20]. In this paper, five regression techniques from supervised learning are employed.

Figure 1 illustrates the flowchart of the optimum model selection operation using the ML algorithms. As shown in Figure 1, all collected data was firstly sorted and standardized. based on needs, so that machine learning algorithms could read and understand it. Then these datasets were divided into two parts, including training and evaluation data. After this division, the training data was entered as inputs to different algorithms to train the machine. After training, the values predicted by the algorithm and the actual outputs of the evaluation section were compared, and the accuracy of the algorithm was estimated. Finally, the algorithm with the highest accuracy level was selected as the optimal algorithm to continue the process.

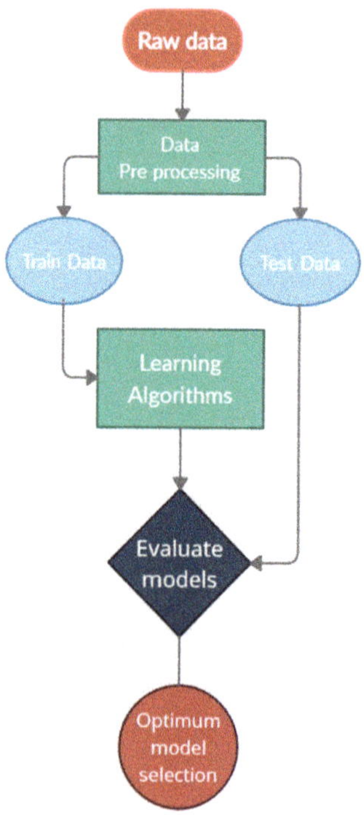

Figure 1. Optimum model selection flowchart.

3. Studied Site

Zenouz kaolin mine is located near Zenouz city, approximately 15 km North of Marand city of East Azerbaijan, Iran. Given that the Zenouz kaolin mine is one of the biggest in the area, it contributes significantly to the local economy. The key factor in the amount of revenue is transportation costs. In addition, kaolin is one of the important components of the ceramic, insulation, and tile industries and has a crucial impact on the entire production process. Therefore, the success and longevity of this mine is important ensure the survival of client enterprises.

The mainstay of innovation today are research and development operations using advanced technology that raise the added value of an industry sector. This study's goal was to provide industries, that have primary costs of raw material production which are tied to transportation expenses, with a rapid, effective, persuasive, and useful technique.

Zenouz kaolin mine is the largest kaolin mine in the Middle East, producing approximately 1,700,000 tonnes of raw kaolin and supplying nearly 70% of the kaolin in the region. This mine includes five working zones. Each zone has its own individual characteristics and processing plant demands. The mining method in this mine is open-pit mining, and kaolin is extracted by blasting, loaded by various types of excavators, and hauled by trucks to the processing plant and low-grade stockpiles.

Figure 2 shows the location of different zones and stockpiles.

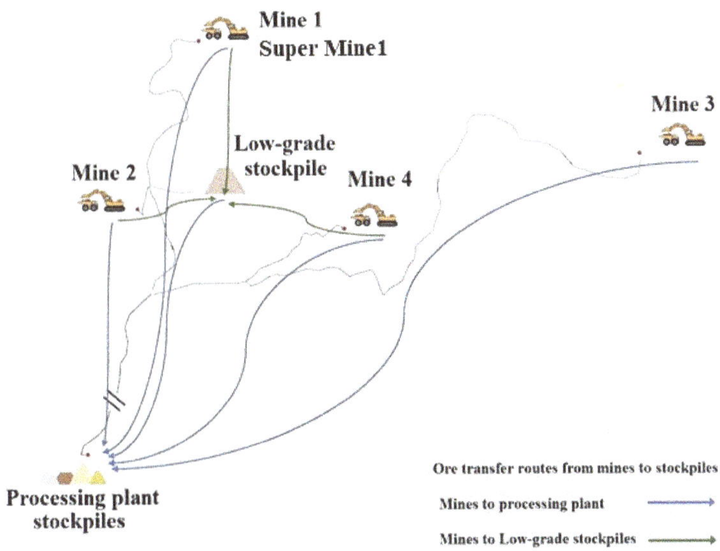

Figure 2. Working zones, low-grade stockpile and processing plant stockpile locations.

Data Collecting:

By collecting five years of records, from May 2017 to May 2021, in seven different divisions, the 1975 data were regarded as machine learning input data. The data were converted into numerical data to transform descriptive data into something understandable and agreeable to machines. These seven categories and their related numerical forms are presented as follows:

3.1. Month

This information was taken into account because the amount of minerals hauled varied by month. Therefore, numbers 1 to 12 were allocated to the data for January to November, as shown in Table 1. On the other hand, due to the fact that months are cyclic variables, the values were encoded by transforming the data into two dimensions using a sine and cosine transformation.

Table 1. Months and their related numerical values.

Month	Encoded Data	Month	Encoded Data	Month	Encoded Data
January	1	May	5	September	9
February	2	August	6	October	10
March	3	June	7	November	11
April	4	July	8	December	12

3.2. Weather Condition

Weather conditions influence the operation of the hauling systems because operators and equipment perform differently in various weather conditions. Since weather conditions affect the amount of ore haulage, related data on this factor were collected and divided into five situations, shown in Table 2.

Table 2. Weather conditions and related encoded data.

Weather Condition	Encoded
Cloudy	(1,0,0,0,0)
Foggy	(0,1,0,0,0)
Rainy	(0,0,1,0,0)
Snowy	(0,0,0,1,0)
Sunny	(0,0,0,0,1)

3.3. Season

According to experimental observations, the amount of mineral transportation varied in different seasons. Hence, this parameter was also analyzed for better consideration as training data and is presented in Table 3.

Table 3. Seasons and related encoded data.

Season	Encoded
Spring	(1,0,0,0)
Summer	(0,1,0,0)
Fall	(0,0,1,0)
Winter	(0,0,0,1)

3.4. Weekday

Due to several spatial and temporal constraints, truck drivers' weekend driving behavior is expected to differ considerably from their weekday driving style. Thus, the weekdays were also considered and evaluated in Table 4.

Table 4. Weekdays and related encoded data.

Weekday	Encoded
Monday	0
Tuesday	1
Wednesday	2
Thursdays	3
Friday	4
Saturday	5
Sunday	6

3.5. Number of Trucks

In Zenouz kaolin mine, two models of trucks, Sahand-WD615 and Mercedes-Benz-OM335, are used, and the carrying capacity of each is 26 tons on average. The number of trucks that haul minerals from different zones to stockpiles was also considered as analyzable data for machine learning.

3.6. Routes

Zenouz mine complex includes six loading spots and two delivery points (see Figure 2). Regarding the distances of these zones from the stockpiles, and considering the production plan, this parameter was separated into nine divisions, shown in Table 5.

3.7. Loader Types

Since different types of excavator load trucks are utilized, the efficiency of these machines was investigated. Four types of excavators were used as loaders at the studied site, which were taken into account as part of the input data. Table 6 displays these loaders, as well as the numerical data associated with them.

Table 5. Routes and related abbreviations and encoded data.

Route	Abbreviation of Routes	Encoded
LG-stockpile to plant	LGP	(1,0,0,0,0,0,0,0,0)
Mine 1 to plant	M1P	(0,1,0,0,0,0,0,0,0)
Mine 1 to LG-stockpile	M1LG	(0,0,1,0,0,0,0,0,0)
Mine 2 to plant	M2P	(0,0,0,1,0,0,0,0,0)
Mine 2 to LG-stockpile	M2LG	(0,0,0,0,1,0,0,0,0)
Mine 3 to plant	M3P	(0,0,0,0,0,1,0,0,0)
Mine 4 to plant	M4P	(0,0,0,0,0,0,1,0,0)
Mine 4 to LG-stockpile	M4L	(0,0,0,0,0,0,0,1,0)
Super Mine1 to plant	SP	(0,0,0,0,0,0,0,0,1)

Table 6. Types of excavators and their related encoded data.

Loader Type	Encoded
Hyundai 250	(1,0,0,0)
Hyundai 320	(0,1,0,0)
Komatsu 200	(0,0,1,0)
Komatsu 220	(0,0,0,1)

Table 7 shows examples of the collected data and Table 8 illustrates final table after converting the data to numerical data.

Table 7. Sample of collected data.

Row	Month	Weather Condition	Season	Weekday	No. of Tucks	Routes	Loader	Hauled Ore (tonne)
50	5	Sunny	Summer	Monday	6	M1 to plant	Hyundai 320	382.800
51	5	Sunny	Summer	Thursday	12	M1 to plant	Hyundai 320	1131.310
52	5	Sunny	Summer	Friday	18	M1 to plant	Hyundai 320	2129.650
53	5	Sunny	Summer	Saturday	18	M1 to plant	Hyundai 320	2277.940
54	5	Sunny	Summer	Wednesday	14	M4 to plant	Hyundai 320	2036.88

Table 8. Encoded value of the data presented in Table 7.

Row	Month	Weather Condition	Season	Weekday	No. of Tucks	Routes	Loader	Hauled Ore (tonne)
50	5	4	2	1	6	1	1	382.800
51	5	4	2	4	12	1	1	1131.310
52	5	4	2	0	18	1	1	2129.650
53	5	4	2	2	18	6	1	2277.940
54	5	4	2	6	14	1	1	2036.88

4. Data Pre-Processing

4.1. Important Data

Weekdays had insignificant impacts on model learning and creation basis and had low impact rates on the learning process (see Figure 3) and this, along with the unpredictability of the weather in the long term, meant the inclusion of weekdays in the continuation of modeling was omitted. Five parameters were used as input data: season, month, number of trucks, routes, and loader types. The data was processed and validated through efficient techniques to train the machine properly.

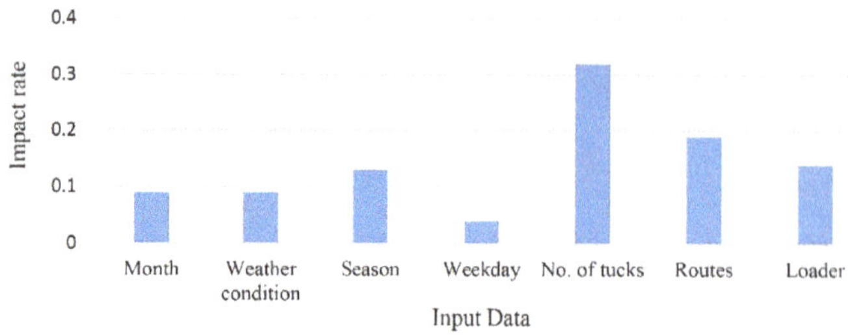

Figure 3. Impact rate of each parameter on the hauled ore.

4.2. Time Series Validator

K-fold cross-validation effectively partitions the data into K chunks, K-1 of which formed the training set R, and the last chunk served as the validation set V. Cross-validation was iterated through all combinations of assignments of chunks to R and V. This procedure was repeated for all K choices for the validation set and the performance of the model from the K runs was averaged in [21]. However, this method chose data randomly and neglect the sequence of time. So, in this paper, another effective method, namely the time series splitter, was used to split both training and validation data. Figure 4 shows how this method ran.

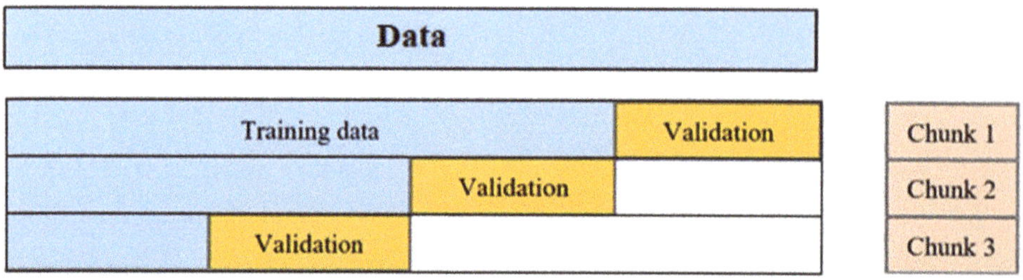

Figure 4. Time series validator performance.

4.3. R^2 Score

The R^2 coefficient (Equation (1)) represents the proportion of variation in the model's predicted result, based on its features and real data [22].

$$R^2_{(\text{ytrue, ypred})} = 1 - \frac{\sum(\text{ytrue} - \text{ypred})^2}{\sum(\text{ytrue} - \bar{\text{y}})^2} = \frac{RSS}{TSS} \quad (1)$$

In which R^2 is the coefficient of determination, *RSS* is the sum of squares of residuals, and *TSS* is the total sum of squares.

5. Modeling

After collecting data, excluding insufficient data, and processing the data, 1580 and 395 data points were imported into the machine as training and test data, respectively. In ML, dozens of unique algorithms perform specialized purposes, including, regression, clustering, and classification. The amount of hauled ore was continuous data; therefore, regression methods that deliver a continuous type of data was selected in this paper. The following sections describe the validation of the five algorithms.

5.1. Linear Regression (LR)

LR is a linear approach for modelling the relationship between scalar response and one or more explanatory variables. In LR, the relationships are modeled using linear predictor functions, the unknown model parameters of which are estimated from the data [21]. By running the algorithm on the processed input data, a model with 74% accuracy was achieved. Figure 5 shows the data's real and predicted values from numbers 50 to 150. The algorithm parameters for this method were fit intercept = True, normalize = deprecated, copy_X = True, n_jobs = None.

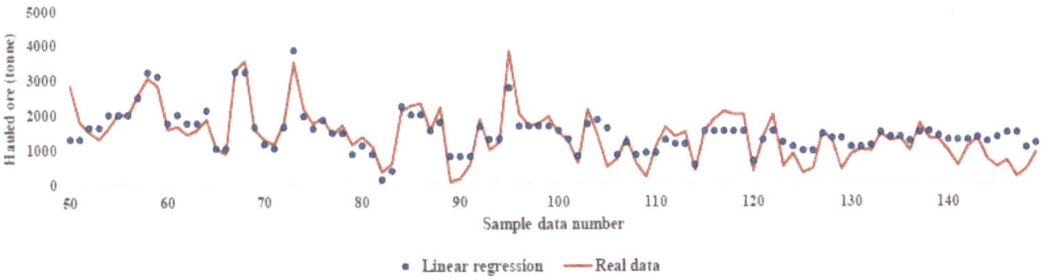

Figure 5. Comparison between real data and prediction of linear regression algorithm.

5.2. Decision Tree Regression (DTR)

Decision tree learning algorithms are based on heuristics, such as a greedy approach, where the tree is constructed gradually, and locally optimal decisions are made at the construction of each node [21]. By attempting this algorithm, predicted data were fitted to real data with 72% accuracy. Figure 6 compares real data, and decision tree regression algorithm predicted data. Parameters used for this algorithm were splitter = 'best', max_depth = 3, min_samples_split = 2, min_samples_leaf = 1.

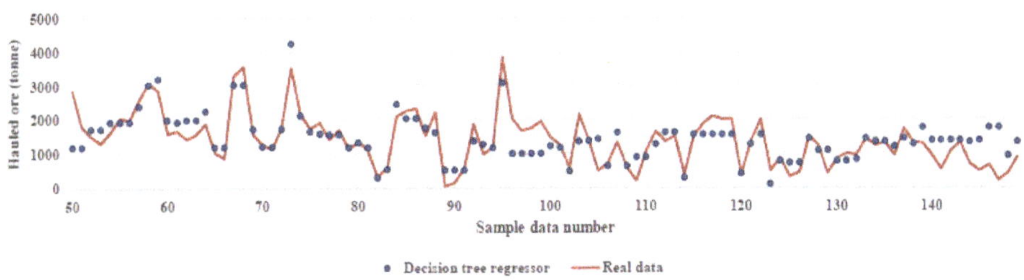

Figure 6. Comparison between real data and prediction of decision tree regression algorithm.

5.3. K-Nearest Neighbors Algorithm (KNN)

K nearest neighbour algorithm is a supervised learning technique used to classify or predict new data points, based on the relationship to nearby data points [23]. Actual and predicted values using the KNN algorithm are shown in Figure 7. The accuracy of the KNN prediction was 78%. Also, the parameters of learning the algorithm to be modified were: n_neighbors = 5, weights = 'uniform', algorithm = 'auto', leaf_size = 30, n_neighbour = 7.

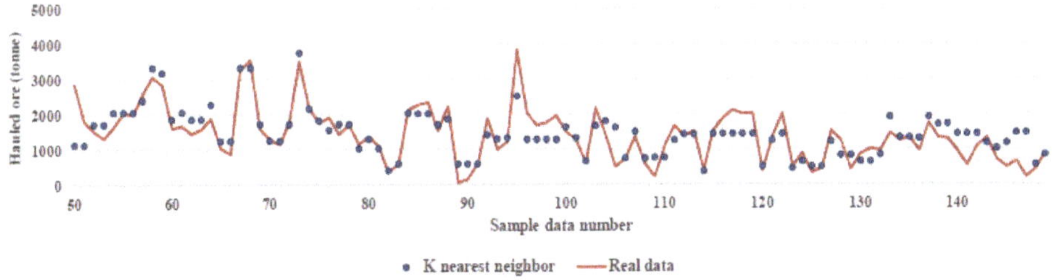

Figure 7. Comparison between real data and prediction of K nearest neighbour algorithm.

5.4. Random Forests (RF)

RF is a regressor consisting of a collection of decision trees. The prediction of the RF is obtained by a majority vote over the predictions of the individual trees, and, also, RF generally outperformed decision trees' performance [21] with the implementation of this algorithm. Figure 8 shows the difference between the actual and predicted values by random forest regression algorithm for the obtained accuracy of 83%.

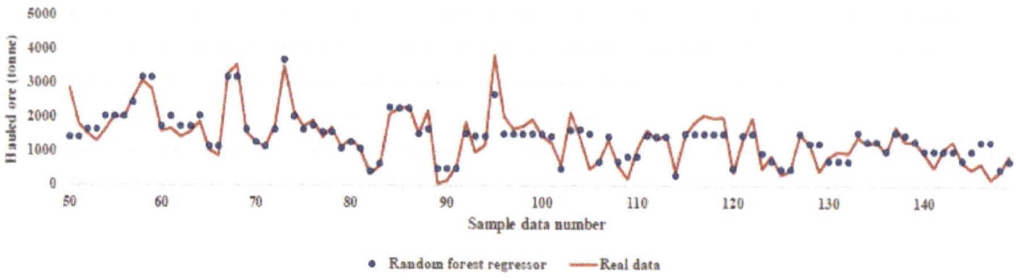

Figure 8. Comparison between real data and prediction of the random forest regression algorithm.

5.5. Gradient Boosting Decision Tree

Rather than selecting combinations of binary questions at random (like RF), gradient boosting selects binary questions that improve prediction accuracy for each new tree. This works because mistakes incurred with the training data are recorded and then applied to the next round of training data. At each iteration, weights are added to the training data, based on the results of the previous iteration. A higher weighting is applied to instances incorrectly predicted from the training data, and those correctly predicted receive less weighting. The training and test data are then compared, and errors are again logged in order to inform weighting at each subsequent round [23]. Figure 9 shows that the gradient boosting algorithm could predict the data with 85% accuracy and the parameters were learning_rate = 0.19, max_depth = 3, n_estimators = 35.

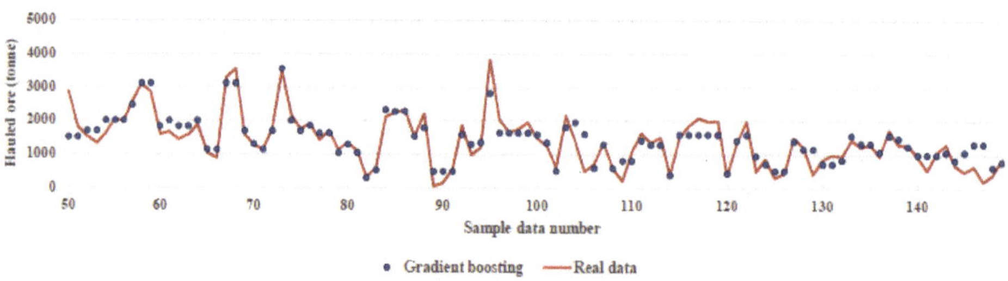

Figure 9. Comparison between real data and prediction of gradient boosting regression algorithm.

6. Model Selection

The gradient boosting algorithm was chosen as the best among the investigated algorithms. With 85% accuracy, this algorithm was used for the rest of the study after measuring the implemented algorithms to achieve an optimal model using the R^2 score formula. Each algorithm's efficiency is depicted in Figure 10.

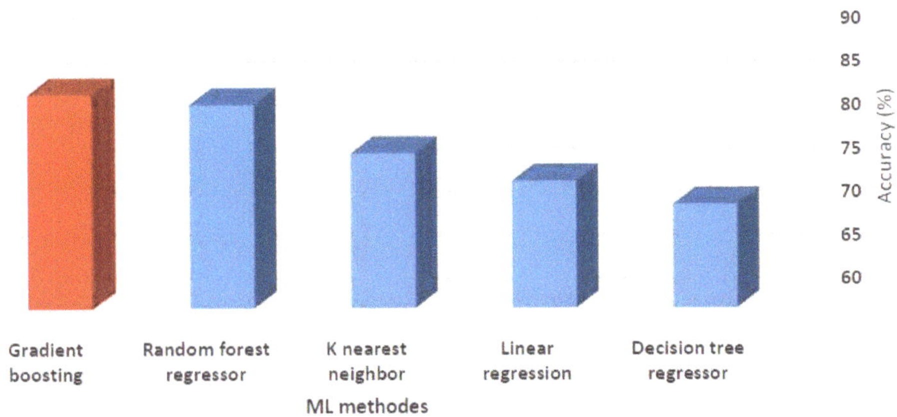

Figure 10. Implemented algorithms accuracy in percent.

7. Ore Transport Schedule

A Mine Planning team calculates the required monthly ore production from each mine based on the processing plant's required monthly feed. Table 9 shows the calculated monthly amount of ore from different zones.

Table 9. Ore annual haulage scheduling (ktonne).

Month	SUP. to P	M1 to P	M2 to P	M3 to P	M4 to P	M1 to LG	M2 to LG	M4 to LG	LG to P	Total Ore to the Plant
1	0	60	15	0	0	15	0	0	0	75
2	12	65	17	0	0	20	0	0	0	94
3	15	65	17	14	0	20	0	0	0	111
4	15	65	17	14	25	20	10	5	0	136
5	15	65	17	0	33	30	0	5	0	130
6	15	65	17	0	33	30	0	5	0	130
7	10	65	17	0	33	30	10	5	0	125
8	8	65	17	0	33	30	0	5	0	123
9	0	65	17	0	33	25	0	5	0	115
10	0	65	17	0	33	25	8	3	0	115
11	0	65	17	0	20	25	0	3	0	102
12	0	60	15	0	0	16	0	3	10	85
Annual ore to the plant										1341

While there are some limitations, simultaneous loading in more than three working zones is not possible. As a result, working days for different zones were planned according to Figure 11. Table 10 shows the required daily ore quantity to cover the processing plant's annual demand depending on this plan.

The algorithm used 1258 individual scenarios after measuring the daily required ore amount. As a result, the minimum difference between the predicted and required data values was calculated, and the optimal fleet was selected, based on related items to this data. According to Figure 12, the machine anticipated a Hyundai 320 excavator and 19 trucks as the ideal fleet in April to transport ore from Mine 1 to the plant's stockpiles.

The most suitable fleet was selected using 11.322 scenarios as input data for five loading points and two mineral discharge stockpiles. Table 11 shows the best loader and number of trucks that different zones should use over the 12 months.

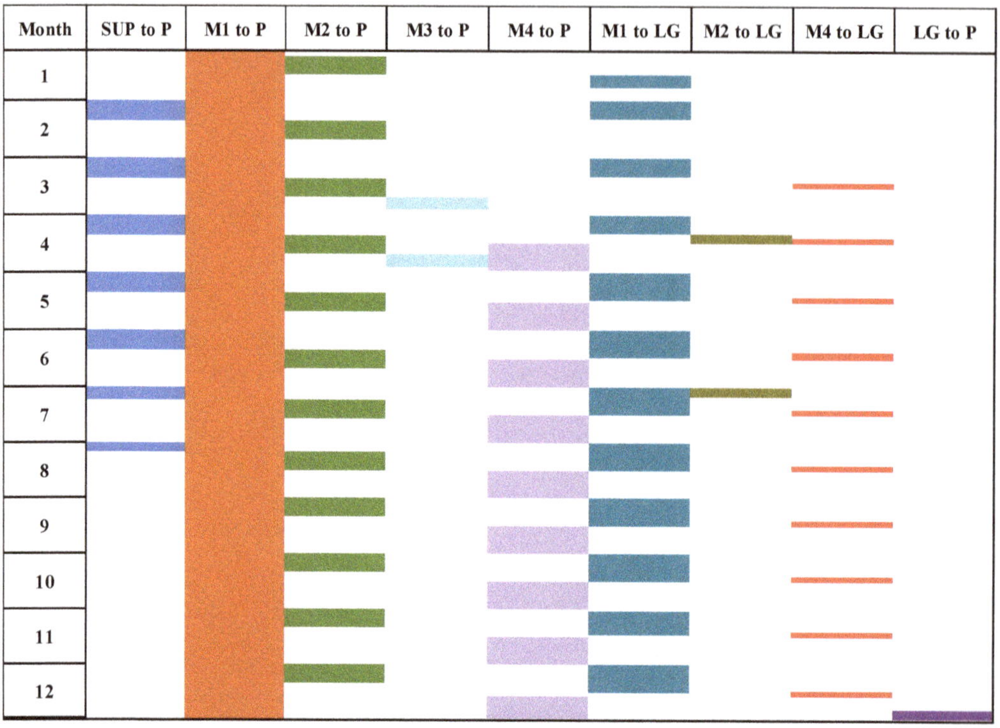

Figure 11. Working zones daily ore hauling schedule plan.

Table 10. Daily ore production schedule.

Month	SUP. to P	M1 to P	M2 to P	M3 to P	M4 to P	M1 to LG	M2 to LG	M4 to LG	LG to P
1	0	2000	1500	0	0	2143	0	0	0
2	1200	2097	1700	0	0	2000	0	0	0
3	1500	2097	1700	2000	0	2000	0	0	0
4	1500	2097	1700	2000	1667	2000	2000	1667	0
5	1500	2097	1700	0	2200	2000	0	1667	0
6	1500	2097	1700	0	2200	2000	0	1667	0
7	1429	2167	1700	0	2200	2000	2000	1667	0
8	1600	2167	1700	0	2200	2000	0	1667	0
9	0	2167	1700	0	2200	1667	0	1667	0
10	0	2167	1700	0	2200	1667	2000	1000	0
11	0	2167	1700	0	1333	1667	0	1000	0
12	0	2000	1500	0	0	1067	0	1000	2000

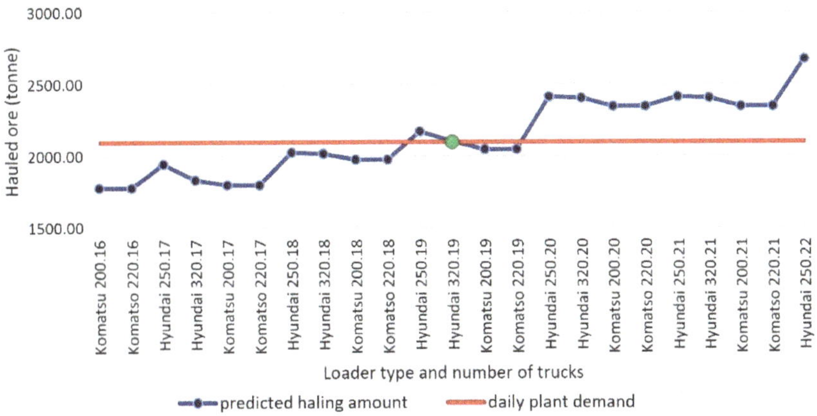

Figure 12. Optimum loader type and number of trucks selection.

Table 11. Optimum fleet to supply processing plant demands.

	Sup. to Plant		Mine1 to Plant		Mine2 to Plant		Mine3 to Plant		Mine4 to Plant		Mine1 to LG		Mine2 to LG		Mine4 to LG		LG to Plant	
Month	**Loader**	**No. of Trucks**	**Loader**	**No. of Trucks**	**Loader**	**No. of Trucks**	**Loader**	**No. of Trucks**	**Loader**	**No. of Trucks**	**Loader**	**No. of Trucks**	**Loader**	**No. of Trucks**	**Loader**	**No. of Trucks**	**Loader**	**No. of Trucks**
1			H250	17	H320	10					H250	11						
2	H320	9	H250	18	H320	11					H250	11						
3	H320	11	H320	19	H250	11	K200	12			K220	12						
4	K200	12	H320	19	H320	11	K200	12	K200	10	K220	12	K220	11	K200	11		
5	K200	12	H320	19	H250	11			H320	13	K220	12			H320	12		
6	K200	12	H320	19	H250	11			H320	13	K220	12			H320	12		
7	H320	11	H320	19	H250	11			K200	13	H250	11	K200	11	H320	12		
8	K220	10	H320	19	H250	11			K200	13	H250	11			H320	12		
9			K220	21	K200	12			K200	13	H250	10			H320	12		
10			H250	19	H320	11			H320	13	K220	10	K220	11	K220	9		
11			K220	18	H320	11			K200	10	H250	10			K200	8		
12			H320	16	K200	10			K200	6	H250	9			K200	8	H320	17

8. Conclusions

According to estimations, mineral transportation costs cover a large share of the operating costs and are becoming a challenge in mining management. So, implementing optimization in this operation can minimize the loss of capital costs, reduce the final price of the mineral, and increase profitability. In this paper, the ML method was used as an innovative approach to simulate operations, and was executed in the Zenouz kaolin mine to optimize fleet selection. Consequently, the Gradient Boosting Regressor, an excellent algorithm, was chosen and taught by various operational and conditional data to fit and predict the most beneficial fleet. Finally, the best daily required fleet to supply ore transportation to stockpiles was obtained by matching the processing plant ore demands and predicted values and finding the minimum difference between these values. As a result, the suggested fleet reduced truck queuing and excavators' idle times, which account for a considerable portion of energy consumption and capital wasting.

Author Contributions: Conceptualization, P.N.; methodology, P.N. and Y.P.; software, P.N. and Y.P.; validation, P.N., Y.P. and F.M.K.; formal analysis, P.N.; data curation, P.N.; writing—original draft preparation, P.N. and F.M.K.; writing—review and editing, Y.P. All authors have read and agreed to the published version of the manuscript.

Funding: This research received no external funding.

Data Availability Statement: The data can be obtained by sending an email to pouya.nobahar@gmail.com.

Conflicts of Interest: The authors declare no conflict of interest.

References

1. Zhao, Y.; Sun, B.; Liu, S.; Zhang, C.; He, X.; Xu, D.; Tang, W. Identification of mining induced ground fissures using UAV and infrared thermal imager: Temperature variation and fissure evolution. *ISPRS J. Photogramm. Remote Sens.* **2021**, *180*, 45–64. [CrossRef]
2. Zheng, J.; Fu, H.; Li, W.; Wu, W.; Yu, L.; Yuan, S.; Kanniah, K.D. Growing status observation for oil palm trees using Unmanned Aerial Vehicle (UAV) images. *ISPRS J. Photogramm. Remote Sens.* **2021**, *173*, 95–121. [CrossRef]
3. Puniach, E.; Gruszczyński, W.; Ćwiąkała, P.; Matwij, W. Application of UAV-based orthomosaics for determination of horizontal displacement caused by underground mining. *ISPRS J. Photogramm. Remote Sens.* **2021**, *174*, 282–303. [CrossRef]
4. Li, Z. A methodology for the optimum control of shovel and truck operations in open-pit mining. *Min. Sci. Technol.* **1990**, *10*, 337–340. [CrossRef]
5. Lizotte, Y.; Bonates, E. Truck and shovel dispatching rules assessment using simulation. *Min. Sci. Technol.* **1987**, *5*, 45–58. [CrossRef]
6. Hashemi, S.; Sattarvand, J. Application of ARENA Simulation Software for Evaluation of Open Pit Mining Transportation Systems—A Case Study. In Proceedings of the 12th International Symposium Continuous Surface Mining—Aachen 2014, Aachen, Germany, 1 October 2014; Springer: Cham, Germany, 2015; pp. 213–224. [CrossRef]
7. Temeng, V.A.; Otuonye, F.O.; Frendewey, J.O. A Nonpreemptive Goal Programming Approach to Truck Dispatching in Open Pit Mines. *Miner. Resour. Eng.* **1998**, *7*, 59–67. [CrossRef]
8. Mena, R.; Zio, E.; Kristjanpoller, F.; Arata, A. Availability-based simulation and optimization modeling framework for open-pit mine truck allocation under dynamic constraints. *Int. J. Min. Sci. Technol.* **2013**, *23*, 113–119. [CrossRef]
9. Topal, E.; Ramazan, S. A new MIP model for mine equipment scheduling by minimizing maintenance cost. *Eur. J. Oper. Res.* **2010**, *207*, 1065–1107. [CrossRef]
10. Gu, Q.; Lu, C.; Guo, J.; Jing, S. Dynamic management system of ore blending in an open pit mine based on GIS/GPS/GPRS. *Min. Sci. Technol.* **2010**, *20*, 132–137. [CrossRef]
11. Cox, W.; French, T.; Reynolds, M.; While, L. A Genetic Algorithm for Truck Dispatching in Mining. *EPiC Ser. Comput.* **2018**, *50*, 78–93. [CrossRef]
12. Ahangaran, D.K.; Yasrebi, A.B.; Wetherelt, A.; Foster, P. Real-time dispatching modelling for trucks with different capacities in open pit mines. *Arch. Min. Sci.* **2012**, *57*, 39–52. [CrossRef]
13. Upadhyay, S.P.; Askari-Nasab, H. Simulation and optimization approach for uncertainty-based short-term planning in open pit mines. *Int. J. Min. Sci. Technol.* **2018**, *28*, 153–166. [CrossRef]
14. Baek, J.; Choi, Y. Deep neural network for predicting ore production by truck-haulage systems in open-pit mines. *Appl. Sci.* **2020**, *10*, 1657. [CrossRef]
15. Moradi-Afrapoli, A.; Upadhyay, S.; Askari-Nasab, H. Truck dispatching in surface mines—Application of fuzzy linear programming. *J. S. Afr. Inst. Min. Metall.* **2021**, *121*, 1–8. [CrossRef]
16. Mohtasham, M.; Mirzaei-Nasirabad, H.; Askari-Nasab, H.; Alizadeh, B. A multi-objective model for fleet allocation schedule in open-pit mines considering the impact of prioritising objectives on transportation system performance. *Int. J. Min. Reclam. Environ.* **2021**, *35*, 709–727. [CrossRef]
17. Yeganejou, M.; Badiozamani, M.; Moradi-Afrapoli, A.; Askari-Nasab, H. Integration of simulation and dispatch modelling to predict fleet productivity: An open-pit mining case. *Min. Technol. Trans. Inst. Min. Metall.* **2021**, *131*, 67–79. [CrossRef]
18. Mohtasham, M.; Mirzaei-Nasirabad, H.; Askari-Nasab, H.; Alizadeh, B. Truck fleet size selection in open-pit mines based on the match factor using a MINLP model. *Min. Technol. Trans. Inst. Min. Metall.* **2021**, *130*, 159–175. [CrossRef]
19. Upadhyay, S.P.; Tabesh, M.; Badiozamani, M.M.; Moradi Afrapoli, A.; Askari-Nasab, H. A simulation-based algorithm for solving surface mines' equipment selection and sizing problem under uncertainty. *CIM J.* **2021**, *12*, 36–46. [CrossRef]
20. Judith Hurwitz, D.K. *Machine Learning for Dummies*; For Dummies: New York, NY, USA, 2018.
21. Shalev-Shwartz, S.; Ben-David, S. *Understanding Machine Learning: From Theory to Algorithms*; Cambridge University Press: Cambridge, UK, 2013; Volume 9781107057.
22. Raschka, S.; Mirjalili, V. *Python Machine Learning*; Packt Publishing: Birmingham, UK, 2019; Volume 69.
23. Theobald, O. *Machine Learning for Absolute Beginners*; Scatterplot Press: London, UK, 2017.

mining

MDPI

Article

The Road to the Mine of the Future: Autonomous Collaborative Mining

Javier Ruiz-del-Solar [1,2]

[1] Advanced Mining Technology Center (AMTC), University de Chile, Santiago 8370451, Chile; jruizd@ing.uchile.cl

[2] Department of Electrical Engineering, Universidad de Chile, Santiago 8370451, Chile

Abstract: The automation of mining mobile equipment is a topic of considerable interest, as it has the potential to significantly reduce the number of accidents and implement the so-called zero-entry mining concept, which would eliminate the need for any human presence on the mine site. Nevertheless, the current state of robotics and automation technology does not yet meet the requirements for the implementation of fully autonomous operations in mines. Autonomous mining equipment continues to operate under the supervision of humans, and a considerable number of mining equipment has not yet been automated. This indicates the necessity of identifying novel strategies to increase the safety of mining operations through the utilization of robotics and automation technologies. One potential solution to address this challenge is to increase the involvement of humans in autonomous mining operations. This could entail integrating human decision-makers into the decision-making loops of autonomous mining equipment. To this end, we propose the paradigm of autonomous collaborative mining, wherein humans and autonomous machines work together in a collaborative manner to increase the safety and efficiency of mining operations. We analyze the enabling factors required to implement this paradigm and present the case of autonomous loading using LHDs based on the autonomous collaborative mining paradigm.

Keywords: mining automation; autonomous vehicles; shared autonomy; field robotics; collaborative robotics; autonomous loading

1. Introduction

The use of automation technologies is critical in today's mining industry because it makes it possible to increase the safety of the operations as well as their productivity and efficiency [1–3]. Moreover, the use of automated systems allows minimizing waste and reducing the environmental impact of mining activities [2].

Automated equipment can be static (e.g., conveyor belts) or mobile (e.g., trucks), and it is the latter type of equipment that is the focus of the analysis in this paper. Automated mobile equipment can have different degrees of autonomy, ranging from pure remote control to full autonomy. Mobile equipment (vehicles) with full autonomy are of great interest because they have the potential to implement the so-called zero-entry mining concept [2,4], where no human is required to enter the mine site, thus drastically reducing the number of accidents.

Groundbreaking projects and initiatives related to mining automation began in Canada [5–7]. In [8], one of the early implementations of industrial automation is reviewed, emphasizing that the use of automation in day-to-day operations provides flexibility and convenience for operators. Shortly thereafter, the first commercially available solutions for autonomous navigation of LHD (Load–Haul–Dump) machines were developed. This groundbreaking work was followed by developments for autonomous underground equipment in Australia (e.g., [9–11]) and Scandinavia (e.g., [12,13]), and autonomous trucks for surface mining in Japan and the United States. The first semi-autonomous LHDs and the first autonomous trucks were used in production in Chile, at Codelco's El Teniente and Gaby mines in 2004 and 2008, respectively.

Following this groundbreaking work, several developments have been carried out by various OEMs (Original Equipment Manufacturers), and many automation systems are in production in various mines around the world. It is interesting to review the current state of automation of mobile mining equipment in the mining industry. As already mentioned, in the case of underground mining, special attention has been given to the automation of LHD vehicles because they operate in high-risk areas [14], but also because the mine production directly depends on the operation of these vehicles, which move the material along the production areas of the mines. The current standard is that semi-autonomous LHDs, which can haul and dump autonomously but are tele-operated during loading, are already in commercial use [15,16]. The same technology used to automate the LHDs has been applied to automate the low-profile trucks that operate at the haulage level of underground mines. In the case of open pit mines, the automation of the haul truck systems has received the most attention. Today, several mines around the world are operating with autonomous haul trucks [17,18], and given the reported benefits in terms of reduced operating and maintenance costs and increased productivity [2,17,18], many other operations are implementing this technology. Similarly, autonomous drills have recently attracted the interest of mining companies, and they are increasingly being used in open pit operations, also demonstrating benefits in terms of increased safety and cost reduction [2,19].

In all of the above cases of automated mobile equipment—trucks, LHDs and drills—they operate under human supervision for safety reasons, meaning that an operator remotely monitors the operation of equipment.

It is relevant to mention that, in addition to trucks, LHDs and drills, there are dozens of other types of mobile equipment/vehicles used in mines that have not yet been automated. Mining includes unit operations related not only to production but also to mine development (i.e., construction), where high-risk activities are carried out using a wide variety of mobile equipment. For example, in [14], it is shown that out of 59 unit operations in a typical block/panel caving mine, only 6 are fully automated worldwide (10%), 8 are teleoperated (14%), and 45 are still manually operated (76%). This means that we are still a long way from having a zero-entry mine [2], i.e., a fully autonomous mine with no people in the mine. This implies that we still need to find ways to increase the safety of mining operations through the use of robotics and automation technologies. In this paper, we argue that having a way to deal with this challenge is using collaborative human–machine systems.

The recent literature in the field of mining automation analyzes the current status, trends, and challenges of mining equipment automation. However, these studies have not clearly identified the potential contributions of collaborative human–machine systems in facilitating the transition to the zero-entry mine concept. A recent review of equipment and operations automation in mining is presented in [2], where several papers describing advances in underground and surface mining operation are discussed. The review focuses on the impact of automation on the safety and efficiency of mining operations, highlighting the successful implementation of autonomous truck systems in surface mining and describing the development of automation in underground mining and its challenges, particularly limitations in communication and localization, which hinder the development and deployment of fully autonomous systems. The review further discusses the potential for machine learning and AI to play a critical role in enhancing future mining automation systems. It also underscores the significance of collaborative human–robot work environments, stating that "as operations begin integrating automation, there will inevitably be a mix of both robotic and human workforce". However, the review does not offer specific implementation strategies or propose any operational frameworks or paradigms to address this issue. In [20], the current state and perspectives of autonomous truck systems in surface mining are presented. The paper reviews the state of autonomous truck technology and identifies several challenges, including a shortage of qualified personnel, a need for interoperability, price issues, and employee resistance to the new technology. However, the paper does not identify the need for cooperative systems. In [21], a review of automation in the mining industry is presented, addressing the following three areas: automation technology, system engineering and management processes around automation, and the role of human factor engineering in automated and semi-automated systems. Notably, the review observes that "full autonomous solutions capable of providing the complete range of ore production capabilities remain limited in industry," and that "most projects are deficient in a holistic approach that integrates process change and the re-engineering of operational systems". The review further examines the role of human factor engineering in automated and semi-automated systems. However, it does not place significant emphasis on the relevance of cooperative autonomous systems or on the proposal of methodologies for addressing them. In [22], a survey is presented on the use of automation and AI technology in surface mining, providing a view of on some of the engineering problems, technological innovations, robotic development, and automation efforts encountered in the mining industry. The survey identifies opportunities for automation across the spectrum in most facets of open-pit mining; however, it does not analyze the technical challenges nor the need of cooperative systems. In [23], the automation of mining equipment in surface and underground mining is examined from a robotics perspective. The study identifies the automation of various equipment using different robotics and automation technologies, as well as the challenges associated with these implementations. Among the identified challenges is the implementation of automation projects as "islands of automation", which underscores the need for greater integration. However, the study does not specifically address the need for cooperative systems. Another review of the current status of automation in the mining industry is presented in [24], identifying benefits such as "enhanced capital and operational productivity as well as improved safety and less use of water, energy, and consumables". The review also states that, although autonomous equipment is used in many underground and surface mining processes, about 3% of all mobile equipment used in mining is truly autonomous. The review further notes the surge of interest within the industry concerning the concept of the zero-entry mine while acknowledging the considerable time and technological advancement required before the widespread implementation of fully autonomous mining operations. It identifies the capacity of robots to effectively maintain other robots as a significant technical challenge. However, the review does not address the potential role of human–robot collaborative systems as a transitional step in this process. In [25], a recent review of robotic automation in mining automation is presented. The review describes real-world industrial implementations and presents a comprehensive collection of performance evaluation methods. Additionally, it presents timelines illustrating the evolution of automation systems for pertinent equipment, including autonomous earthmoving equipment, drilling rigs, impact hammers, and autonomous haulage systems. The review also showcases future trends, including the use of artificial intelligence, data network challenges, multi-machine cooperation, and equipment electrification. However, the role of human–robot collaborative systems is not mentioned as a future trend.

In order to address the challenge of increasing the safety of mining operations through the use of robotics and automation technologies, this paper analyzes the current state of automation in mining and presents a path to the mine of the future. The urgent and immediate goal must be to reduce the risks associated with mining operations by eliminating exposure to unsafe environments, particularly the production area. It is proposed that the future mine will be a safe environment in which humans and autonomous machines will work collaboratively. To this end, the paradigm of autonomous collaborative mining is proposed wherein humans and autonomous machines work together in a collaborative manner. The paper presents a case study of the autonomous loading of minerals using LHDs. Finally, some conclusions and outline future work in this area are drawn.

It is important to clarify that this paper presents the Autonomous Collaborative Mining paradigm at a conceptual level and that its main components still need to be further developed and implemented. Our goal is to motivate the need to develop automation solutions for the mining industry using this paradigm, which emphasizes the need for collaborative automation solutions.

2. Key Challenges in Mining Automation

2.1. Current State of Mining Automation

A common misconception is that mining automation, understood as the automation of static and mobile equipment, has been remarkably successful. This view holds that the majority of equipment utilized in underground and open mines can operate either autonomously or teleoperated, and this would allow, as a subsequent step, the drastic reduction in the number the humans inside the mine sites and the implementation of the zero-entry mining concept, where no human is required to enter the mine site. Nevertheless, we contend that this is not the case. A significant number of mobile equipment still require automation, and the manner in which current automated mobile equipment operates needs to be improved. Additionally, when different automated equipment is operated together as a fleet of heterogeneous autonomous and teleoperated equipment, coordination and optimization still needs to be addressed. To illustrate the present state of mining automation, we may consider the case of block/panel caving mines, which are massive underground mining operations.

Block/panel caving is a mass mining method employed in hard rock mining, whereby "gravity is used in combination with internal rock stresses to fracture and break the rock mass into pieces that can be handled by miners" [26]. The considerable scale of block/panel caving operations entails a high degree of mechanization across a range of operational activities. This facilitates enhanced productivity and safety within the operational framework. The high level of mechanization inherent to these operations facilitates the implementation of new technologies and improvements to existing ones in a manner that is more straightforward than in other types of mining operations. The automation of these activities plays a significant role in addressing the safety and production challenges posed by evolving rock mass conditions. By reducing personnel exposure to risk situations and enhancing control and management of operations, automation can help mitigate potential hazards and improve overall safety and efficiency. In a block/panel mine, there are over 50 distinct units and auxiliary operations related to the processes of mine operation (production), mining preparation, equipment and infrastructure maintenance, development of civil works, and auxiliary support. The authors of [14] provide a comprehensive examination of the automation status for the various operations inherent to block/panel caving mines. These findings are summarized in Table 1.

Table 1. Level of automation of different unit and auxiliary operations according to the mining processes and subprocesses for a block/panel caving mine. Details of the different unit and auxiliary operation can be obtained from [14].

Process	Subprocess	Number of Operations	Level of Automation		
			None	Remote Control	Autonomous
Preparation	Horizontal Development	24	21	2	1
	Installation of network and IT infrastructure	4	4	0	0
	Vertical Development	2	2	0	0
Operation (Production)	Undercut	4	4	0	0
	Production	5	1	2	2
	Ore Pass	2	0	2	0
	Ore Pass Transport	4	0	1	3
	Infrastructure	1	1	0	0
Auxiliary Activities	Support	7	6	1	0
	Maintenance	5	5	0	0
	Supply	1	1	0	0

As evidenced in Table 1, a total of 59 operations were identified, comprising 16 mining operations, 30 mine preparation operations, and 13 auxiliary/support activities. Of these operations, only six are fully automated worldwide, representing 10% of the total. Eight are teleoperated (14%) and 45 are still manually operated (76%). The six autonomous operations are as follows: ventilation (in horizontal development), LHD hauling (in production), LHD dumping (in production), rail hauling (in ore pass transport), conveyor belt hauling (in ore pass transport), and CAEX hauling (in ore pass transport). The eight teleoperated operations are as follows: LHD ore extraction (in horizontal developments), gallery cleaning (in horizontal developments and in ore pass), LHD loading (in production), fixed rock breaker (in production, in ore pass, and in ore pass transport), and street cleaning (in support). These numbers demonstrate that there are significant opportunities for the automation of mining equipment and that the majority of efforts have concentrated on the automation of the equipment used in production.

In order to enhance the safety of underground operations, it is essential to direct attention to the processes of development and auxiliary activities. In [14], this aspect is also analyzed. The study defines a risk index for the different operations and determines that 43 operations have a high-risk index (74%), 1 operation has a medium risk index (2%), and 14 operations have a low-risk index (24%). This indicates that there is a need for automation of these high-risk operations in order to reduce the level of worker exposure.

Other underground mining methods as well as open pit mining methods employ distinct mining processes, subprocesses, and unit operations. However, in all cases, the majority of the automation efforts have been focused on the automation of a select few unit operations, predominantly utilizing autonomous haul trucks, LHDs, and drills, along with a few additional pieces of teleoperated equipment. In light of the aforementioned, it is evident that the immediate goal should be to have mines where humans are situated at a safe distance from risky areas, and in which humans and autonomous and teleoperated machines work in an efficient and collaborative manner, instead of having a mine with no humans. In the next subsections, we argue that the current robotic technology does not allow for the full automation of the multiple unit processes and tasks required in mining.

2.2. Challenges in Robotics Technology for Autonomousvehicle Navigation

Mining is one of the few industries where autonomous mobile equipment, i.e., vehicles, are regularly used (e.g., autonomous haul trucks in open pit mines). As a result, there are few examples of other industries from which to draw technological developments. One potential source of inspiration could be the autonomous vehicle/automotive industry, where billions of dollars have been invested, but where satisfactory results have yet to be achieved. The development of general-purpose autonomous vehicles is worthy of analysis.

Although the first autonomous vehicles/cars were developed in Japan and in Germany in the early 1980s [27,28], the idea of having a massive adoption of autonomous vehicles started after the success of the second DARPA (Defense Advanced Research Projects Agency) Grand Challenge (2005) in which five vehicles were able to complete the challenge [29,30]. This success inspired companies like Google to pursue the development of this technology. Subsequently, billions of dollars have been invested in these developments. Nearly every major automotive manufacturer has initiated its own autonomous vehicle project, and numerous new companies and joint ventures have been established with this objective [28]. Nevertheless, this technology remains insufficiently mature to be used in most cities [31,32]. Consequently, following two decades of development and substantial investment, the application of this technology remains largely confined to a few select cities, including San Francisco and Phoenix, and states (e.g., Arizona and California in the United States). The technology is not yet sufficiently robust. It relies on the use of extensive LIDAR-based scans of the routes (high-density sensor maps), requiring a constant human supervision of the vehicles from central station. Furthermore, it only works under normal traffic conditions. In conclusion, the technology remains in an immature state: (1) it is not plug-and-play solution, as it requires a previous scan of the routes, (2) it does not work without human supervision, and (3) it is not operational under heavy or under unstructured traffic conditions [32]. The reasons for this lack of success can be attributed to two primary factors. First, the current state of robotics technology is not yet sufficiently robust to effectively deal with changing environmental conditions. Second, it remains challenging to accurately anticipate and respond to the myriad of potential scenarios that can arise on a populated road. Learning algorithms fail to make decisions in rare cases because there are no data to train the algorithms in these cases (out-of-distribution problem [33]).

This last point is significant because it may be assumed that the most recent AI algorithms, including convolutional neural networks, transformers, reinforcement learning, and diffusion models, provide a solution to the autonomous driving problem. While AI can undoubtedly assist in addressing the aforementioned challenges, it is imperative to acknowledge the existence of a fundamental limitation inherent to the majority of AI algorithms: the out-of-distribution (OOD) problem. An AI algorithm is trained on a specific distribution of examples or cases, and its efficacy is contingent upon its ability to generalize to similar examples or cases. If the examples or cases differ from those used for training, an out-of-distribution (OOD) problem arises, preventing the AI algorithm from inferring or reasoning correctly. This phenomenon occurs in autonomous vehicles when they encounter situations not included in their training data. Such scenarios are not uncommon when driving in real-world settings, particularly in complex urban environments.

A further DARPA Challenge, the DARPA Subterranean (SubT) Challenge, which is more closely aligned with mining, was recently launched with the objective of developing innovative technologies that can enhance underground operations. "The SubT Challenge program explored new approaches to rapidly map, navigate, search, and exploit complex underground environments such as human-made tunnel systems, urban underground, and natural cave networks" [34]. The main scientific and technical outcomes of this four-year endeavor are summarized in [35]. Advancements were observed in the domains of perception and SLAM (Simultaneous Localization and Mapping) under underground conditions, as well as in multi-robot operation and multi-sensor integration. However, with regard to decision-making, all teams still include humans in the main decision-making loops (human supervisor): "For most teams, the role of the human supervisor included commanding high-level missions, providing manual intervention when needed, deciding when to deploy communication nodes, coordinating the next system deployments with the pit crew, monitoring incoming sensor streams, reviewing artifact detections for viability, and sending artifact reports to the DARPA command post" [35]. This reinforces the continued importance of humans in ensuring the successful operation of robotic systems. In fact, one of the primary reported insights emphasizes the pivotal role of humans in system performance: "The Human Teammate Is (Often) the Critical Performance Limiter" (Insight #6 in [35]).

In conclusion, the current state of robotics technology is insufficient for full autonomous navigation (i.e., no human intervention) in complex, real-world environments. To address this challenge, it is necessary to integrate human supervision and, in many cases, incorporate human decision-makers into the decision-making loops.

2.3. Continuous Operation of an Automated Task Requires Human Intervention

Autonomous systems cannot operate without human intervention for long periods of time. As an example, let us analyze the case of the most widely used household robot in the world, the robot vacuum cleaner. A robot vacuum cleaner can vacuum autonomously, but it requires a human being to place it in the room to be cleaned, making sure that it has access to the entire place where it must perform the task for which the human must open the doors to the rooms where the cleaning must be performed and, if the house has more than one floor, take the robot to the corresponding floor. In addition, the vacuum cleaner has a waste container that must be emptied by a human being when it is full, and if water is used for cleaning, it must be filled by a human being. Moreover, after days/weeks of operation, maintenance and repair of the robot vacuum cleaner must also be performed by a human being. In summary, the robot vacuum cleaner can operate autonomously for short periods of time, but its medium- and long-term operation depends on tasks such as placement, supply, maintenance, and repair that only a human being can perform.

Similarly, if we think of any equipment that has to work inside a mine, in a shift regime, and perform tasks such as installing of infrastructure (meshes, bolts, etc.), drilling, blasting, transporting material, etc., it can operate autonomously for limited periods of time, but undoubtedly humans will be required to bring the equipment to its place of operation and provide it with the necessary supplies. Moreover, depending on the task, humans will supervise the safe operation of the equipment and evaluate the results obtained (e.g., mesh installation). In the medium and long term, humans will also perform maintenance and repair tasks, most of the time, inside the mine facilities.

In conclusion, the performance of tasks by autonomous equipment inside a mine requires human involvement.

2.4. Autonomous Equipment Confinement and Operational Interference Management

In underground mining operations, as well as in some surface mining setups, the number of interactions between mobile autonomous equipment and human personnel is minimized by isolating them from each other to achieve a safe operation. This is achieved by confining mobile autonomous equipment to designated mine areas, where human operators and manual equipment are prohibited. The isolation of autonomous equipment is facilitated by a confinement system that should encompass the following functionalities [36]:

1. It must restrict unauthorized entry of equipment and people to the operating area of the autonomous systems, which is referred to as the autonomous operation zone (AOZ).

2. It must be able to detect any unauthorized entry of people to the AOZ or any transgression of the system security.
3. It must prevent unplanned departure of autonomous equipment in operation from the AOZ.
4. It must provide controlled access for autonomous equipment to the AOZ.
5. It must provide controlled and secure access for operators and manual equipment to the AOZ, when they are required to address operational interferences.

The confinement of autonomous mobile equipment may have a notable impact on system efficiency and production. As an example, the segregation of autonomous LHDs allows for the operation to be conducted in a safe manner at the production level. However, this has a detrimental impact on production, as each instance of human access to production areas disrupts the operation of autonomous equipment. It is important to consider that human operators must enter the AOZ on a regular basis to address operational interferences, such as resolving draw point hang-ups and performing secondary reduction tasks. Currently, operations implement a static confinement policy, whereby all autonomous LHDs cease functioning while human workers are present in the production area. Consequently, operations are entirely halted at the production level during the periods that human workers are required to be inside this level.

According to [36], the replacement of static confinement policies with dynamic confinement policies may prove an effective means of reconciling the competing considerations of safety and efficiency.

3. The Autonomous Collaborative Mining Paradigm

The mine of the future must be a safe environment in which humans work in areas that are free from hazardous conditions. In light of the current limitations of available robotics technology, it is not yet feasible to implement fully autonomous operation of mines. Consequently, some tasks will be conducted by autonomous mobile equipment, while many others will be performed by human teleoperation from nearby positions or from operating centers, inside or outside the mine, or directly by human operators. Furthermore, humans will continue to be essential for other activities, including equipment placement, supply provision, task evaluation, maintenance, and repair.

In this context, the key idea of the Autonomous Collaborative Mining paradigm is that humans and machines interact in a flexible, collaborative, and synergistic manner to achieve safer and more efficient mine operations. This is achieved through the use of various complementary technologies, including robotics, automation, IT, and machine learning. This paradigm is aligned and inspired in the new trends of collaborative robotics which is a "new form of robotic automation built to work safely alongside human workers in a shared, collaborative workspace" [37]. According to the Association for Advancing Automation, "the ability to work collaboratively with humans greatly expands the potential applications of robotic automation" [37].

The Autonomous Collaborative Mining paradigm includes the following components, which can be implemented individually or in combination:

- *Autonomous Operation under Human Supervision*
- *Shared Autonomy*
- *Dynamic Confinement and Online Monitoring*
- *Augmented Reality to Enhance Human Capabilities*

Furthermore, it is essential to consider the enabling factors that may facilitate or impede the implementation of the paradigm.

Before describing the main components of the paradigm, it is important to remember that the paradigm and its components are presented in this paper at a conceptual level.

They still need to be further developed and implemented. Our goal is to motivate and encourage the community to adopt the ideas behind this paradigm to further develop systems based on it, which may combine some or all of its components.

3.1. Autonomous Operation Under Human Supervision

In consideration of the present state of robotics technology, automated operations, defined as operations carried out using autonomous equipment, will continue to require human supervision for safety reasons, as is the case in the autonomous vehicle industry.

Furthermore, it is also important to note that in many unit and auxiliary operations related to mine development, installation of infrastructure, cleaning, support, maintenance and supply, in which the equipment is installed and used for a fixed period of time, an increase in the level of automation of an operation does not entail the complete withdrawal of the personnel from the mine. Rather, it entails a withdrawal of the personnel from the work zone, with high associated risks, to a safer zone. This is mainly because even after an operation is automated, human operators are still required to prepare, monitor, supervise, evaluate, and complete the operation. This aspect is frequently overlooked in the design of automated mining operations. Some of the tasks that will still be carried out by humans when a manual mining operation is automated include [14].

- Inspection of the work area and the operational variables present in the environment.
- Verification of the state of the equipment to be used.
- Transportation and installation of the equipment in the area of operation.
- Administration of supplies required in the operation.
- Execution of the unitary operation.
- Uninstallation and removal of the equipment from the work area.
- Quality assurance (QA) of the result of the operation.

Figure 1 illustrates the continued involvement of humans in the operations of autonomous units, including preparation, withdrawal, quality assurance, and supervision (either locally or remotely).

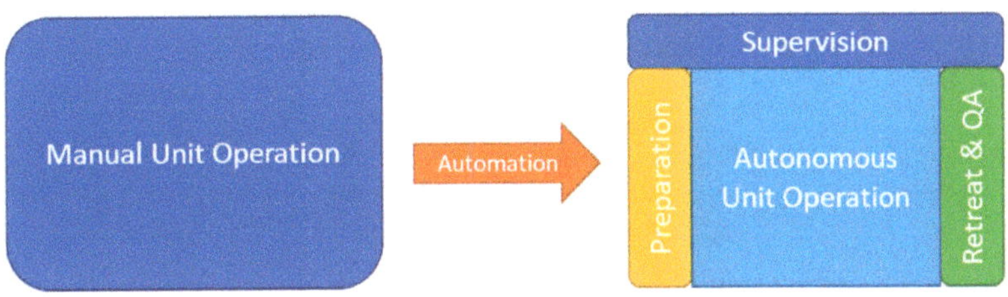

Figure 1. Diagram of the modification of a unit operation produced by its automation.

Consequently, within the autonomous collaborative mining paradigm, humans remain a presence, assuming a variety of roles. One such role is that of supervision of autonomous operations.

3.2. Shared Autonomy

The concept of *shared autonomy* provides an alternative to full autonomy, whereby a human operator can provide situation awareness, decision-making support and other assistance to an autonomous system [38]. In the context of mining, where the operation of autonomous mobile equipment requires human supervision for safety reasons, shared autonomy may be a more suitable option than full autonomy, as it could lead to more reliable and safer operations.

Let us explain how shared autonomy works and describe its distinguishing features in comparison to alternative automation modes. The following formalization is employed: as shown in Equation (1), the orders that the mobile equipment (machine) receives in timestep t, $u(t)$, are contingent upon a function $h_c(t)$, which combine the human orders, $u_h(t)$, and the autonomous system orders, $u_a(t)$, depending on the context, c. The context in which these orders are given may be dependent upon a number of factors, including the task completeness, the environment, the human operator activity, or other human-related factors.

$$u(t) = h_c(u_h(t), u_a(t),\ t) \tag{1}$$

If we apply this definition to the levels of autonomy shown in Table 2, we see that in the cases of *manual operation*, *remote control*, and *teleoperation* without assistance, $u(t) = u_h(t)$. Conversely, in the case of *autonomy*, $u(t) = u_a(t)$. In the case of *semi-autonomy*, $u(t) = u_h(t)$ or $u(t) = u_a(t)$ depending on the task at hand. For instance, in the context of a semi-autonomous LHD, $u(t) = u_h(t)$ for loading and $u(t) = u_a(t)$ for hauling and dumping. In the case of *teleoperation with assistance*, $u(t)$ is a combination of $u_h(t)$ and $u_a(t)$, as defined when the system is designed. In this case, the assistance provided to the human operator may be static, or it may be contingent upon the operator's internal state or the external environment. For instance, the assistance level may increase when the operator is fatigued or when the machine is in proximity to other vehicles or humans. Finally, in the case of *shared autonomy*, $u(t)$ is a combination of $u_h(t)$ and $u_a(t)$, with the autonomous system determining the specific manner and timing of this combination. Thus, $u(t)$ is controlled by the autonomous systems of transferred to the user. For instance, in the case that the autonomous system request assistance from a human operator, $u(t) = u_a(t)$ transitions to $u(t) = u_h(t)$ while the human is providing assistance and then reverts to $u(t) = u_a(t)$ when the autonomous system regains control. Naturally, we also have the case in which the human does not assume complete control of the machine. In such case, the order received by the machine is a combination of $u_h(t)$ and $u_a(t)$. What is relevant in the shared autonomy case is that the autonomous system is the one that decides how and when the human orders and its own orders are combined. However, it is important to note that in cases where the human supervisor detects unsafe or incorrect behavior, they can take full control of the machine.

Table 2. Levels of autonomy of a mobile system operating in a mine.

Manual Operation	**A Human Operates the Equipment**
Remote Control or line-of-sight Teleoperation	The equipment is teleoperated from a distance of a few meters, with the operator observing its operation directly.
Teleoperation	The equipment is teleoperated from an operating center, inside or outside the mime.
Teleoperation with Assistance	The teleoperation process include the use of assistance whereby the operator receives some form of assistance from an artificial agent or algorithm.
Semi-Autonomy	The equipment performs certain tasks teleoperated and certain tasks autonomously.
Shared Autonomy	Adaptively, or on demand, it is decided whether the equipment operates autonomously or in a teleoperated manner.
Autonomy	The equipment operates autonomously.

Thus, we propose that within the autonomous collaborative mining paradigm, shared autonomy may be regarded as a viable alternative to full autonomy in instances where the latter is not readily attainable. To illustrate this proposition, we present a case study in Section 4, wherein the application of shared autonomy to the task of loading from draw points using LHD is proposed.

It is worth noting that some current fleet management systems for mining equipment are already incorporating the concept of shared autonomy, although not necessarily by that name.

3.3. *Dynamic Confinement and Online Monitoring*

As previously stated, autonomous mobile equipment is confined to specific autonomous operation zones (AOZs) for reasons of safety. The presence of human operators or manual equipment is prohibited in these zones. This so-called static confinement results in inefficiencies due to the necessity for autonomous equipment to halt operations each time human operators must enter the production area. A better solution to this suboptimal approach to confinement is the implementation of dynamic confinement policies. In contrast to the static confinement paradigm, the AOZ is updated in real time as the operation progresses under the dynamic confinement paradigm. The dimensions of the AOZ are contingent upon the prevailing operational status and are not necessarily determinable in advance. Key elements for implementing dynamic confinement policies are (1) the use of sensors embedded within the environment (e.g., laser barriers) that indicate the mobile equipment the limits of permitted areas and act as virtual gates which can be dynamically activated (see examples in [36]) and (2) the real-time provision of positional data from all mobile equipment and human operators within the AOZ to a central system. This central system is responsible for defining the AOZ at any given moment and generating alarms when any agent makes an unauthorized movement, such as an unauthorized entry or departure from the AOZ.

In order to illustrate this idea, Figure 2 presents a conceptual diagram of a central system that allows the determination of the poses of multiple agents on an AOZ. The color ellipses and arrows represent the estimated position of the different agents—humans, autonomous systems, teleoperated systems, and manually operated systems—and their direction of movement. The position of each agent is estimated based on the reported position of the agent, with variance (represented by the size of the ellipse) added to account for errors in the estimation of the own position by the agent (self-localization error) and time delays in communication. Additionally, static sensors (S) estimate the position of some agents within their field of view and provide supplementary observations.

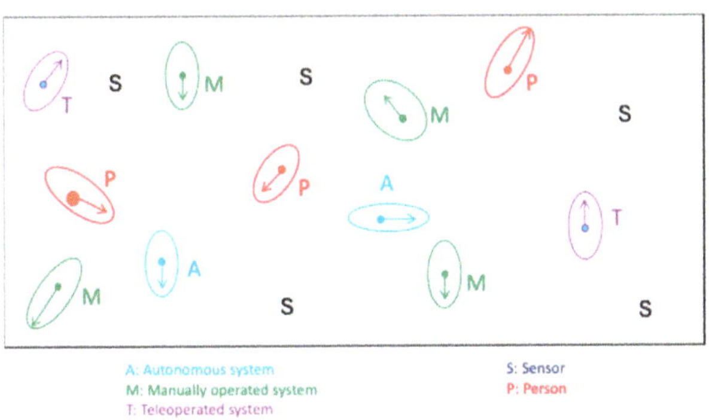

Figure 2. Conceptual diagram of a system that allows the determination of the poses of multiple agents on an AOZ. Color ellipses and arrows represent the variance in the positions of the different agents (humans, autonomous systems, teleoperated systems, and manually operated systems) and their direction of movement. The static sensors (S) allow the direct observation of the agents in the environment (see main text for details).

The implementation of dynamic confinement policies not only enhances operational efficiency but also provides a greater degree of operational flexibility. This enables the coexistence of automated equipment and human personnel within the same designated areas, with the establishment of virtual boundaries delineating their respective domains.

For example, in [36], simulations are carried out to assess the performance of different confinement policies in a panel caving mine with 16 production drives. The results show a reduction in the operational losses (interference time) from a 24% in the case of using a static confinement policy to 5–7% in the case of using a dynamic confinement policy, converting most of this time into effective time and having a better utilization of the semi-autonomous LHDs. The increase in effective time results in a 35–47% increase in production.

3.4. Augmented Reality to Enhance Human Capabilities

The term *augmented reality* (AR) is defined as the provision of visual interfaces that superimpose computer-generated digital information onto the perceived physical world. This enables the provision of real-time insights, visualizations, and instructions in a direct and immediate manner within the user's field of vision using devices such as smartphones, special glasses, and other wearables. There are numerous applications for this technology, including (i) the provision of information about parts and components during an inspection process of facilities or equipment, wherein the system guides the operator to the specific locations that need to be inspected and identifies components and defects, (ii) equipment maintenance or repair, wherein the operator is able to follow visual instructions superimposed on the equipment part to ensure precise execution of maintenance procedures, and (iii) remote support, whereby the onsite operator can stream their view of the equipment to a remote operator (expert) who can provide instructions or annotate data directly in the field of view of the onsite operator. The information provided to the operator may be obtained from annotations (notes, comments, and markings) made previously during a planning stage. Alternatively, the information may be provided online by an AI assistant or by an expert placed in another location.

The application of AR in a mining context has the potential to enhance the capabilities of operators, providing them with more effective tools and multi-purpose functionality. A given operator is able to perform a variety of tasks, including repairs on different types of equipment, with the assistance of AR-based systems. One can envisage a scenario in which an operator is situated inside a mine facility and is required to either initialize or repair a piece of mine equipment, despite lacking complete knowledge of the task at hand. In such a situation, a virtual assistant can furnish the operator with the necessary information through the utilization of an AR interface, thereby enabling the operator to complete the task successfully.

AR can also be used to enhance the visualization employed by a teleoperator when remotely controlling equipment. In this case, supplementary data, such as the distance to objects and the identities of object components and parts, can be superimposed upon the visualization displayed to the operator. In addition, *telepresence*, defined as teleoperation using special sensing and display technologies that enable the human operator to feel present at the remote location despite the operator's physical absence [39], also enhances the quality of remote control tasks. *Virtual reality* (VR) also offers a variety of applications in the mining industry, including operator training and teleoperation. In the latter case, the teleoperation task can be conducted in a virtual environment generated using data from sensors monitoring the environment.

In the context of autonomous collaborative mining, the use of AR, VR and telepresence technologies facilitates an enhanced interaction between human operators and the equipment to be controlled or inspected. But more importantly, it allows for the enhancement of human operators' capabilities, transforming them into versatile operators capable of performing a diverse range of tasks. This is relevant to achieve the goal of reducing the number of humans in potentially hazardous working areas, as multi-purpose human operators allow for a reduction in the required on-site personnel.

It should be noted that the use of AR, VR, and telepresence technologies requires an appropriate design in order to avoid mental fatigue of the operator/user. Mental workload must be considered in the design of these interfaces, and in general human factors must always be considered in the design of any computer interface. The proposed idea is not to use these interfaces permanently, but only when needed to solve specific tasks.

3.5. Integration

The various components of the paradigm can be used in an integrated fashion. Figure 3 shows a conceptual diagram that integrates the different components and technologies. In this diagram, machines with different levels of autonomy and humans using different types of interfaces work together in an AOZ using dynamic confinement. All machines are subject to human supervision for safety and efficiency reasons; at any time, a supervisor can take full control of a particular machine if unsafe or incorrect behavior is detected. Of course, supervisors can also detect incorrect human behavior and take corrective actions.

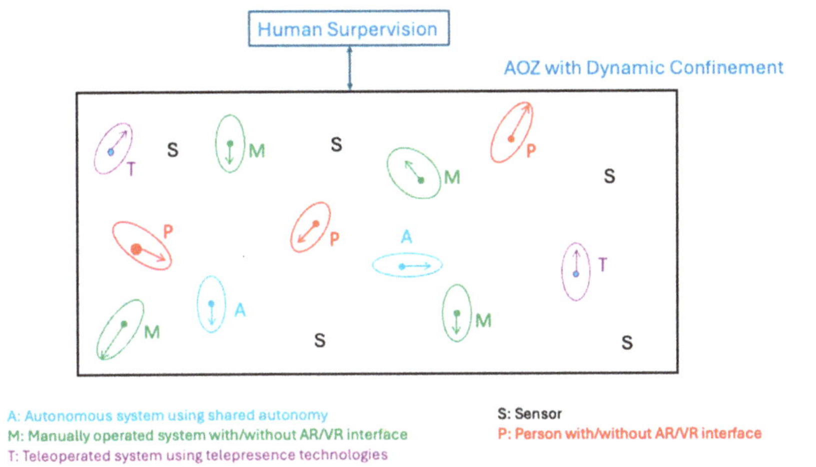

Figure 3. Conceptual diagram of a system integrating the different components of the paradigm. The colored ellipses and arrows represent the variance in the positions of the different agents and their direction of movement. The static sensors (S) allow the direct observation of the agents in the environment (see main text for details).

Machines can be operated either autonomously using shared autonomy, remotely using telepresence technologies, or manually. People within the AOZ can be operating the machines or performing other tasks such as sampling, inspection, or repair. In all cases, they can use AR or VR interfaces to augment their capabilities, as needed. The use of dynamic confinement allows for a safe interaction between the various machines and the people inside the AOZ.

3.6. Safety of Human–Machine Interactions

One of the most important ideas of the autonomous collaborative mining paradigm is to increase collaboration between humans and machines, which often includes closer human–machine physical interaction. This closer interaction requires the development of safety regulations, procedures, and protocols to prevent any possible collision or similar incidents that could put people in danger. In addition, proper technology, such as surveillance systems that monitor the activity of humans and machines, should be developed to avoid risky situations. The safety procedures and protocols to be implemented must consider potential system failures, algorithm errors, communication problems, and human factors. Human factors analysis is especially important because humans make errors and do not always follow the rules and safety protocols.

As mentioned earlier, the mine environments and unit operations exhibit significant variability, as does the employed mobile equipment. Therefore, safety procedures and protocols must be meticulously designed for each distinct application of collaborative automation systems that operate under the proposed paradigm. This issue is of paramount importance and must be addressed prior to any physical implementation of the paradigm.

For instance, in [36], the dynamic confinement of semi-autonomous LHDs operating in a panel/block caving mine is proposed at a conceptual level. This dynamic confinement is implemented using sensors that act as virtual gates (e.g., laser barriers), allowing a high level of segregation for both people and equipment and registration areas for control access. In the context of a practical implementation of this concept, it is imperative to ensure the redundancy of sensors in the event of failure, as well as the incorporation of safety protocols that address scenarios such as network and sensor failures, malfunction of the software controlling the LHDs, and human errors.

3.7. Enabling Factors

The advancement of the autonomous collaborative mining paradigm requires the development of interoperability standards, new regulations, the adaptation of the mining designs and procedures to facilitate the use of autonomous and teleoperated equipment, the development of new sensors to be placed in the mines, high-bandwidth and low-latency data networks, and a workforce trained in the use of the new technologies. These factors, among others, must be considered to ensure the successful implementation of the autonomous collaborative mining paradigm. In order to achieve this, it is necessary to coordinate the efforts of mining companies, original equipment manufacturers (OEMs), research and development institutions, technological companies, and regulatory bodies.

To illustrate the significance of the aforementioned enabling factors, the lack of interoperability standards for autonomous mining vehicles hinders the development of new modules and products by non-OEM entities, thereby slowing down the development of new solutions. For instance, in light of the recent advancements in artificial intelligence (AI), particularly in the domains of machine and deep learning, there arises an opportunity to leverage novel methodologies wherein AI can enhance the functionality of existing automation technologies. This may include the development of enhanced object and human detection algorithms, more advanced optimization algorithms for navigation, more sophisticated sensing methodologies, more advanced learning algorithms, etc. However, the application of AI-based solutions developed at research and development centers or at technology companies to commercial mining mobile equipment is hindered by two main factors: the lack of interoperability standards and the presence of contractual restrictions that prohibit any intervention with the equipment. In order to effectively address these challenges, it is essential to foster collaborative initiatives among OEMs, mining companies, and technology developers.

4. Case Study: Autonomous Loading with LHDs

To date, there are no commercial autonomous loading systems for LHD that can be used in in block/panel caving mines and sublevel stoping mines due to the complexity of the loading process; loading fragmented rock from draw points is a complex task due to many factors, including bucket–rock interaction forces that are difficult to model, moisture that increases cohesion forces, and the possible presence of boulders.

In order to address this challenge, the authors in [40] proposed an autonomous loading system based on the autonomous collaborative mining paradigm. The system is designed to integrate all the relevant tasks required for ore loading, including rock pile identification, LHD positioning in front of the ore pile, charging and excavating into the ore pile, pull back, and payload weighing. The system employs the concept of shared autonomy whereby, in cases where the loading process cannot be completed autonomously, the LHD's controller is able to detect this situation and request assistance from a human operator.

4.1. On the Complexity and Requirements of the Loading Process

Loading of fragmented rock involves a sequence of steps: As the LHD's operator approaches the draw point, a quick assessment of the condition of the rock pile is made. The LHD's bucket is tilted down until the tip is pressed against the ground. The machine is then commanded to charge/attack the rock pile. Once the LHD has made contact, it buries the bucket into the rock pile and the bucket is tilted and lifted as it advances through the fragmented rock. When sufficient penetration is achieved, the bucket is fully retracted, and the vehicle is withdrawn with a reverse motion. The operator then shakes the bucket briefly and assesses whether or not enough ore has been loaded. If necessary, a new loading maneuver is performed.

The step in which the machine is being controlled to dig into the rock pile is herein referred to as the excavation process. During excavation, bucket–rock interaction forces affect the bucket motion through the material and ultimately define the amount of ore loaded. Inter-particle forces in granular material and bucket–rock interactions have been studied previously, but their complexity has led only to stochastic modeling [41]. As a result, an analytical approach to bucket motion control is not possible, and it is highly unlikely that these interactions can be accurately replicated to develop a system in simulation. Therefore, only a real-world environment can provide the appropriate conditions for developing and experimenting with an excavation method. Furthermore, fragmented rock is much more difficult to load than piled sand or gravel. This is exacerbated by the environmental conditions of underground mining. Depending on the mining method, the column of blasted rock can exert large compressive forces on the open face of the extraction point. Humidity can also increase cohesion forces, making the whole process more difficult [42]. In addition, large rocks may be present at the draw point. For this reason, the operator must be aware of the presence of large rocks as he approaches the draw point, and, depending on their size, may be forced to change the goal of the maneuver from filling the bucket to just loading one large rock. If the rock is too big to be loaded, the task is aborted. Furthermore, other conditions, such as narrow tunnels, uneven or unprepared terrain, and/or significant amounts of water and mud near the draw point, can make the loading process very challenging.

An autonomous system should be prepared to deal with all of these situations, in addition to performing certain processing and steps that a human operator performs with while driving the LHD, such as detecting a sufficiently accurate enough position of the rock pile before charging at it, making sure not to hit the tunnel while pulling back from the rock, and estimating whether the bucket is full enough.

4.2. Autonomous Loading Based on the Autonomous Collaborative Mining Paradigm

An autonomous loading routine can be thought of as a sequence of several steps that perform specific actions. Here, *rock pile identification* is the process of finding the location and status of the draw point (whether it is suitable for loading or not). *Positioning* refers to orienting the LHD machine to ensure that a forward thrust will end in a collision with the rock pile. *Charging* is the step that considers lowering the bucket and accelerating towards the pile. *Excavation* is the machine control that performs the digging action. *Pull back* comprises the backward movement of the vehicle and bucket shake. Finally, *payload weighing* estimates the amount of ore loaded. Most of these steps may fail due to the highly variable environment in underground mining. For example, (i) the on-board sensors may not accurately locate the rock pile, (ii) the machine positioning may not avoid colliding with a tunnel wall, (iii) the collision between the machine and the rock pile may not be detected, or (iv) the bucket of the LHD may become stuck during excavation. In any of these cases, human intervention would be required to resolve the problem. In fact, most advanced mining automation systems for mining operations require human interaction with automated mobile machines, at least to monitor/supervise a fleet of machines. Since large operations have a fleet of vehicles, it is common for one or more operators to monitor/supervise and assist multiple semi-autonomous machines, as each machine does not require constant attention. For these reasons, it is imperative that this human interaction be taken into account in the design phase of an autonomous loading system and thus included in the formulation of its behavior.

Taking these considerations into account, an autonomous loading process with these characteristics is presented in [40]. Figure 4 shows the proposed state machine that models the autonomous loading process. Two distinct assistance states are defined: *positioning assistance*, wherein the operator assumes control of the machine to position it in front of the rock pile, and *loading assistance*, wherein the operator assumes control of the LHD to perform the remaining steps of the loading maneuver up until *payload weighing*. *Positioning assistance* is called up either when the *rock pile identification* fails to characterize the rock or if *positioning* is unable to place the LHD to start the charging procedure. Once the LHD is positioned correctly, the LHD's controller regains control of the LHD in order to commence *charging* the pile. *Loading assistance* is called up if *charging*, *excavation*, or *pull back* cannot be completed successfully. Following the successful completion of the pullback step, the LHD's controller regains control of the LHD in order to complete the *payload weighing* step.

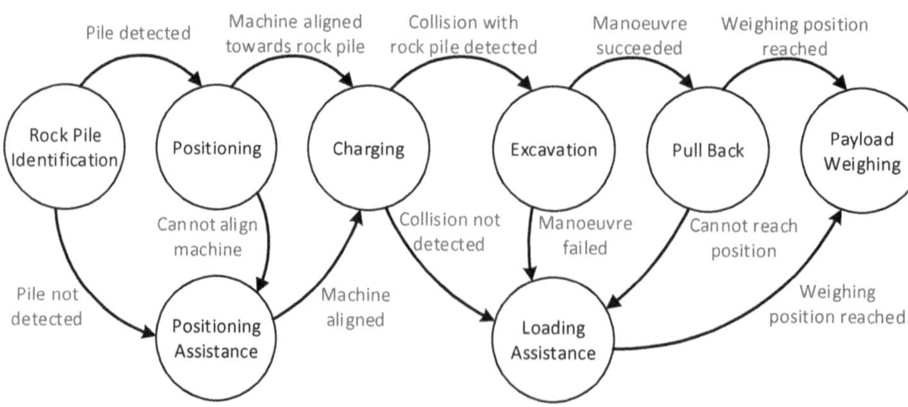

Figure 4. Finite state machine formulation for the autonomous loading with human operators in the loop. Image taken from [40].

In Algorithm 1, a pseudocode is shown that implements the state machine of the loading process.

Algorithm 1: Autonomous loading process.

```
current_state ← autonomous_rock_pile_identification();
if current_state == pile_not_detected then
  | current_state ← positioning_assistance();
else
  | current_state ← autonomous_positioning();
  | if current_state == cannot_align_machine then
  |   L current_state ← positioning_assistance();
if current_state == machine_aligned then
  L current_state ← charging();
if current_state == collision_with_rock_pile_detected then
  | current_state ← excavation();
  | if current_state == manoeuvre_failed then
  |   | current_state ← loading_assistance();
  | else
  |   | current_state ← pull_back();
  |   | if current_state == weighted_position_reached then
  |   |   | current_state ← payload_weighing();
  |   | else
  |   |   L current_state ← loading_assistance();
else
  L current_state ← loading_assistance();
if current_state == weighted_position_reached then
  L current_state ← payload_weighing();
```

4.3. Validation in Real Mining Environments

The excavation algorithm of the autonomous loading system was first validated for six months in a sublevel stoping copper mine in northern Chile. In these experiments, an LHD model LF11H from the German supplier GHH with a bucket of 11-ton capacity was used. The main sensors used were two SICK LIDARs model LMS511-10100 (see more details of the experiments in [40]). The rock pile identification module was validated using data obtained during the same tests.

At the end of the validation period, a total of 30 complete loading trials were performed. Similar to manual operation, multiple loading attempts were required to fill the bucket during each trial (manual operation of the LHD in sublevel stopes typically requires two or three attempts to achieve a proper bucket fill factor, as the procedure is performed using line-of-sight radio control with limited visibility behind the LHD). A total of 81 loading attempts were made to complete the 30 full loads. Of these attempts, 7 failed (early collision trigger) and 74 were successful, resulting in a 91% success rate of the loading and collision detection methods. For the 30 completed loads, an average of 3.6 attempts were required to obtain a bucket full of material, with an average fill factor of 90% (see Table 3 in [40] for more details).

After this first validation in a real mine, the system was improved and adapted for use in Room and Pillar mines and subjected to a full validation (all steps) again, during six months, in a potash mine located in Germany. In these experiments, an LHD model SLP-14H from the German supplier GHH with a bucket of 14-ton capacity was used. The main sensors used were two SICK LIDARs model LMS511-10100 and one 3D Ouster LIDAR model OS2-64. To evaluate the proposed autonomous loading method, eight consecutive loads of material (trials) were performed. Then, eight consecutive loads of material were performed by an expert operator in the same sector. In each trial, the amount of material loaded and the loading time were recorded for comparison. In seven out of eight trials, the autonomous LHD was able to complete the loading process (88% of the time). In one trial, the LHD generated an alarm and aborted at the pullback stage. The average weight achieved by the autonomous loading system across all trials was 11.3 tons, 79.25% of the average weight achieved by manual loading. The details of this validation are described in [3].

It must be noted that in Room and Pillar mines, there is a wide variety of muck piles, and working under the principle of shared autonomy makes the system more flexible to deal with this diversity; thus, due to the very dynamic nature of the muck piles used in Room and Pillar mines, the here-described autonomous loading system is the first one to be successfully validated in this kind of mine.

Since November 2024, the system is being fine-tuned to achieve the production KPI defined by the customer, and there is a good chance that the system will be adapted by this mine.

The reader is referred to [40] and [3] for details of the technical implementation of the technology and how the validation process was carried out.

5. Conclusions

The future mine must be designed to provide a safe environment for humans, situated at a safe distance from potentially hazardous areas. The automation of mining operations plays a pivotal role in achieving this objective. In this context, we put forward the autonomous collaborative mining paradigm as a means of enhancing the automation of mining operations by incorporating human–robot collaborative mechanisms. This paradigm entails a flexible, collaborative, and synergistic interaction between humans and machines, with the objective of achieving safer and more efficient mining operations. The paradigm posits that autonomous operations still require human involvement, particularly in supervisory roles. Furthermore, shared autonomy is regarded as a potential alternative to full autonomy in instances where the latter is not a viable option. Dynamic confinement policies facilitate the physical coexistence of automated equipment and humans within the same working areas. Furthermore, the utilization of augmented reality (AR) technologies allows for the enhancement of human operators' capabilities, thereby transforming them into versatile operators capable of performing a diverse range of tasks. This is a crucial aspect in achieving the objective of minimizing the number of humans in potentially hazardous working areas, as multi-purpose human operators allow for a reduction in the number of personnel required.

The autonomous collaborative mining paradigm represents not an end in itself, but rather a means of working towards the mine of the future. The implementation of the paradigm necessitates the consideration of pertinent enabling factors, including the development of interoperability standards, new regulations, the adaptation of the mining designs and procedures to facilitate the use of autonomous and teleoperated equipment, the development of new sensors to be placed in the mines, high-bandwidth and low-latency data networks, and the training of a workforce in the use of new technologies. The implementation of the paradigm cannot be addressed by a single institution; it requires the coordinated efforts of a number of stakeholders, including mining companies, original equipment manufacturers (OEMs), research and development institutions, technological companies, and regulatory bodies. This represents a significant challenge, but it is nevertheless a worthwhile endeavor to begin working towards the development and implementation of the paradigm.

It is notable that the autonomous collaborative mining paradigm has a direct application in space mining. The processes involved in mining on a planet or an asteroid can be conceptualized as analogous to those employed in the mining of distant locations, whether underground or in a remote desert. This indicates that the automation technologies currently utilized in terrestrial mining can be effectively applied in space mining. In this context, some key elements of the autonomous collaborative mining paradigm such as shared autonomy, human–machine collaboration, and general-purpose operators may be considered.

Funding: This work was funded by the Chilean National Research Agency ANID under project grant Basal AFB230001.

Data Availability Statement: No new data were created or analyzed in this study. Data sharing is not applicable to this article.

Acknowledgments: This work was funded by the Chilean National Research Agency ANID under project grant Basal AFB230001.

Conflicts of Interest: The author declares no conflicts of interest.

References

1. Ali, D.; Frimpong, S. Artificial intelligence, machine learning and process automation: Existing knowledge frontier and way forward for mining sector. *Artif. Intell. Rev.* **2020**, *53*, 6025–6042. [CrossRef]
2. Long, M.; Schafrik, S.; Kolapo, P.; Agioutantis, Z.; Sottile, J. Equipment and Operations Automation in Mining: A Review. *Machines* **2024**, *12*, 713. [CrossRef]
3. Cárdenas, D.; Loncomilla, P.; Inostroza, F.; Parra-Tsunekawa, I.; Ruiz-del-Solar, J. Autonomous detection and loading of ore piles with Load-Haul-Dump machines in Room & Pillar mines. *J. Field Robot.* **2023**, *40*, 1424–1443. [CrossRef]
4. GMG Zero Mine Entry Initiative. Available online: https://gmggroup.org/zero-entry-mining/ (accessed on 24 November 2024).
5. Baiden, G. Multiple LHD teleoperation and guidance at Inco limited. In Proceedings of the International Mining Conference, Kalgoorlie, Australia, 5–8 July 1993.
6. Labonte, F.; Cohen, P. Perceptual aspects of mining equipment teleoperation. In Proceedings of the 6th Canadian Symposium on Mining Automation, Montreal, QC, Canada, 16–19 October 1994.
7. Vagenas, N.; Scoble, M.; Baiden, G. A review of the first 25 years of mobile machine automation in underground hard rock mines. *CIM Bull.* **1997**, *90*, 57–62.
8. Poole, R.A.; Golde, P.V.; Baiden, G.R.; Scoble, M. A review of INCO's mining automation efforts in the Sudbury Basin. *CIM Bull.* **1998**, *91*, 68–74.
9. Scheding, S.; Dissanayake, G.; Nebot, E.; Durrant-Whyte, H. Slip modelling and aided inertial navigation of an LHD. In Proceedings of the International Conference on Robotics and Automation, Albuquerque, NM, USA, 25 April 1997; Volume 3, pp. 1904–1909.
10. Madhavan, R.; Dissanayake, M.W.M.G.; Durrant-Whyte, H.F. Autonomous underground navigation of an LHD using a combined ICP-EKF approach. In Proceedings of the IEEE International Conference on Robotics and Automation, Leuven, Belgium, 16–20 May 1998; Volume 4, pp. 3703–3708.
11. Scheding, S.; Dissanayake, G.; Nebot, E.M.; Durrant-Whyte, H. An experiment in autonomous navigation of an underground mining vehicle. *IEEE Trans. Robot. Autom.* **1999**, *15*, 85–95. [CrossRef]
12. Makela, H.; Lehtinen, H.; Rintanen, K.; Koskinen, K. Navigation system for LHD machines. *IFAC Proc. Vol.* **1995**, *28*, 295–330. [CrossRef]
13. Mäkelä, H. Overview of LHD navigation without artificial beacons. *J. Robot. Auton. Syst.* **2001**, *36*, 21–35. [CrossRef]
14. Salvador, C.; Mascaro, M.; Ruiz-del-Solar, J. Automation of Unit and Auxiliary Operations in Block/Panel Caving: Challenges and Opportunities. In Proceedings of the MassMin2020—The 8th International Conference on Mass Mining, Santiago, Chile, 9–11 December 2020.
15. Sandvik to Automate New LHD Fleet at Codelco's El Teniente Copper Mine. 2021. Available online: https://im-mining.com/2021/02/16/sandvik-to-automate-new-lhd-fleet-at-codelcos-el-teniente-copper-mine/ (accessed on 24 November 2024).
16. Larsson, J.; Appelgren, J.; Marshall, J. Next generation system for unmanned LHD operation in underground mines. In Proceedings of the Annual Meeting and Exhibition of the Society for Mining, Metallurgy & Exploration (SME), Phoenix, AZ, USA, 28 February–3 March 2010.
17. The Rise of Autonomous Trucks in the Mining Industry. 2024. Available online: https://www.mixtelematics.com/us/resources/blog/autonomous-trucks-in-mining/ (accessed on 24 November 2024).
18. Autonomous Haulage Systems—The Future of Mining Operations. 2024. Available online: https://www.futurebridge.com/industry/perspectives-industrial-manufacturing/autonomous-haulage-systems-the-future-of-mining-operations/ (accessed on 24 November 2024).

19. What's New in Automated Drilling and Blasting? Mining Technology. 2024. Available online: https://www.mining-technology.com/features/mining-automated-drilling-and-blasting/?cf-view (accessed on 24 November 2024).
20. Voronov, Y.; Voronov, A.; Makhambayev, D. Current state and development prospects of autonomous haulage at surface mines. In *E3S Web of Conferences*; EDP Sciences: Les Ulis, France, 2020; Volume 174, p. 01028.
21. Rogers, W.P.; Kahraman, M.M.; Drews, F.A.; Powell, K.; Haight, J.M.; Wang, Y.; Baxla, K.; Sobalkar, M. Automation in the Mining Industry: Review of Technology, Systems, Human Factors, and Political. *Min. Metall. Explor.* **2019**, *36*, 607–631.
22. Leung, R.; Hill, A.J.; Melkumyan, A. Automation and Artificial Intelligence Technology in Surface Mining: A Brief Introduction to Open-Pit Operations in the Pilbara. *IEEE Robot. Autom. Mag.* **2023**, *2023*, 2–21. [CrossRef]
23. Marshall, J.; Bonchis, A.; Neobot, E.; Scheding, S. Robotics in Mining. In *Springer Handbook of Robotics*; Siciliano, B., Khatib, O., Eds.; Springer: Cham, Switzerland, 2016; pp. 1549–1576, ISBN 978-3-319-32552-1.
24. Cucuzza, J. The status and future of mining automation: An overview. *IEEE Ind. Electron. Mag.* **2021**, *15*, 6–12.
25. Du, H.; Chan, L.; Tong, J.; Raad, R.; Naghdy, F.; Guo, Q.; Yu, Y.; Islam, M.R.; Tubbal, F.; Ros, M.; et al. Industrial Progress of Robotic Automation in Mining Applications: A Survey. *Min. Metall. Explor.* **2025**. [CrossRef]
26. Pourrahimian, Y.; Askari Nasab, H. *An Overview of Block Caving Operation and Available Methods for Production Scheduling of Block Cave Mines*; Report Number: MOL Research Report Two 2009/2010; Mining Optimization Laboratory, Universidad de Alberta: Edmonton, AB, Canada, 2010.
27. Delcker, J. *The Man Who Invented the Self-Driving Car (in 1986)*; Politico: Arlington, VA, USA, 2018.
28. The Future of Autonomous Vehicles, Future Agenda. 2020. Available online: https://www.futureagenda.org/future-autonomous-vehicles-finalreport/ (accessed on 24 November 2024).
29. Thrun, S.; Montemerlo, M.; Dahlkamp, H.; Stavens, D.; Aron, A.; Diebel, J.; Fong, P.; Gale, J.; Halpenny, M.; Hoffmann, G.; et al. Stanley: The Robot That Won the DARPA Grand Challenge. In *The 2005 DARPA Grand Challenge*; Springer Tracts in Advanced Robotics; Buehler, M., Iagnemma, K., Singh, S., Eds.; Springer: Berlin/Heidelberg, Germany, 2007; Volume 36. [CrossRef]
30. DARPA Grand Challenge. 2005. Available online: https://en.wikipedia.org/wiki/DARPA_Grand_Challenge_(2005) (accessed on 24 November 2024).
31. Autonomous Vehicles Are Coming, but Slowly. Special Report: Hands of the Wheel. *Economist*, 14 April 2023.
32. Kaszas, D.; Roberts, A. Comfort with varying levels of human supervision in self-driving cars: Determining factors in Europe. *Int. J. Transp. Sci. Technol.* **2023**, *12*, 809–821. [CrossRef]
33. Shen, Z.M.; Liu, J.; He, Y.; Zhang, X.; Xu, R.; Yu, H.; Cui, P. Towards out-of-distribution generalization: A survey. *arXiv* **2021**, arXiv:2108.13624. [CrossRef]
34. DARPA Subterranean (SubT) Challenge. 2024. Available online: https://www.darpa.mil/program/darpa-subterranean-challenge (accessed on 24 November 2024).
35. Chung, T.; Orekhov, V.; Maio, A. Into the Robotic Depths: Analysis and Insights from the DARPA Subterranean Challenge. *Annu. Rev. Control Robot. Auton. Syst.* **2023**, *6*, 477–502.
36. Espinoza, J.; Mascaro, M.; Morales, N.; Ruiz-del-Solar, J. Improving productivity in block/panel caving through dynamic confinement of semi-autonomous Load-Haul-Dump machines. *Int. J. Min. Reclam. Environ.* **2022**, *36*, 552–573. [CrossRef]
37. What Are Collaborative Robots? A Fast-Growing Robotics Technology with Major Market Potential. 2024. Available online: https://www.automate.org/robotics/cobots/what-are-collaborative-robots (accessed on 24 November 2024).
38. Selvaggio, M.; Cognetti, M.; Nikolaidis, S.; Ivaldi, S.; Siciliano, B. Autonomy in Physical Human-Robot Interaction: A Brief Survey. *IEEE Robot. Autom. Lett.* **2021**, *6*, 7989–7996. [CrossRef]
39. Sheridan, T.B. Teleoperation, telerobotics and telepresence: A progress report. *Control Eng. Pract.* **1995**, *3*, 205–214. [CrossRef]
40. Tampier, C.; Mascaro, M.; Ruiz-del-Solar, J. Autonomous Loading System for Load-Haul-Dump (LHD) Machines Used in Underground Mining. *Appl. Sci.* **2011**, *11*, 8718.
41. Erikson, J.M.; Mueggenburg, N.W.; Jaeger, H.M.; Nagel, S.R. Force distributions in three-dimensional compressible granular packs. *Phys. Rev. E* **2002**, *66*, 040301. [CrossRef]
42. Herminghaus, S. Dynamics of wet granular matter. *Adv. Phys.* **2005**, *54*, 221–261. [CrossRef]

mining

MDPI

Article

Autonomous and Operator-Assisted Electric Rope Shovel Performance Study

Ali Yaghini [1,*], Robert Hall [2] and Derek Apel [1]

[1] Department of Civil and Environmental Engineering and the School of Mining and Petroleum Engineering, University of Alberta, Edmonton, AB T6G 2R3, Canada

[2] Department of Mining Engineering and Management, University of South Dakota, Vermillion, SD 57069, USA

* Correspondence: yaghini@ualberta.ca

Abstract: Automation has been changing the mining industry for the past two decades. Material handling is a critical task in a mining operation, and truck-shovel handling systems are the primary method for surface mining. Mines have deployed autonomous trucks, and their positive impact on both production and safety has been reported. This paper aims to study the extent to which autonomous and operator-assisted loading units could improve different aspects of a mining operation. Four different levels of automation ranging from operator-assisted swing and return to fully autonomous for a shovel were considered. A discrete event simulation model was developed and verified using detailed data from a shovel monitoring system. Later, the developed model was deployed to assess how each of the proposed technologies could improve productivity and efficiency. Results show that up to a 41% increase in production can be achieved. Both mining companies and equipment manufacturers can use the methodology and results of this study for future decision-making and product development.

Keywords: electric rope shovel; automation; performance monitoring; discrete event simulation; surface mining

1. Introduction

Loading and hauling materials, typically referred to as a haulage system, are critical in any mining activity. In a surface mining operation, one of the prominent types of such systems consists of shovels and trucks as loading and hauling equipment. Material movement accounts for 35% to 60% of operation cost in surface mining [1,2]. Therefore, minor improvements in this system could create significant savings in operational costs and increase the production of the mine.

Shovels, as reliable and long-lasting loading units, play a crucial role in the haulage system [3], and it is expected that minor variations in their operation would have a significant influence on the system outcome.

Previous research shows that the performance of an electrical rope shovel is influenced to a great extent by machine condition, operators practices and the muck-pile characteristics [4–6]. Automation offers the potential to eliminate human errors and variability caused by human factors and provide consistent performance and optimized output. Automation of loading units has been an active area of research, mainly focused on excavators and wheel loaders [7].

Mines use a fleet management system to optimize their hauling system and maximize their production while feeding the processing plant with the targeted rate and predefined grade blending constraints [8,9]. Although even the most sophisticated fleet management algorithms are trying to reach their goal by finding the best routes and optimizing trucks and shovels assignment, it is proved that in practice, the actual results deviate from the predicted ones [10].

Deficiencies, particularly variations in the hauling system operation, can be seen as a significant reason for this deviation in which the human factor plays an important role [11,12]. Different levels of automation ranging from tele-remote, operator-assisted, and driver-less equipment have been deployed to reduce operator errors, remove issues related to human behaviour such as variability as well as increase equipment utilization [13–17].

Although the complex nature of mining activities and uncertainties associated with them is difficult to capture using traditional analysis methods [18], discrete event simulation (DES) has been proven as a tool to model, assess, and investigate operational scenarios in the mining industry [19,20].

This paper investigates how different levels of automation of loading units could improve both shovel and mine production and how other mining activities are influenced by it. This paper also describes how a detailed simulation model can be developed to capture the loading operational uncertainties. This model uses the key performance activities of shovel operators as an input to demonstrate their influences on other activities in a mine and, more importantly, mine productivity.

2. Background

2.1. Automation in Mining

Automation is generally defined as the intelligent operation of equipment using technology without direct human contribution [21]. The mining industry has been considered a slow industry in adopting and implementing new technologies [22]. The combination of a long time for new technology development -between seven to ten years-, the often short-term financial focus, and the risk of failed technology are contributors to the mining industry's slow adoption of technology [23].

The initial purpose of automation in the mining industry was to protect the health and safety of mineworkers, but it was also considered for improvement in productivity by minimizing human error and eliminating poor driving behaviours of tired and distracted workers [24,25]. In addition, the reduction in lost time due to shift change, breaks, etc., was viewed as how automation could improve productivity.

Complexities related to the automation of loading units have encouraged researchers to aim for different levels of automation. Roberts, Duff and Corke [26] suggested the following steps toward automation:

(1) Manual operation: The operator physically presents in the operator's cab and performs the operation.
(2) In-sight teleoperation: The operator is located in the machine's vicinity and performs all tasks using a remote control.
(3) Tele-remote operation: The operator controls the machine from a remote location using a remote control and video and audio feedback.
(4) Assisted tele-remote operation: the machine control system performs many tasks, and the operator intervenes via tele-remote control needed.
(5) Fully autonomous: The machine operators autonomous.

Operator assistance technologies such as path planning and navigation [27,28] and collision detection and avoidance [29] have been developed mainly for Load-Haul-Dump (LHD) units by researchers. Dunbabin and Corke [30] automate the digging cycle of an electric rope shovel, proposed techniques to detect and avert dipper stall, and the on-line estimation of the dipper fill factor. At the time of writing this paper, there are several operator assistance systems offered by electric rope shovel manufactures that mainly prevent inadvertent mishandling of the machine.

Automation in mining has had many implications ranging from an automated valve to a whole fleet of autonomous haulage trucks. Increased production hours and lower employee and associated costs are among the main benefits reported by researchers [22]. The higher productivity and reduced costs might make some previously uneconomical mines profitable.

2.2. Simulation

Simulation has been deployed to understand and mimic real-world systems and assess different scenarios. Early attempts to use simulation in mining date back to 1950 [31]. Since then, simulation has been used by researchers and professionals to address problems such as truck fleet sizing, dispatch, and allocation, short-, medium-, and long-term mine planning and efficiency improvement. [32–34]

DES in particular has been used by researchers to model different aspects of mine operations. DES decomposes a system into a set of logically separate but sequential processes and events. Each event occurs at a specific instant in time and changes the state of the system. The conventional approach to model shovels in DES, in the absence of details on specific activities, i.e., each pass payload, dig, swing and return times, is to obtain truckload information from truck onboard monitoring systems and use shovel bucket capacity to estimate the number of passes it took to load the truck. Then, the average loading cycle time could be calculated by dividing the total loading time from a dispatch software and the estimated number of passes. It can be argued that this approach relies on data from sources other than the shovel itself. It is believed that this leads to a wide range of estimations and assumptions made throughout the calculations that reduce the value of the results. Additionally, by obtaining loading time from dispatch software, this approach ignores shovels' other activities such as propel and walk that could lead to an oversimplified model.

Data from a shovel onboard monitoring system provides details such as digging, swinging and returning times, each individual pass and total truckload and many other KPIs gathered directly from shovel operation. These details could be used to model shovels more precisely and create an understanding of where improvements can be attained.

3. Approach

3.1. Discrete Event Simulation Model

A DES model was developed to assess the influence of different levels of automation on shovel, truck, and mine production and performance [33]. The developed DES model evaluates the truck haulage system and shovel loading units and ignores the processing plant. It is assumed that the quality and density of materials throughout the simulation are consistent. The equipment mean time between failure and time to repair are used to model equipment failures. The impact of truck driver practices and weather changes is ignored in this model.

The DES model is developed based on historical data from a currently active surface mine. The operation name cannot be revealed due to confidentiality considerations. The mine employs five shovels, three waste and two ore. The shovel monitoring database contains records for a fleet of P&H4100 shovels. A fleet of Cat 793C trucks haul materials to two crushers and a waste dump based on a prescheduled plan.

Truck fleet size and their availability strongly influence shovel production and utilization. As the main focus of this study is on the loading aspect of the operation, a fleet of 60 trucks is considered for this operation to meet the required production in the base model.

The DES model is developed in Rockwell ARENA software (Rockwell Automation, Milwaukee, WI, USA). ARENA has been successfully used by researchers in mining, operation research and other industries for the purpose of studying the behavior of complex systems. The software uses SIMAN [35] processor and simulation language. The following five steps are carried out to develop the model:

(1) ARENA Input Analyzer software is used for all input parameters to calculate and draw their histogram. Then, best-fitted distributions are selected based on the sum of square error criteria calculated by the software (Appendix A).
(2) A mine road network needs to be developed. A VBA macro is used to build the network in ARENA.

(3) The fitted distributions and production schedule are stored in an MS Excel file. Before each run and in the check state, ARENA reads all those parameters and uses them to run the model.
(4) The mine DES model is build using ARENA modules (Figure 1). Trucks and shovels are modelled as entities and resources, respectively. ARENA transporter is used to model truck interactions. Record and write modules are used to save KPIs.
(5) Total production, utilization of the equipment, queue length, and time at shovels are used as KPIs to evaluate and compare different operational scenarios. As a convention, the DES model is set up to run for 30 shifts of 24-h. It is also set up for 25 replications to reach the required half-width for the production within a 95% confidence interval of 20,000 tons.
(6) Results are written in an MS Excel file for the data analysis steps.

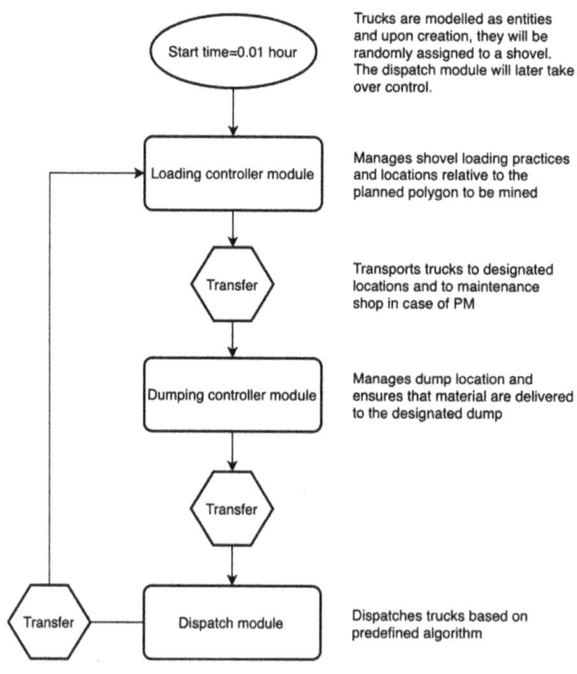

Figure 1. Flow of the mine operation DES.

Additionally, to model loading activities, a shovel submodule is developed. Figure 2 presents the sub-module flow chart. The sub-module uses the number-of-passes-to-load distribution to assign the number of buckets to each truck. Then, for each pass cycle, its specific distribution is used to calculate the bucket load and dig, swing and return times. This will repeat for the assigned number of buckets to the truck.

If the addition of the bucket load to the truckload will cause it to exceed the maximum allowed capacity, that pass load will be discarded. The shovel will be tasked to take an extra pass to reach the minimum allowable load for the truck to be dispatched if it has not yet. This procedure complies with the author's field observation that operators dump their bucket load as soon as the system notifies them about the excessive truck final load considering their current bucket load is added. In this situation, it is common for the operator to dig another bucket load with less material. After the loading process is completed, a propel time will be calculated from its distribution and added to the total cycle time.

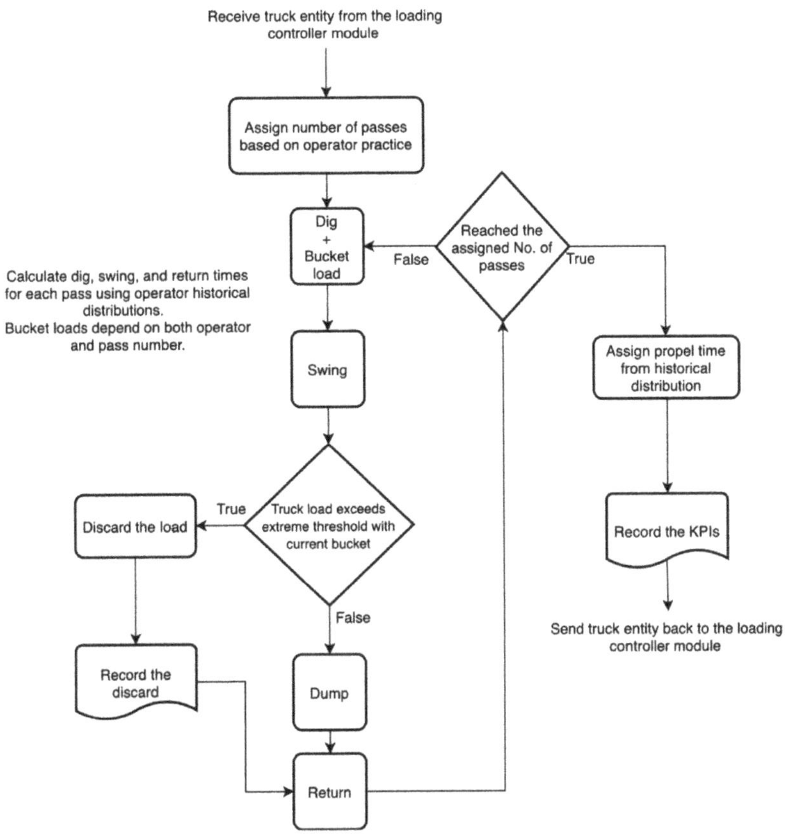

Figure 2. Electric rope shovel simulation sub-module.

3.2. Operational Scenarios

In order to study the extent to which autonomous and operator-assisted loading units could improve different aspects of a mining operation. Four different levels of automation, ranging from operator-assisted swing and return to fully autonomous for a shovel, are considered and compared against a manual fleet as the benchmark.

Automating the dig, swing, and return part of a loading activity can be seen as the initial steps toward automating electric rope shovels. Complexities in the dig section of a loading activity suggest that their automation could occur separately from swing and return. Thus, here in the first autonomous scenario, a swing and return system is studied. The second scenario evaluates adding the dig-assisted function to the system.

The third scenario investigates how a fully autonomous shovel can improve the shovel and mine production and performance. Furthermore, as previously noted in an autonomous LHD study [29], the most significant advantage of automation could be seen in fewer errors and higher consistency. Hence, the fourth scenario explores what benefits could be gain by a hypothetical, optimal autonomous electric rope shovel. Table 1 summarizes automation scenarios.

Table 1. Automation scenarios.

Automation	Practice		
	Swing and Return	**Dig**	**Propel (Positioning and Movement)**
Level 1	Automated	Operator	Operator
Level 2	Automated	Automated	Operator
Level 3	Automated	Automated	Automated
Level 4	Automated *	Automated *	Automated *

* Improved, compared to the base automated scenario.

3.2.1. Manual

The base scenario represents a manually operated fleet of shovels. Operating data from a shovel monitoring system is used to calculate and fit distributions to critical indicators of operator performance (i.e., dig, swing, and return times, number of buckets to load a truck, each pass load, and propel time). The shovel monitoring system database includes 364,655 truck loading cycle records. For each loaded truck, in addition to the truck capacity and final load, other KPIs, including digging, swinging and returning times and type of material, are recorded.

Tables 2 and 3 present average and standard deviation of main KPIs for the fleet of shovels. It should be noted that regular breaks and shift change downtimes are included in the DES to capture lost production hours due to those events (Appendix A).

Table 2. Manual fleet key activity statistics.

Activity	Statistics		
	Mean	**Standard Deviation**	**Distribution**
Dig time (s)	15.1	4.15	NORM (15.1, 4.15)
Swing time (s)	7.02	2.6	LOGN (7.02, 2.6)
Return time (s)	8.2	2.8	NORM (8.2, 2.8)
Propel time (s)	22.6	14.4	1 + LOGN (21.7, 16)
No. of bucket	3.17	0.41	DISC (3,0.83,0.967,4,1,5)

Table 3. Bucket load statistics for the fleet of operators.

Pass No.	Load Statistics (tons)		
	Mean	**Standard Deviation**	**Distribution**
Pass 1	105	21.8	11 + WEIB (98.4, 4.57)
Pass 2	105	17.8	11 + 129 × BETA (8.85, 3.32)
Pass 3	97.7	18.2	11 + 129 × BETA (7.35, 3.6)
Pass 4	69.2	27.8	11 + 128 × BETA (1.94, 2.33)
Pass 5	80.9	30.1	TRIA (11, 99, 139)

3.2.2. Automation Level 1 and 2. Swing-, Return- and Dig-Assisted System

A swing and return assisted system is studied in the first automation scenario. The second scenario evaluates adding the dig-assisted function to the system.

To capture the extent to which the proposed automation influences mine productivity and operational KPIs, statistics for the fleet average in the manual DES model are replaced with the best practices among operators [The best operator is selected using the ORS index suggested by authors (for more information, please refer to Modeling the influence of electric shovel operator performance on mine productivity, Yaghini, A., Hall, R)]. This is a reasonable assumption as any proposed ideal automation technology needs to outperform its human counterpart. Tables 4 and 5 presents the best operator KPIs.

Table 4. Best operators loading statistics.

Activity	Statistics		
	Mean	**Standard Deviation**	**Distribution**
Dig time [1] (s)	13.5	3.3	NORM (13.5, 3.3)
Swing time [2] (s)	6.3	2.66	LOGN (6.3, 2.66)
Return time [2] (s)	7.3	2.4	NORM (7.3, 2.4)
Propel time (s)	11.3	8.06	1 + LOGN (10.3, 7.17)
No. of buckets [1]	3.03	0.2	DISC (3, 0.97, 1, 4)

[1] Dig time and No. of buckets are used for the second scenario in the DES model. [2] Swing and return times are used for both scenarios in the DES model.

Table 5. Best operator bucket load statistics [1].

Pass No.	Load Statistics (tons)		
	Mean	**Standard Deviation**	**Distribution**
Pass 1	108	16.2	11 + 129 × BETA (10.5, 3.49)
Pass 2	110	15.8	11 + 128 × BETA (12.7, 4.25)
Pass 3	98.4	16.9	11 + WEIB (92.1, 6.58)
Pass 4	48.9	21.4	17 + WEIB (54, 2.43)

[1] Bucket loads are used for the second scenario.

Figure 3 shows how the best operators' truckload ratio and dig and swing times compare with the fleet values. It can be seen that the best operator not only spends less time performing the activities but are also more consistent. Additionally, the best operator tends to load trucks slightly higher than 100% with noticeably better consistency (5% standard deviation compared to the fleet average 9%).

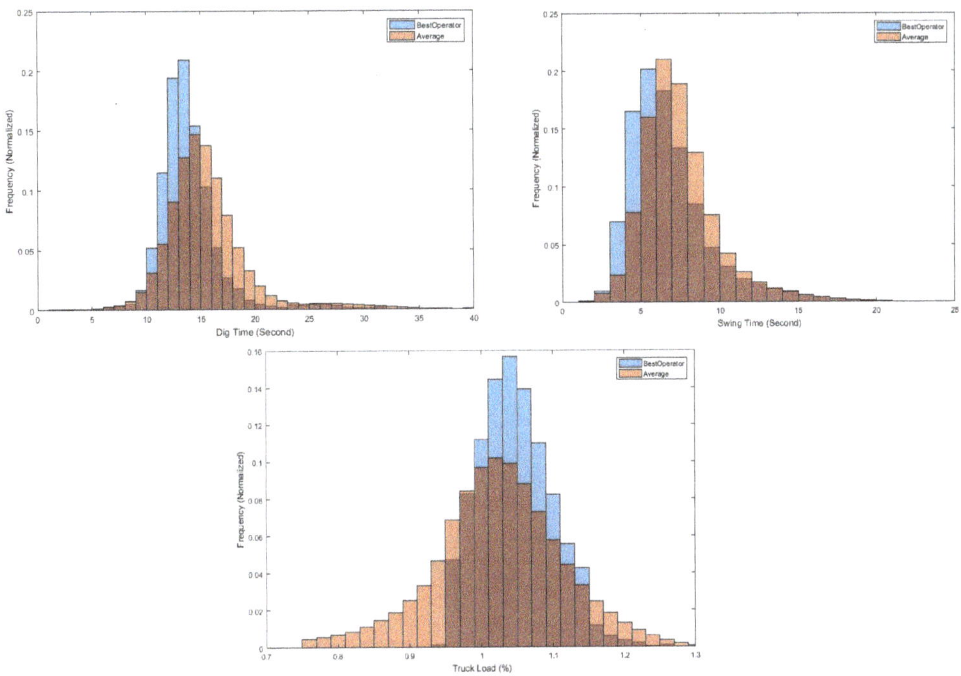

Figure 3. Best operator statistics vs. fleet average.

3.2.3. Automation Level 3 and 4. Full and Improved Autonomous Shovel

It is expected that a full autonomous shovel could load a truck without the need for human intervention. Hence, as argued for the automation level 1 and 2, statistics from loading activities best practices are used to model an autonomous shovel performance in the DES model. In addition to digging, swing, and return times presented in Table 3, it is assumed that shovels load trucks consistently with three buckets. Each load estimated from the distribution belongs to the first pass in Table 5.

Furthermore, as previously noted in an autonomous LHD study [29], the most significant advantage of automation could be seen in fewer errors and higher consistency. A scenario with 10% and 50% hypothetical improvements in the average and standard deviation of loading activities is examined (except for the shovel bucket load as the 10% increase would cause it to exceed its maximum capacity). This scenario could be interpreted as an ideal automation implementation. Table 6 presents the main statistics used for the proposed system.

Table 6. Hypothetical improved autonomous shovel key activity statistics.

Activity	Statistics		
	Mean	Standard Deviation	Distribution
Dig time (s)	12.15	1.65	NORM (12.15,1.65)
Swing time (s)	5.67	1.33	NORM (5.67,1.33)
Return time (s)	6.57	1.2	NORM (6.57,1.2)
Propel time (s)	10.17	4.03	NORM (10.17, 4.03)
Bucket Load (ton)	108.8	8.2	NORM (118.8, 8.2)
No. of bucket (no)	3	N/A	CONST (3)

4. Results and Discussion

4.1. DES Model

4.1.1. Verification of the Simulation Model

Before the developed simulation model results can be used and interpreted, the model needs to be verified. The first step is to monitor the model and reflect the mine operation visually. Afterward, shovel-loaded tonnages and delivered materials to the dump locations are used to verify the model. Figure 4 shows a summary of material moved during the run time for all DES models. It can be seen that all loaded materials are delivered to their designated dump locations.

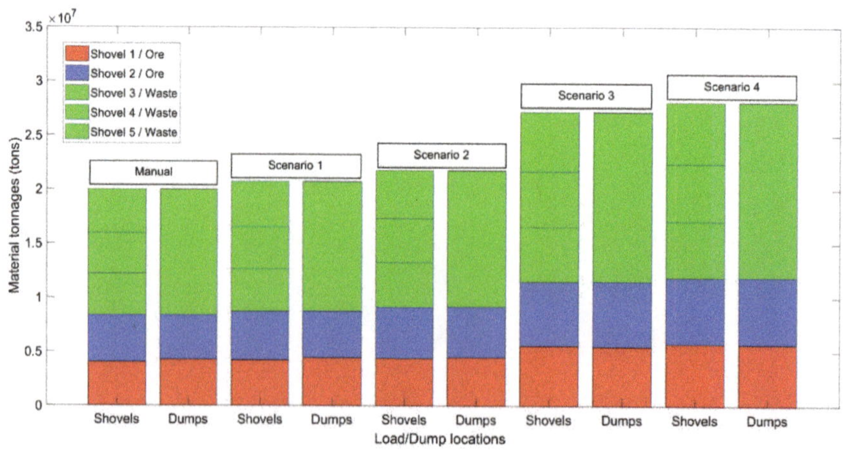

Figure 4. Material moved during the simulation period from shovels to dump locations.

4.1.2. Shovel Production

Table 7 shows the total production of the shovel fleet during the simulation period and compares automated scenarios with the base (i.e., manual) scenario. While implementing an assisted swing and return system in the first scenario has resulted in 3.6% improvement, it has been the automated dig function in the second scenario that improved production by 8.8%, both compared to the manual case. This could be directly linked to the fact that in the automated dig scenario, trucks are being loaded with more optimum bucket loads, and hence, not only their average final load has lower variability, but also the number of passes required to fill them and total loading time are decreased on average. Figure 5 shows these changes.

Table 7. Shovel fleet production during the simulation period.

	Shovel Production (Million tons)						
	1	**2**	**3**	**4**	**5**	**Total**	**% Difference [1]**
Manual fleet	4.04	4.33	3.84	3.74	4.01	19.97	0.0%
Scenario 1	4.21	4.51	3.93	3.86	4.19	20.69	+3.6%
Scenario 2	4.41	4.75	4.13	4.05	4.40	21.73	+8.8%
Scenario 3	5.55	5.96	5.02	5.15	5.51	27.19	+36.1%
Scenario 4	5.74	6.16	5.17	5.31	5.70	28.09	+40.6%

[1] Compared to the manual fleet.

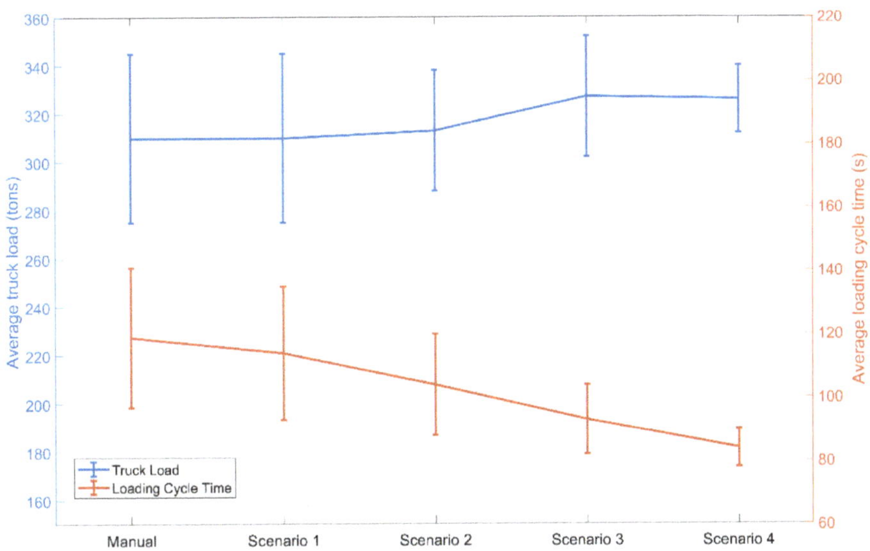

Figure 5. The average and standard deviation of total loading cycle time and a truck load.

As expected before, automation resulted in an astounding 36.1% improvement in production. Reduced propel time and number of buckets to load a truck collectively with the higher bucket load could be regarded as main operational contributors. Furthermore, eliminating the shift changes and lunch break times (a total of 1.5 h per 12-h shift including one shift change and two half an hour breaks) has a significant role in this achievement. It can be seen from Table 8 that those lost production hours can be accounted for almost half of the automation production gain compared to the manual scenario.

Table 8. Influence of shift change and break times on operational KPIs.

Case	Loading Time (s)	Truckload (tons)	Scheduled Utilization (%)	Production (tons × 10^6)
	Mean ± std	Mean ± std	Fleet average	Total
Scenario 3	92.6 ± 10.9	326 ± 25	83.4	27.2
Scenario 3 with break time	92.5 ± 10.9	323 ± 24.8	83.6	23.8

In addition to their influence on production, loading cycle times could affect other aspects of the operation. Figure 6 presents the average number of trucks and their waiting time at loading locations. It can be seen that there is a lower wait time for all non-manual scenarios. While all truck fleet operational parameters (i.e., spotting, backing, and dump times, speeds, capacity, and dispatch algorithm) were the same for all scenarios, it is the shovel loading practices that could be solely considered as a primary contributor to the decrease in truck wait times. Additionally, the extra trucks assigned to active shovels when a shovel is out of service, either due to operator breaks or shift change, could also result in a higher number of trucks in a queue at loading locations in non-autonomous scenarios. Minimizing the wait time improves productivity and reduces operational costs and could be seen as a significant opportunity to reduce the air pollution caused by truck idling.

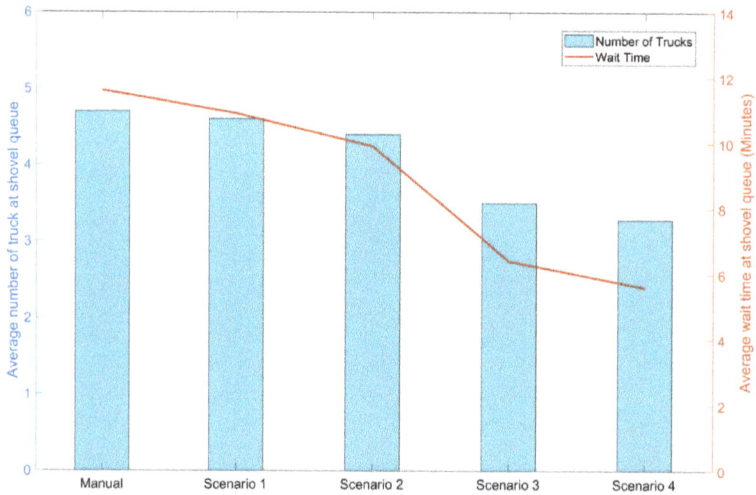

Figure 6. The average number of trucks in the queue and their wait time at loading locations.

Truck fleet size and availability play a significant role in a shovel's utilization of availability. As the direct result of decreased loading times and improved efficiency of shovels, trucks have become the bottleneck of the operation. Hence, shovels have to wait for trucks to become available, and the average shovel fleet utilization dropped by 10% in the improved autonomous scenario compared to the manual. It could be argued that the autonomous and improved autonomous shovel production would have been even more if the truck fleet size were increased. Furthermore, if increasing the number of trucks is not an option for the operation, a scenario with fewer shovels can be considered to meet the same production level.

5. Conclusions

This paper presented a detailed assessment of different levels of automation for an electric rope shovel and their potentials to improve mining loading and haulage system. Four hypothetical levels of automation were considered for shovels, and the influence of each proposed level on both shovel and mine productivity was evaluated. It was found that in addition to increased production for shovels, wait times at loading locations could also be reduced. The increased production for the autonomous scenarios was almost equally due to the elimination of breaks and shift time changes and the assumption of more efficient loading using automation.

It is envisioned that mine professionals can use the proposed simulation approach to help them assess the extent to which automation could improve their activities.

Future work is needed to investigate whether the automation technologies could improve overall machine health by preventing human errors (i.e., extra stresses on machine frame, or events such as boom jacks). Furthermore, how trucks and their health could be influenced by higher and more consistent payloads with better load distribution needs to be studied.

Author Contributions: All authors contributed equally to this paper. All authors have read and agreed to the published version of the manuscript.

Funding: This research received no external funding.

Data Availability Statement: Data is private and not available to public.

Conflicts of Interest: The authors declare no conflict of interest.

Appendix A

After removing outliers and zero values, Arena input analyzer were used to fit distributions on the input data. Table A1 represent mean, standard deviation and the best-fitted distribution on truck input parameters. In the simulation model, based on engineering judgment, a minimum and maximum value were set for all parameters to make sure a sampled number always falls within an acceptable range.

Table A1. Truck input parameters.

Parameter	Mean	Standard Deviation	Distribution
Empty speed (km/h)	35.7	10.7	NORM (35.7, 10.7)
Loaded speed (km/h)	18.8	7.52	5 + LOGN (18.8, 7.52)
Spotting time (s)	37	27.1	1 + LOGN (37.8, 39.1)
Backing time (s)	15.6	10.9	1 + LOGN (14.8, 11.5)
Dumping time (s)	47.9	21.2	NORM (47.9, 21.2)

Table A2 represent mean-time-between-failure (MTBF) and mean-time-to-repair (MTTR) for shovels.

Table A2. BRO input parameters.

Parameter	Distribution
Shovel MTBF	10 × WEIB (130, 0.76)
Shovel MTTR	LOGN (11.39, 14.73)

References

1. Chaowasakoo, P.; Seppälä, H.; Koivo, H.; Zhou, Q. Improving fleet management in mines: The benefit of heterogeneous match factor. *Eur. J. Oper. Res.* **2017**, *261*, 1052–1065. [CrossRef]
2. Hustrulid, W.A.; Kuchta, M.; Martin, R.K. *Open Pit Mine Planning and Design*, 3rd ed.; CRC Press: Boca Raton, FA, USA, 2013; Available online: https://www.crcpress.com/Open-Pit-Mine-Planning-and-Design-Two-Volume-Set--CD-ROM-Pack/Hustrulid-Kuchta-Martin/p/book/9781466575127 (accessed on 1 July 2020).
3. Awuah-Offei, K. Energy Efficiency in Cable Shovel Operations. In *Green Energy and Technology*; Springer: Berlin/Heidelberg, Germany, 2018; pp. 147–157. [CrossRef]
4. Osanloo, M.; Hekmat, A. Prediction of Shovel Productivity in the Gol-e-Gohar Iron Mine. *J. Min. Sci.* **2005**, *41*, 177–184. [CrossRef]

5. Singh, S.P.; Narendrula, R. Factors affecting the productivity of loaders in surface mines. *Int. J. Min. Reclam. Environ.* **2006**, *20*, 20–32. [CrossRef]
6. Vukotic, I.; Kecojevic, V. Evaluation of rope shovel operators in surface coal mining using a Multi-Attribute Decision-Making model. *Int. J. Min. Sci. Technol.* **2014**, *24*, 259–268. [CrossRef]
7. Hemami, A.; Hassani, F. 2017 An Overview of Autonomous Loading of Bulk Material. In Proceedings of the 2009 International Symposium on Automation and Robotics in Construction (ISARC 2009), Online, 24–27 June 2009. [CrossRef]
8. Hauck, R.F. A real-time dispatching algorithm for maximizing open-pit mine production under processing and blending requirements. In *Proceedings, Seminar on Scheduling in Mining, Smelting, and Steelmaking;* Canadian Institute of Mining and Metallurgy: Montreal, QC, Canada, 1973.
9. Matamoros, M.E.V.; Dimitrakopoulos, R. Stochastic short-term mine production schedule accounting for fleet allocation, operational considerations and blending restrictions. *Eur. J. Oper. Res.* **2016**, *255*, 911–921. [CrossRef]
10. Vasquez Coronado, P.P. Optimization of the Haulage Cycle Model for Open Pit Mining Using a Discrete-Event Simulator and a Context-Based Alert System. 2014. Available online: https://s3.amazonaws.com/academia.edu.documents/38121100/APCOM002001.pdf?AWSAccessKeyId=AKIAIWOWYYGZ2Y53UL3A&Expires=1510953242&Signature=UXSbDOTWIbsrPKxKNfVYou%2FDSSg%3D&response-content-disposition=inline%3Bfilename%3DOptimization_of_open_pit_haulage_c (accessed on 1 July 2020).
11. Horberry, T.; Burgess-Limerick, R.; Steiner, L.J. *Human Factors for the Design, Operation, and Maintenance of Mining Equipment;* CRC Press: Boca Raton, FA, USA, 2010.
12. Yaghini, A.; Pourrahimian, Y.; Hall, R. The human factor analysis and classification system (HFACS); a review and lessons from other industries. In *CIM MEMM 2017;* The Chartered Institute of Marketing: Saskatoon, Canada, 2017.
13. Dadhich, S. Assisted Tele-Remote Control of Wheel Loaders in Underground Mining. 2016. Available online: http://www.diva-portal.org/smash/get/diva2:1033000/FULLTEXT01.pdf (accessed on 1 July 2020).
14. Fisher, B.S.; Schnittger, S. Autonomous and Remote Operation Technologies in the Mining Industry Benefits and Costs. BAeconomics Pty Ltd. 2012. Available online: http://www.baeconomics.com.au/wp-content/uploads/2010/01/Mining-innovation-5Feb12.pdf (accessed on 1 July 2020).
15. Hanowski, R.J.; Wierwille, W.W.; Dingus, T.A. An on-road study to investigate fatigue in local/short haul trucking. *Accid. Anal. Prev.* **2003**, *35*, 153–160. [CrossRef]
16. Lewis, M.N.; Werner, J.; Sambirsky, B. Capturing Unrealized Capacity. *CIM Bulletin* **2004**, *97*, 57–62.
17. Roberts, J.; Corke, P. 1997 Automation of underground truck haulage. In Proceedings of the Fourth International Symposium on Mine Mechanisation and Automation, Newcastle, NSW, Australia, 3–4 December 1997.
18. Komljenovic, D.; Nour, G.A.; Popovic, N. An approach for strategic planning and asset management in the mining industry in the context of business and operational complexity. *Int. J. Min. Miner. Eng.* **2015**, *6*, 338. [CrossRef]
19. Zeng, W. A simulation model for truck-shovel operation. 2018. Available online: http://ro.uow.edu.au/theses1/270/ (accessed on 1 July 2020).
20. Upadhyay, S.P. Simulation Optimization of Mine Operations for Uncertainty Based Short Term and Operational Planning in Open Pit Mines. Available online: https://era.library.ualberta.ca/items/bfbf37ab-e3bb-4b8b-912c-84ec394443c7/view/609e54f9-db7f-4ce3-8c9e-7438d26373d2/Upadhyay_Shiv_P_201612_PhD.pdf (accessed on 1 July 2020).
21. Sheridan, T.B. *Humans and Automation: System Design and Research Issues;* Wiley: Hoboken, NJ, USA, 2022.
22. Bellamy, D.; Pravica, L. Assessing the impact of driverless haul trucks in Australian surface mining. *Resour. Policy* **2011**, *36*, 149–158. [CrossRef]
23. Bartos, P.J. Is mining a high-tech industry? *Resour. Policy* **2007**, *32*, 149–158. [CrossRef]
24. Dudley, J.J.; McAree, P.R. Why the mining industry needs a reference architecture for automation initiatives. In Proceedings of the 2013 IEEE/ASME International Conference on Advanced Intelligent Mechatronics, Wollongong, NSW, Australia, 9–12 July 2013; pp. 1792–1797. [CrossRef]
25. Parreira, J. An Interactive Simulation Model to Compare an Autonomous Haulage Truck System with a Manually-Operated System. Ph.D. Thesis, University of British Columbia, Vancouver, BC, Canada, 2013.
26. Roberts, J.M.; Duff, E.S.; Corke, P.I. Reactive navigation and opportunistic localization for autonomous underground mining vehicles. *Inf. Sci.* **2002**, *145*, 127–146. [CrossRef]
27. Gustafson, A. Dependability Assurance for Automatic Load Haul Dump Machines. 2011. Available online: http://www.diva-portal.org/smash/record.jsf?pid=diva2:991068 (accessed on 1 July 2020).
28. Larsson, J. Unmanned Operation of Load-Haul-Dump Vehicles in Mining Environments. 2011. Available online: http://www.diva-portal.org/smash/record.jsf?pid=diva2:512493 (accessed on 1 July 2020).
29. Larsson, J.; Broxvall, M.; Saffiotti, A. An evaluation of local autonomy applied to teleoperated vehicles in underground mines. In Proceedings of the 2010 IEEE International Conference on Robotics and Automation, Anchorage, AL, USA, 3–7 May 2010; pp. 1745–1752. [CrossRef]
30. Dunbabin, M.; Corke, P. Autonomous Excavation Using a Rope Shovel. In *Field and Service Robotics;* Springer: Berlin/Heidelberg, Germany, 2006; Volume 25, pp. 555–566. [CrossRef]
31. Sturgul, J.R. Discrete mine system simulation in the United States. *Int. J. Surf. Min. Reclam. Environ.* **1999**, *13*, 37–41. [CrossRef]
32. Ataeepour, N.; Baafi, E.Y. ARENA simulation model for truck-shovel operation in despatching and non-despatching modes. *Int. J. Surf. Min. Reclam. Environ.* **1999**, *13*, 125–129. [CrossRef]

33. Mena, R.; Zio, E.; Kristjanpoller, F.; Arata, A. Availability-based simulation and optimization modeling framework for open-pit mine truck allocation under dynamic constraints. *Int. J. Min. Sci. Technol.* **2013**, *23*, 113–119. [CrossRef]
34. Tarshizi, E.; Sturgul, J.; Ibarra, V.; Taylor, D. Simulation and animation model to boost mining efficiency and enviro-friendly in multi-pit operations. *Int. J. Min. Sci. Technol.* **2015**, *25*, 671–674. [CrossRef]
35. Sturrock, D.T.; Pegden, C.D. Introduction to SIMAN. In Proceedings of the 1990 Winter Simulation Conference Proceedings, New Orleans, LA, USA, 9–12 December 1990; pp. 109–114. [CrossRef]

mining

MDPI

Review

Recent Developments in Path Planning for Unmanned Ground Vehicles in Underground Mining Environment

Abdurauf Abdukodirov * and Jörg Benndorf

Department of Mine Surveying and Geodesy, TU Bergakademie Freiberg, 09599 Freiberg, Germany; joerg.benndorf@mabb.tu-freiberg.de
* Correspondence: abdurauf.abdukodirov@doktorand.tu-freiberg.de

Abstract: The navigation of Unmanned Ground Vehicles (UGVs) in underground mining environments is critical for enhancing operational safety, efficiency, and automation in hazardous and constrained conditions. This paper presents a thorough review of path-planning algorithms employed for the navigation of UGVs in underground mines. It outlines the key components and requirements that are essential for an effective path planning framework, including sensors and the Robot Operating System (ROS). This review examines both global and local path-planning techniques, encompassing traditional graph-based methods, sampling-based approaches, nature-inspired algorithms, and reinforcement learning strategies. Through the analysis of the extant literature on the subject, this study highlights the strengths of the employed techniques, the application scenarios, the testing environments, and the optimization strategies. The most favorable and relevant algorithms, including A*, Rapidly-exploring Random Tree (RRT*), Dijkstra's, Ant Colony Optimization (ACO), were identified. This paper acknowledges a significant limitation: the over-reliance on simulation testing for path-planning algorithms and the computational difficulties in implementing some of them in real mining conditions. It concludes by emphasizing the necessity for full-scale research on path planning in real mining conditions.

Keywords: underground mining; navigation strategies; path planning algorithms; Unmanned Ground Vehicles; Robot Operation System

1. Introduction

In the age of technological advancement, automation has permeated almost all industries, including underground mining [1]. The high risk and challenges inherent in traditional mining have precipitated the introduction of automation measures. The elevated probability of injury in such a harsh environment poses a significant threat to workers' lives [2]. In addition, survey accuracy is affected by worker fatigue, which impacts measurement accuracy and increases survey time.

Underground automation heavily depends on efficient and reliable navigation systems, serving as a key enabler for autonomous operations in complex and hazardous environments [3]. Autonomous navigation is applicable in various environments, including outdoor, indoor, underground mines, and even underwater [4]. Beyond improving precision and reliability, effective navigation reduces the risks associated with manual operations, making it a fundamental component of underground automation.

Compared to other environments, navigation in underground mining presents unique challenges, including the unpredictable nature of underground terrain, such as irregular surfaces, narrow passages, and dynamic obstacles, as listed in Table 1, requiring robust navigation to ensure safety and efficiency [5]. Additionally, the absence of GPS signals necessitates advanced techniques like Simultaneous Localization and Mapping (SLAM) and Dead-reckoning methods [6,7].

Table 1. Unique factors of each environment and their impact on navigation.

Factors	Mine	Indoor	Outdoor
Surface Conditions	Irregular surfaces, narrow passages, and mud	Smooth and structured surfaces	Varied surfaces
Obscurants	Dust, smoke, and gas	Nearly clear	Rain, fog, and snow
GPS Availability	GPS denied	Available with extended setups	Widely available and effective
Obstacle Density	Relatively high	Low	Moderate

Navigation in mining is crucial for enabling autonomous vehicles and systems to perform various tasks efficiently and safely in challenging underground environments. The following are examples:

- For surveying and mapping. Autonomous vehicles equipped with LiDAR and SLAM systems navigate the mine to generate 3D maps for planning and operational purposes [8];
- Transportation. Autonomous haul trucks navigate predefined routes to transport ore, minerals, or waste material between loading and dumping points in underground mines [9];
- Drilling and Blasting Operations. Navigation systems guide autonomous drilling machines to precise coordinates within the mine for efficient drilling [10];
- Inspection and Maintenance. Autonomous robots navigate mine tunnels to detect structural integrity issues and gas leaks [11];
- Search and Rescue Operations (SARs). Navigation enables unmanned vehicles to explore hazardous or collapsed mine areas where human entry is unsafe [12].

Underground navigation in autonomous vehicle applications involves several key steps to ensure safe and efficient operation, as outlined below:

- Localization, where the vehicle determines its position and orientation within its environment using techniques like SLAM or dead-reckoning;
- Perception, which involves sensing and mapping the surroundings using Light Detection and Ranging (LiDAR), cameras, or radar to detect obstacles and terrain features;
- Path planning, which determines the optimal route to reach a destination while avoiding obstacles and adapting to environmental changes;
- Motion control executes the planned path by steering, accelerating, or braking, ensuring precise and smooth movement.

These steps are supported by sensor integration and real-time decision-making to handle dynamic conditions effectively [13].

In the context of autonomous underground navigation, a variety of platforms assume significant roles in enhancing efficiency, safety, and productivity. Unmanned Ground Vehicles (UGVs) and Unmanned Aerial Vehicles (UAVs) are among the key technologies driving automation in underground mining operations. Both of them are essential for surveying, exploration, inspection, and SAR tasks, where human intervention is either constrained or infeasible. UGVs with high power and payload capacity have advantages in long-duration missions and transportation, but they are limited in mobility. UAVs, on the other hand, are good at quick inspections and can reach areas that are difficult to access [14]. Beyond unmanned vehicles, many mining machines, including Load-Haul-Dump (LHD) machines, are being redesigned for full automation, further enhancing the efficiency and safety of underground operations [15].

As one of the crucial components of navigation, path planning might be modified or enhanced with optimization strategies to achieve efficient and accurate navigation in the underground mining environment. While extensive research has been conducted on surface-level path planning [16], underground navigation remains a critical and evolving research area due to its unique environmental challenges and constraints. This paper provides a comprehensive review of recent advancements in underground UGV path planning. The structure of this paper is as follows:

- Section 2 (Requirement Analysis and Methodological Basis for Path Planning in Underground Mining Environments) outlines the key requirements and methodology for successful navigation and path planning of UGVs in underground mining environments;
- Section 3 (Materials) discusses the essential components for UGV operation, including environmental sensors and operational framework;
- Section 4 (Path Planning Framework) presents a general overview of path planning frameworks, algorithms, and optimization strategies, highlighting their respective advantages and limitations;
- Section 5 (Related Works on Underground Mining Path Planning with UGVs) reviews related research, identifies emerging trends, and examines methodological improvements in path planning algorithms, as well as existing research gaps;
- Section 6 (Discussion) analyzes the results of the reviewed algorithms, highlights the main limitations of this study, and proposes directions for future research based on these findings;
- Finally, Section 7 (Conclusion) provides a summary of this research.

The analysis aims to offer insights into current limitations and identify opportunities for future research and development in optimizing autonomous navigation for underground mining applications.

2. Requirement Analysis and Methodological Basis for Path Planning in Underground Mining Environments

The requirements outlined in this section serve as a key consideration for navigation and the evaluation of existing path planning research in underground mining using UGVs. General requirements for underground navigation are as follows:

- UGVs must operate autonomously in GPS-denied environments;
- Performing effectively in low-light, high-dust, and uneven terrain conditions with the help of advanced sensors such as LiDAR, cameras, radars, Inertial Measurement Unit (IMU), and motor encoders for accurate environmental mapping and obstacle detection;
- UGVs must be compatible with the Robot Operating System (ROS) framework, enabling modular communication, multi-sensor integration, and scalability;
- UGVs must reliably detect and avoid both static and dynamic obstacles, generating accurate and efficient paths in real-time while incorporating recovery behaviors to handle unexpected environmental changes;
- Adaptability to environmental complexity, such as dynamic obstacles and sudden changes in tunnel structures, is necessary for reliable underground navigation;
- Energy-efficient operation is also essential to prolong the operational lifespan of UGVs in an underground mine [17].

Path planning requirements are primarily determined by the algorithms employed. The selection of an appropriate algorithm is based on an analysis of environmental and vehicle parameters. To evaluate path planning algorithms and review existing research, the key requirements and evaluation criteria for planned paths are summarized in Table 2.

Table 2. Requirements and evaluation criteria for planned paths in underground mining environments.

Requirements	Evaluation Criteria
Optimal path	Minimal travel time and path length
Smoothness	Spatial and temporal smoothness coefficients [18]
High accuracy and safety	Avoids obstacles and ensures collision-free path
Success rate	Percentage of successful path completions
Computational cost	Processing time and resource consumption
Robustness	Ability to handle uncertainties
Handling narrow tunnels	Minimum passable width

Considering the above-mentioned requirements, this review paper aims to address the following key questions:

- What types of sensors and an operational framework are used in underground mining navigation?
- How can alternative algorithms be compared and classified based on their strengths and application scenarios?
- Which path planning algorithms are most frequently applied in underground mining scenarios?
- What optimization techniques are commonly used to meet the requirements for successful path planning?
- What kinds of experimental methods are employed in the literature to assess how well these techniques reflect real-world mining conditions?

To address these questions, relevant studies were selected from major academic databases based on relevance to underground mine UGV path planning. Algorithms were categorized by type (graph-based, sampling-based, nature-inspired, etc.), and their implementation details, optimization strategies, testing environments, and limitations were comparatively analyzed.

This analysis enables the identification of research gaps between theoretical models and practical deployment challenges in path planning for underground mining UGVs.

3. Materials

3.1. Sensors to Perceive the Environment

Sensors are critical components in underground mining applications, enabling UGVs to perceive their surroundings, detect obstacles, and generate optimal paths in challenging environments. Considering the different operating conditions in underground mines, the selection of suitable sensors depends on the specific navigation requirements of the UGV. The following sensors are frequently employed for navigation purposes in mining operations:

- LiDAR (Light Detection and Ranging);
- Depth Camera;
- Radar;
- IMUs (Inertial Measurement Units);
- Motor encoders.

LiDAR is a widely utilized technique for 3D mapping and obstacle detection in which precise point clouds of the surroundings are generated. Laser scanners, on the other hand, operate similarly but offer high-resolution scans for detailed surface analysis. Depth cameras capture depth information to identify variations in terrain and assist in real-time object detection. Radar is effective in low-visibility conditions such as dust and fog, using radio waves to detect obstacles beyond the limitations of optical sensors. IMUs enhance localization accuracy by tracking motion changes through accelerometers and gyroscopes. In addition, motor encoders, which measure wheel rotation, provide odometry data to estimate the vehicle's displacement and support dead reckoning-based navigation [19]. Each sensor has its strengths and weaknesses in different applications, depending on the environment and accuracy required, as described in Table 3.

Table 3. Sensors' features.

Sensors	Strengths	Weaknesses	Range and Frequency
LiDAR 3D [20]	Works well in low-light environments;	Performance can degrade in dust, smoke or reflective surfaces;	10–300 m; 10–100 Hz
	High accuracy in 3D mapping and obstacle detection.	High power consumption.	
Laser scanner 2D [21]	Works in low-light environments;	Performance may degrade in heavy particulate environments;	0.5–25 m; 10–100 Hz
	Provides precise distance measurements.	Computationally intensive.	
Depth Camera [22]	Provides both color (RGB) and depth (D) information;	Sensitive to reflective and transparent surfaces;	0.2–10 m; 30–90 Hz
	Compact and lightweight	Can be disrupted by dust or fog.	
Radar [23]	Works in harsh environments (dust, smoke);	Difficult to interpret data without additional processing;	0.1–250 m; 10–200 Hz
	Reliable for detecting dynamic objects.	Lower resolution.	
IMU [24]	Small, lightweight, and power-efficient;	Susceptible to external vibrations and sensor noise;	100–1000 Hz
	Provides high-frequency motion data.	Accumulates drift over time.	
Motor Encoder [25]	Simple integration with UGV control systems;	Accumulates drift over long distances (wheel slippage);	10–1000 Hz
	High frequency data.	Not reliable for rough terrain or slippery surfaces.	

For robust underground mine path planning, sensor fusion (combining multiple sensors) is often necessary. A typical UGV utilizes LiDAR or scanners, along with IMUs and motor encoders. Additionally, radar complements LiDAR in harsh conditions, while depth cameras are used for object recognition and enhanced depth sensing [26].

Although sensor types do not directly determine the optimal path planning algorithm, there is an indirect relationship between sensor accuracy and planning performance. Inaccurate or sparse sensor data can lead to uncertain obstacle maps, limiting the planner's ability to generate smooth or efficient paths, particularly in constrained underground environments with narrow corridors and sharp turns. Conversely, more accurate sensing allows planners to exploit available space more effectively, potentially improving path smoothness and reducing the need for overly conservative maneuvers.

In underground navigation, the ROS (Robot Operating System) plays a key role as a middleware, facilitating communication between sensors, data acquisition, and processing in UGVs. It provides a structured framework for integrating multiple sensors, ensuring synchronized data flow for perception, localization, and navigation.

3.2. ROS as Main Operating System

ROS is a flexible framework used primarily in robotics for developing robot software. Despite its name, ROS is not a traditional operating system but rather a middleware framework that provides tools, libraries, and conventions for creating complex and robust robot applications [27]. There are several advantages that ROS can provide, including the following:

- ROS is open source, so it can be augmented by developers;
- It is a modular and flexible framework that allows researchers to customize and adapt it for a wide range of applications;
- Cross-platform compatibility and hardware-agnostic design, which supports diverse robotic systems, from drones and robotic arms to autonomous vehicles;
- Robust built-in libraries for computer vision, motion planning, and navigation;
- Simulation tools, such as Gazebo and RViz, allow researchers to efficiently prototype, test, and refine robots;
- Integration with AI and machine learning technologies extends its usefulness in creating intelligent and autonomous systems [28].

ROS has evolved through two major generations. ROS 1 (2007) laid the foundation for robotics development with modularity and ease of use, being widely adopted for research and non-critical applications. Key versions include ROS Kinetic (2016) and ROS Melodic (2018). However, it faced limitations in real-time performance and scalability. On the other hand, ROS 2 (2017) addressed these issues with parallel processing, secure communication, and real-time support, making it ideal for industrial use. Built on a Data Distribution Service (DDS), ROS 2 ensures robust communication and scalability, with key releases like Foxy (2020) and Humble (2022). Although ROS 1 is still in use, the focus is gradually shifting to ROS 2 as the emerging industry standard [29].

In ROS, a node is a fundamental building block that represents a single process performing a specific task, such as controlling a sensor or executing a robot's movement logic. Nodes communicate with each other using topics, which are named channels for message passing. A node can publish messages to a topic or subscribe to a topic to receive messages, enabling modular and scalable communication between different parts of a robot's system. For instance, a camera node might publish image data to a topic, while another node subscribes to process that data. The ROS file system provides a structured way to organize resources and facilitate development within the ROS framework. It is designed to manage the various components of a robotics project, including nodes, packages, messages, and configurations.

The key components of ROS file systems are as follows:

- Packages are the fundamental unit of software organization in ROS. They contain all the necessary files for a specific functionality, such as nodes, libraries, configurations, and launch files. Each package is self-contained, enabling easy reuse and sharing;
- Metapackages are collections of related packages grouped together under a common purpose, such as ros_base or desktop_full;
- Workspaces are directories where ROS packages are developed and built.

Each package has a standardized directory structure, as outlined below:

- src/: Source code for nodes and other scripts;
- *launch/*: Launch files to start nodes and set parameters;
- *config/*: Robot description configuration files to manage its behavior;
- *msg/* and *srv/*: Definitions for custom message and service types;
- CMakeLists.txt and package.xml: Build system and package metadata files.

ROS defines specific file types for communication between two nodes, as shown in Figure 1:

- Messages: Define data structures for node-to-node communication;
- Services: Specify request-response structures for synchronous interactions;
- Actions: Provide a framework for long-running tasks with feedback.

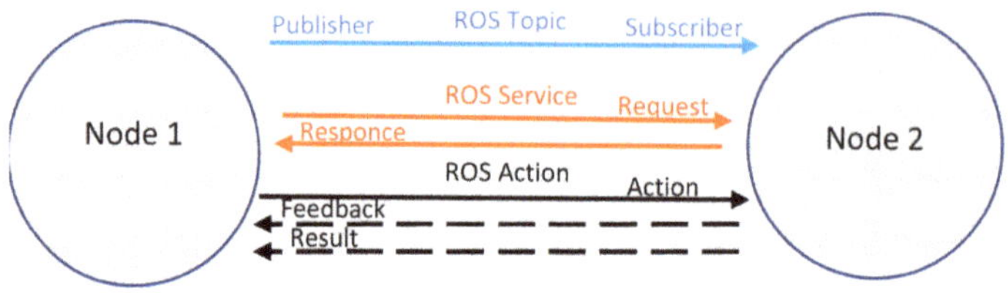

Figure 1. ROS communication types between two nodes.

The ROS file system ensures consistency, modularity, and ease of collaboration. By organizing code and resources into packages, it simplifies sharing, testing, and scaling across projects [30].

The operating system requirements for ROS vary depending on the generation. ROS 1 is primarily designed for Ubuntu Linux, as it leverages Ubuntu's package management system and development ecosystem. In contrast, ROS 2 offers broader compatibility and can be installed on multiple operating systems, including Ubuntu and Windows.

ROS is also ideal for underground and mining applications due to its modular and flexible architecture, enabling customization for navigation, mapping, and inspection in challenging environments. It enables UGVs to perform essential tasks, including SLAM, path planning, and visualization. In addition, it supports multi-sensor integration, such as LiDAR, radar, and cameras, for reliable operation in low-light and dusty conditions. ROS facilitates real-time data processing, remote operation, and multi-robot coordination, enhancing safety and efficiency. Numerous completed and ongoing ROS-based projects focus on automating mining operations and conducting rescue missions in mining disasters. Take, as an example, the successful completion of the DARPA Subterranean Challenge in 2021 [31], which was a robotics competition aimed at pushing the boundaries of autonomous navigation and exploration in complex underground environments. The challenge sought innovative approaches to quickly and effectively map, navigate, and search through difficult subterranean environments such as caves, tunnels, and urban underground structures.

4. Path Planning Framework

Path planning is a fundamental component of autonomous UGV navigation, responsible for determining an optimal or feasible path from a starting position to a destination while avoiding obstacles and considering environmental constraints. Developing a path planning framework is inherently complex, as it encompasses environment representation, localization and perception, path planning algorithms, obstacle avoidance, and motion control. This framework ensures that UGVs can safely and efficiently navigate underground environments, where GPS is unavailable and traditional navigation methods are ineffective [32].

The environment representation plays a crucial role in path planning, as it defines how the UGV perceives and interacts with its surroundings. The various representation methods include grid-based maps, topological maps, point cloud maps, and hybrid maps, each of which offers different advantages depending on the navigation requirements and computational constraints [33]. For localization in a known environment, sensor-based techniques such as Monte Carlo Localization (MCL) and feature-based localization are commonly employed [34]. However, in unknown environments, SLAM is required to build a map while simultaneously determining the UGV's position. Once the UGV has identified its start or current position, it can execute path planning algorithms to navigate efficiently toward the target location. After that, motion control manages the vehicle's movement.

Path planning algorithms can be broadly classified into global and local planning approaches, each serving a specific role in navigation, as described in Figure 2.

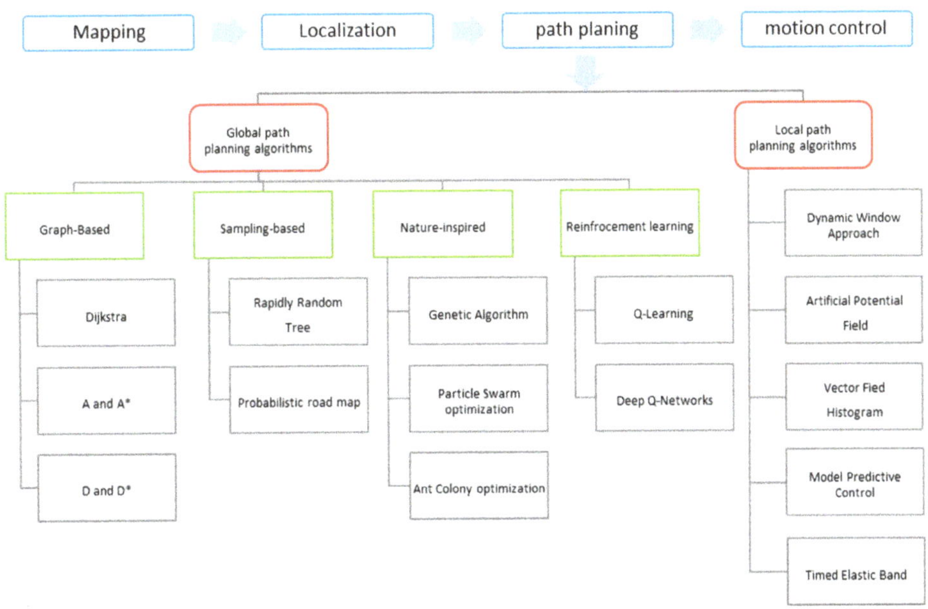

Figure 2. Workflow and types of path planning.

4.1. Global Path Planning Algorithms

Global path planning operates with a pre-existing map and computes a path before execution. These algorithms are suitable for structured environments where a detailed representation of the terrain is available. The global planner gives information about obstacles, the positions of the robot and targets, as well as the environment represented in the map. It creates a global path to reach the target position within this environment. Based on how they model the environment and search strategies for an optimal path, global path planning algorithms are divided into Graph-based, Sampling-based, Nature-Inspired, Reinforcement learning (RL)-based, and Hybrid algorithms [35].

1. **Graph-Based algorithms.** These algorithms model the environment as a graph and find the shortest path using various search techniques and algorithms, including the following:
    - **Dijkstra's algorithm.** It is an exact algorithm designed to determine the minimum-cost path from a source vertex to all other vertices in a directed graph. The algorithm operates by iteratively selecting the closest unvisited node, updating the shortest known distances to its neighboring vertices, and continuing this process until the destination node is reached or all reachable nodes have been explored. Since Dijkstra's algorithm follows a breadth-first search-like approach, it systematically expands the search from the starting node outward. However, this

leads to relatively high time and space complexity, especially in large graphs, as it requires maintaining and updating distance information for multiple vertices throughout the execution [36];

- **A and A* algorithms.** The algorithm, often linked to Dijkstra's approach, identifies the shortest path between nodes in a weighted graph by evaluating all possible routes based on cumulative cost. However, it lacks heuristic guidance, which can make it inefficient in larger environments. The A* algorithm improves upon this by integrating a heuristic function with cost-based search, allowing it to prioritize more promising paths. It evaluates the total estimated cost from the start node and estimates the remaining cost to the goal [37];
- **D* and D* Lite**. The D* algorithm is an incremental path-planning method designed for dynamic environments, where obstacles or terrain conditions may change over time. It initially computes an optimal path from the start position to the goal under the assumption of a static environment. When the robot moves and encounters updated information, D* efficiently recalculates only the affected portions of the path instead of recomputing from scratch, making it suitable for real-time applications in navigation and robotics. D* Lite, a simplified version of D*, follows a similar approach but computes paths in reverse, from the goal to the start. This backward search allows for efficient path updates when obstacles appear or disappear. By maintaining a cost-efficient priority queue and selectively updating affected nodes, D* Lite reduces computational complexity while pursuing a near-optimal solution [38].

2. **Sampling-Based Algorithms.** These algorithms generate a path by randomly sampling points in the environment. Some important sample algorithms are as follows:
    - **Rapidly-exploring Random Tree (RRT).** It is designed to efficiently navigate complex, high-dimensional spaces. It incrementally builds a tree by randomly sampling points within the search space and connecting them to the nearest existing node, effectively exploring feasible paths in environments with numerous obstacles. While RRT is proficient at quickly finding a viable path, it doesn't guarantee optimality [39]. To mitigate this limitation, RRT* was developed as an improvement over the original algorithm. It aims to enhance path quality through iterative refinement, where the tree structure is rewired to explore shorter or more efficient paths as new samples are added. While RRT* improves performance compared to RRT, it does not guarantee exact optimality. Rather, it is asymptotically optimal, meaning it can approach an optimal solution given sufficient sampling and time under certain conditions [40];
    - **Probabilistic Roadmap (PRM).** This method efficiently navigates high-dimensional spaces. It works in two phases: first, the construction phase generates random collision-free nodes and connects them to form a roadmap; and, second, the query phase links the starting and goal positions to the roadmap, using graph search algorithms like Dijkstra's to determine the optimal path. PRM is particularly effective in static environments where multiple queries are needed, as the roadmap can be reused, reducing computational complexity [41].
3. **Nature-Inspired Algorithms.** These algorithms are modeled after biological processes and use mathematical optimization techniques to find the best path. Some examples of these algorithms are as follows:
    - **Genetic Algorithm (GA).** It is an evolutionary optimization technique inspired by the principles of natural selection. It begins with a randomly generated population of candidate solutions, which evolve over multiple iterations through

selection, crossover, and mutation. A fitness function evaluates each candidate, so a parent selection operator favors the best solutions for reproduction. The algorithm applies crossover to combine parent solutions, thus generating children, and then it applies mutation to introduce small variations to these children, generating the mutated children that will become the parents of the next generation. This iterative framework introduces diversity to the search space, continuing until a stopping criterion is reached. GA is widely used in complex optimization problems where traditional methods struggle with large or nonlinear search spaces [42];

- **Particle Swarm Optimization (PSO).** An algorithm that is a population-based optimization technique inspired by the collective movement of birds and fish. It initializes a group of candidate solutions, called particles, which explore the search space by updating their positions based on both their individual best-known solution and the globally best-known solution within the swarm. This dynamic adjustment allows particles to converge toward near-optimal solutions over multiple iterations. PSO is particularly useful for solving high-dimensional and nonlinear optimization problems due to its simplicity and efficiency in navigating complex search spaces [43];
- **Ant Colony Optimization (ACO).** It mimics the foraging behavior of ants to find near-optimal paths in a graph. Artificial ants explore potential routes, depositing virtual pheromones that influence the decisions of subsequent ants. Pheromones evaporate at a certain rate, but over successive iterations, paths with stronger pheromone trails will outpace evaporation and become more favorable, leading the swarm toward convergence and efficient solutions [44].

4. **Reinforcement learning (RL)-based algorithms.** RL-based methods learn optimal paths through trial-and-error interactions with the environment, refining their policies over time based on cumulative rewards, consisting of an approach that does not require a predefined model [45]. Sample algorithms are as follows:
    - **Q-Learning algorithm.** It builds a Q-table where each state-action pair is assigned a value, which is then updated iteratively using the Bellman equation. By maximizing cumulative rewards, the algorithm ensures efficient obstacle avoidance and goal-reaching, making it particularly useful in dynamic or uncertain environments [46];
    - **Deep Q-Networks (DQN).** It employs a Deep Neural Network (DNN) to approximate Q-values for state-action pairs, enabling the robot to learn an optimal policy through continuous interaction with the environment. This approach is particularly beneficial in dynamic, uncertain, or partially observable settings where traditional methods struggle. Additionally, experience replay and target networks stabilize training, mitigating the overestimation of Q-values and improving convergence [47].
5. **Hybrid Algorithms.** These algorithms combine two or more of the mentioned approaches to improve efficiency and robustness. Examples of hybrid algorithms used for path planning of Autonomous Mobile Robots are the A-RRT* and D-lite with RRT* [48].

4.2. Local Path Planning Algorithms

Local path planning algorithms are crucial for UGVs, especially in dynamic environments. The global path planner determines the overall trajectory for vehicle motion but lacks efficient real-time obstacle avoidance. Therefore, direct execution of the global path often results in suboptimal performance. Additionally, as the search space expands, the computational time required for global path planning algorithms increases significantly [49]. Local planning adapts a planned path based on real-time data, ensuring that the robot can navigate effectively while considering its physical limitations and avoiding obstacles. They work within the framework established by the global planner but have a more dynamic role that involves constant adjustments rather than a strictly defined trajectory generation. Among the common local path planning algorithms, the following can be mentioned:

- **Dynamic Window Approach (DWA).** It is a real-time motion planning algorithm that ensures both collision avoidance and adherence to dynamic constraints. It evaluates a range of possible velocities within a short time horizon and selects the trajectory that optimally balances safety, efficiency, and goal direction. By considering the kinematic limitations of the robot and environmental obstacles, DWA enables smooth and reactive navigation in dynamic environments [50];
- **Artificial Potential Field (APF).** In this approach, the robot is influenced by an artificial force field composed of attractive forces, which pull it toward the goal, and repulsive forces, which push it away from obstacles. The robot navigates by following the resultant force vector, aiming for a collision-free path to reach the target. While APF is computationally efficient and straightforward to implement, it can encounter issues such as trapping in local minima, where the robot becomes trapped in a position that is not the goal. Various modifications have been proposed to address these limitations and enhance the effectiveness of this method [51];
- **Vector Field Histogram (VFH).** This approach constructs a two-dimensional Cartesian histogram grid using range sensor data. This grid is continuously updated to reflect the environment of the robot. The algorithm reduces this grid to a one-dimensional polar histogram centered on the current position of the robot, where each sector represents the obstacle density in a specific direction. By analyzing these sectors, VFH identifies obstacle-free paths and determines the most suitable steering direction, allowing the robot to navigate toward its target while avoiding collisions. This method effectively balances reactive obstacle avoidance with goal-oriented navigation [52];
- **Model Predictive Control (MPC).** It is an advanced control strategy that utilizes a dynamic model system to predict and optimize its future behavior over a specified time horizon. In the context of mobile robot local path planning, MPC involves formulating an optimization problem that accounts for the kinematic constraints of the robot, environmental obstacles, and a predefined cost function. At each time step, the controller solves this optimization problem to determine the optimal control inputs, resulting in a trajectory that guides the robot toward its target while avoiding collisions. This process is repeated in a receding horizon manner, allowing the robot to adapt its path in real-time to dynamic changes in the environment. MPC's ability to handle multivariable control problems and incorporate constraints makes it particularly effective for complex path planning tasks in uncertain environments [53];
- **Timed Elastic Band (TEB).** This algorithm optimizes the trajectory of the robot by considering both spatial and temporal constraints. Starting with an initial path, TEB refines it into a time-parametrized trajectory by adjusting the positions and velocities of intermediate points, ensuring adherence to the robot's kinematic constraints and obstacle avoidance. This optimization process allows the robot to navigate efficiently in dynamic environments, responding adaptively to changes while maintaining smooth and feasible motion [54].

Several algorithms, such as ACO, RRT*, A*, and RL, can be utilized for both global and local path planning after appropriate optimization [55].

However, despite the existence of a wide range of path-planning algorithms, their practical implementation often reveals specific limitations in meeting all operational requirements. These limitations are dependent on the testing environment and the characteristics of the UGV. As a result, many algorithms require optimization or fusion with other approaches to enhance efficiency, adaptability, and robustness. Table 4 presents a detailed comparison of the advantages and limitations of each algorithm.

Table 4. Strengths and weaknesses of path-planning algorithms.

Algorithms	Strength	Weakness
Dijkstra	Guarantees the shortest path; Works well in static environments.	Computationally intensive; Inefficiency in dynamic environment.
A*	Fast and Efficient; Deterministic and reliable on structured maps.	Depends on map quality and resolution; Slow in high-dimensional spaces.
D*	Efficient replanning; Handles dynamic obstacles well.	Higher memory usage; Slower with frequent changes.
RRT*	Efficient in high-dimensional spaces; Incremental path improvement.	Paths are not always smooth; Slower in narrow passages.
PRM	Reusable path for repeated queries; Works well in large, open spaces.	Paths are not always smooth; Struggles with dynamic obstacles.
GA	Efficient in high-dimensional spaces; Works well with noisy data.	Slow convergence; Sensitive to parameter selection.
PSO	Simple to implement; Adaptive to dynamic environments.	Prone to local minima; Sensitive to parameter selection.
ACO	Good for multi-agent path planning; Works in dynamic environments.	Computationally intensive; Slower convergence.
Q-Learning	Works with Partial Information; Can Handle Complex Environments.	Higher memory usage; Slow convergence in large spaces.
DQN	Suitable for high-dimensional spaces; Works with Partial Observability.	Requires high computational power; Sensitive to parameter selection.
DWA	Smooth and feasible paths; Computationally lightweight.	Prone to local minima; Struggles with complex environment.
APF	Lightweight, fast obstacle avoidance; Simple, best for reactive navigation.	Prone to local minima; Struggles with dynamic obstacles.
VFH	Efficient real-time obstacle avoidance; Handles noisy sensor data well.	Prone to local minima; Oscillations in Narrow Spaces.
MPC	Optimizes performance; Generates smooth and efficient paths.	Computationally intensive; Limited real-time application.
TEB	Generates smooth, time-optimal path; Dynamic Obstacle Avoidance.	Computationally intensive; Sensitive to parameter selection.

Global path-planning algorithms differ significantly in efficiency and adaptability based on the environment and constraints. Graph-based methods, such as A* and Dijkstra, work well in structured environments but are computationally expensive, while sampling-based approaches, like RRT* and PRM, offer better scalability in high-dimensional spaces but may require additional smoothing. Nature-Inspired Algorithms, like GA and ACO, are useful for solving complex optimization problems, but may suffer from slower convergence. RL techniques, including DQN, provide adaptability in dynamic and unknown environments, but require computational resources.

Local path planning algorithms also offer common and distinct advantages and drawbacks. DWA, APF, and VFH are prone to local minima; however, they are good at real-time obstacle avoidance. Moreover, DWA, MPC, and TEB ensure smooth and optimized paths, but MPC and TEB demand high computational power. TEB is excellent for dynamic trajectory adjustments but requires fine-tuned parameters. In the field of practical robotics, a universal algorithm that is best suited for all applications is not available. Consequently, there is a preference for hybrid methods that integrate multiple classical or RL-based approaches, particularly in the context of efficient path planning. Alternatively, the existing algorithms may be modified through the implementation of optimization strategies [56].

In terms of using path planning algorithms in underground mines, the evaluation criteria mentioned in Section 2. The Requirement Analysis and Methodological Basis for Path Planning in Underground Mining Environments will be selected based on key operational and environmental requirements, including the optimality of the path, computational demand, robustness, and obstacle avoidance. Additionally, the variability in excavation methods influences the requirements for these algorithms.

4.3. Optimization Strategies in Underground Path Planning

To improve path quality, safety, and computational efficiency in underground environments, numerous optimization strategies are employed within path planning algorithms. These techniques address specific issues such as jagged trajectories, redundant nodes, slow convergence, or collision risks. The strategies reviewed in the literature can be broadly categorized into several functional groups:

- **Curve and Spline-Based Path Smoothing** [57]. Path smoothing techniques, such as Gaussian filtering, B-spline curves, and Bessel curves, are widely used to convert discrete waypoints into smooth, continuous trajectories, which are essential for navigating narrow underground tunnels. As one of the common smoothing technique, Gaussian filtering applies a weighted average to a set of neighboring points using a Gaussian kernel [58]. The filtered position P_i' is computed as shown in Equation (1):

$$P_i' = \sum_{j=-k}^{k} w_j \cdot P_{i+j} \tag{1}$$

  where P_i' is the filtered waypoint, w_j are the weights from the Gaussian kernel, and $2k + 1$ is the size of the window. This method reduces noise and sharp transitions in the path, making it more navigable for UGVs in confined underground environments;
- **Search Space or Node optimization**. These techniques aim to reduce computational load and improve planning efficiency by refining how nodes are generated or selected. These strategies include key node selection, adaptive step size expansion, and articulation-angle-based expansion. Each method seeks to avoid oversampling in open areas and prioritize exploration in more constrained regions. One example is adaptive step size expansion [59], which adjusts the growth rate of the tree based on environmental feedback or remaining distance to the goal. A common form to calculate the step size for node expansion is shown in Equation (2):

$$\delta_{\text{step}} = \min\left(\delta_{\text{max}}, \alpha \cdot d_{\text{goal}}\right) \tag{2}$$

  where δ_{step} is the current expansion step size, δ_{max} is a predefined upper limit, α is a scaling factor, and d_{goal} is the Euclidean distance to the goal. This approach helps maintain planning accuracy near the target while speeding up search in free space;
- **Evolutionary optimization.** This category encompasses bio-inspired and probabilistic methods, such as the pheromone updating model, entropy increase strategy, and ant retreat strategy, which enhance path planning by optimizing path quality and computational efficiency.

A representative mechanism is the pheromone updating rule used in ACO [60]. After each iteration, the pheromone levels τ_{ij} on the path between nodes i and j are updated to reflect the quality of the solutions found. The update equation is calculated as Equation (3):

$$\tau_{ij} \leftarrow (1-\rho) \cdot \tau_{ij} + \Delta\tau_{ij} \quad (3)$$

where:

- τ_{ij} is the pheromone value on edge (i, j);
- ρ is the evaporation rate ($0 < \rho < 1$);
- $\Delta\tau_{ij}$ is the amount of pheromone deposited, often defined by Equation (4):

$$\Delta\tau_{ij} = \begin{cases} \frac{Q}{L}, & if\ edge\ (i,j)\ is\ in\ the\ best\ path \\ 0, & otherwise \end{cases} \quad (4)$$

Here, Q is a constant, and L is the length of the path. This mechanism reinforces efficient paths while allowing less optimal ones to fade, balancing exploration and exploitation in the search process.

5. Related Works on Underground Mining Path Planning with UGVs

The automation of underground mining vehicles has been the focus of research and industry for several decades. However, path planning for autonomous navigation remains a critical challenge and an active area of investigation. Despite several studies have been conducted on the review of path planning algorithms on the ground [16,35], a limited number of studies have successfully addressed optimal path planning for mining vehicles and robotic applications in underground environments.

Effective path planning allows UGVs to traverse complex, unstructured terrains while ensuring efficiency, safety, and adaptability. Researchers have explored various techniques, ranging from classical graph-based methods to modern metaheuristic and RL-based approaches, each aiming to optimize path efficiency, obstacle avoidance, and computational feasibility. These approaches have been validated in real underground mines, controlled indoor environments, and virtual simulation frameworks implemented in environments such as MATLAB, Python, and Gazebo. Table 5 reviews key contributions by researchers in the field, categorizing existing studies based on algorithmic approaches, testing strategies, and demonstrated advantages in underground mining applications.

Table 5. Analysis of various path planning techniques for UGVs in underground mine.

Author	Path Type	Basic Algorithms	Optimization Strategy	Test Field	Advantages	Application
[61]	Global	A*	Gaussian filtering method; Quadratic programming method.	MATLAB	Safety; Smoother path.	LHD in mining transportation
[62]	Global	A*	Expanding nodes by articulation angle; Collision threat cost.	Indoor; C++.	Enhanced search efficiency; Collision free.	LHD in mining transportation
[63]	Global Local	A* + APF	Exponential function weighting; Cubic spline interpolation; Repulsion potential field correction factor.	MATLAB; Indoor.	Smoother path; More robust; Guaranteed safety.	Transportation in coal mine
[64]	Global Local	A* + DWA	Key node selection strategy; Clamped-B spline; Fuzzy control.	MATLAB; Indoor.	Safety; Smoother path; Optimal path.	Vehicle in mine transportation
[65]	Global Local	A* + DWA	-	Indoor	Shortest path; Smoothest path.	Mobile robot in inspection
[66]	Global Local	A* + DWA	Floyd Algorithm; B-Spline curves.	MATLAB; Indoor.	Smoother path; Safer.	Inspection robot in coal mine

Table 5. *Cont.*

Author	Path Type	Basic Algorithms	Optimization Strategy	Test Field	Advantages	Application
[67]	Global	VFH-A*	Bezier interpolation;	MATLAB; Mine.	More efficient search; Smooth path; Fast calculation speed.	LHD in mining operation
[68]	Global Local	Dijkstra-ACO	MAKLINK lines. Shifting locally; Symmetric polynomial curve.	Mine	Smooth path	Robot in SAR
[69]	Global	Dijkstra-PSO	PSO-based optimization	MATLAB	Shorter path; Safety.	Mine mapping and inspection
[70]	Global	Dijkstra	3D Environment-based adaptation	MATLAB	Feasible path	Survey robot in mine exploration
[71]	Global	D*	Danger-aware cost function	MATLAB	Safer path; Reduced planning time and cost.	Downhole robot in mine detection
[72]	Global	RRT*	Vectorized map; Optimal tree reconnection.	Python	Shorter path; Smoother path.	LHD in mining operations
[73]	Global	RRT*	Line corner	MATLAB; Outdoor.	Smooth path; Better in narrow passages with tight turn;	Robotic excavator loader
[74]	Global	RRT-PRM	Fan-shaped goal orientation; Adaptive step size expansion; Third-order Bessel curve.	MATLAB; Indoor.	Higher success rate; Smoother path; Shorter path length.	Inspection robot in Mining
[75]	Global	RRG [1]	Ray tracing method; Next-Best View.	Gazebo	Shorter path; Fast calculation.	Mobile Robot in Mine rescue
[76]	Global	ACO	Retreat-punishment strategy; Serial number and Cartesian coordinate methods.	MATLAB	Consumes less resources	Coal mine robot
[77]	Local	ACO	Membrane computing	Simulation	Faster convergence; Better robustness.	Mobile robot
[78]	Global	ACO	16-directional 24-neighbourhood ant search approach; Ant retreat strategy.	MATLAB	Improved search efficiency; U-shaped trap solution; Shorter path.	Rescue robot in Mine
[79]	Global	ACO	Annealing algorithm; Entropy increase strategy.	MATLAB	Strong in robustness; High convergence speed.	Mining robot in SAR
[80]	Global Local	ACO + TEB	Pheromone updating model;	MATLAB; Indoor.	Enhanced search capabilities; Faster convergence.	Drilling robot in mining rockburst
[81]	Global	BFA [2]-PSA [3]	Metaheuristic optimization	MATLAB	Fast convergence speed; Better Robustness	Rescue robot in coal mine
[82]	Global Local	AFSA [4] + DWA	Improved genetic algorithm Adaptive trajectory evaluation function	MATLAB; Gazebo; Indoor.	Shorter path; Smoother path.	Patrol robot in mine safety inspection
[83]	Global	QLearning	Even gray model; Multi-attribute intelligent.	MATLAB	Smoother convergence; Shorter path;	UGV in SAR
[84]	Local	Skeleton-Based	Thinning algorithm	Mine	High robustness; Stable;	LHD in mining operation
[85]	Local	APF	Velocity and acceleration fields; Global potential field line; Genetic Trust Region Algorithm	MATLAB; Indoor	Dynamic collision avoidance; Escape local minima.	Mobile robot in coal mine rescue

[1] Rapidly Random Graph; [2] Bacterial Foraging algorithm; [3] Polyclonal selection algorithm; [4] Artificial fish swam algorithm.

A thorough analysis of the data presented in Table 5 indicates that several underground mining operations are being targeted by UGV navigation systems. The majority of research in path planning focused on mine transportation, which is primarily carried out using LHD. Additionally, mine exploration and inspection tasks with mobile robots equipped with specialized sensors emerged as a significant area of research interest. SAR is another critical topic for underground mining operations, and numerous researchers are exploring optimal path planning algorithms for successful rescue operations after disasters. It is notable that coal mines are considered to be high-risk environments due to their potential for accidents.

Despite the limited research in this area, existing studies have employed all types of environment representation and search strategy models. The most prevalent techniques employed in underground mining environments include classical approaches such as graph-based and sampling-based algorithms, as illustrated in Figure 3a, which is based on an analysis of the existing research.

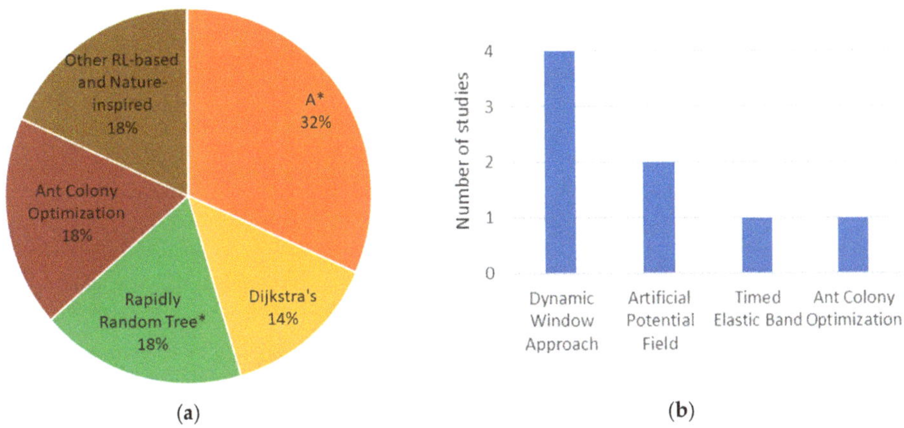

Figure 3. Percentage and number of existing path planning algorithms that are being investigated by researchers for the underground mining environment (**a**) Global path planning; (**b**) Local path planning.

It is illustrated that the A* algorithm is the most prevalent global path planning algorithm in the field, with approximately one in three studies utilizing this approach. This method incorporates a range of optimization strategies, including Gaussian filtering, quadratic programming, expanding and selecting nodes, collision threat cost, exponential function weighting, splines, and other features. These techniques ensure safe, smooth, and optimal paths for UGVs in mining transportation and inspection tasks [61–66]. Graph-based algorithms have also demonstrated a high level of efficiency in hybrid contexts. For example, the utilization of A* with VFH employing Bazier interpolation and Dijkstra's with ACO and PSO employing curve strategies have been shown to enhance search efficiency, smoothness, and accelerated calculation in mining operations and SAR [67–69]. Dijkstra's algorithm also demonstrated strong performance in path planning within 3D environments [70], representing 14% of the analyzed studies.

Sampling-based methods have gained significant attention, with the RRT accounting for 18% of all research in the field. Optimized versions incorporating vectorized map, optimal tree reconnection, and line corner for narrow passages have demonstrated superior path smoothness and shorter path lengths in operating mining transports [72,73]. A hybrid RRT*—PRM model integrating fan-shaped goal orientation, adaptive step-size expansion, and Bessel curve has been shown to enhance success rates and smoothness for mining inspection robots [74]. Similarly, the incorporation of ray-tracing and next-best-view methods into the RRG has been applied to mine rescue operations, achieving shorter path generation and faster computational speed [75].

Nature-inspired approaches have been frequently explored, with ACO accounting for approximately 18% of applications for underground mining navigation and rescue applications. Studies have employed retreat-punishment mechanisms, serial numbering and Cartesian coordinates, along with a 16-directional, 24-neighborhood ACO approach. Other implementations incorporated techniques such as the ant retreat strategy, annealing algorithms, pheromone updating models, and entropy-increasing strategies to reduce consumption costs, overcome U-shaped obstacle traps, and enhance convergence speed, efficiency, and robustness [76,78–80].

The last 18% of the analyzed research is primarily divided between various types of nature-inspired and RL-based algorithms. For example, the hybrid approaches BFA combined with PSA, tested when solving the traveling salesman problem, and AFSA with an improved genetic algorithm, demonstrated fast convergence, better robustness, and shorter paths in SAR applications [81,82]. Additionally, RL algorithms, such as Q-learning, combined with the Even Gray Model and Multi-Attribute Intelligent Decision-Making, demonstrated smoother convergence and high robustness in SAR operation [83].

Unlike global path planning, local path planning involves fewer algorithms. As illustrated in Figure 3b, the DWA is the most favorable method for adjusting the kinematics of UGVs. In the second place stands APF, which could be modified with global potential field line and genetic Trust Region strategies to overcome local minima and dynamic collision in coal mine rescue tasks [85]. Moreover, membrane computing-based ACO enhances robustness and convergence speed [77], while skeleton-based thinning algorithms provide high stability and robustness [84]. Additionally, the TEB method was also utilized for local path planning.

The test environments used in the reviewed studies highlight the progression from simulation-based validation to real-world implementation. Many algorithms have been tested in MATLAB and Python simulations, providing controlled conditions for evaluating path efficiency, convergence, and obstacle avoidance. However, some studies extend their validation to indoor lab environments, where methods have been tested in structured, tunnel-like settings, enabling real-time sensor fusion and navigation performance analysis. The most reliable assessments come from real underground mining experiments. Only a few algorithms have been deployed to evaluate practical feasibility, adaptability to real-world constraints, and system robustness.

6. Discussion

The analysis of various path planning algorithms for underground mining UGVs highlights significant trends and challenges in achieving efficient, safe, and adaptive autonomous navigation. Most existing research is focused on transportation, inspection, and SAR operations in mines. The comparison of graph-based, sampling-based, nature-inspired, RL-based, and local path planning approaches demonstrates that each method has unique advantages and limitations depending on the operational environment and constraints.

Graph-based approaches, such as A* and Dijkstra's, remain widely used due to their deterministic nature and best performance in structured environments. However, these algorithms often struggle with real-time adaptability, particularly in dynamic underground environments where sudden obstacles and unpredictable conditions are common. To overcome these drawbacks, especially for the A* collision threat cost and repulsion potential field correction factor, strategies are used. Furthermore, A* typically produces paths that can be jagged or not smooth, especially in complex environments. Techniques like cubic spline interpolation and B-spline curves provide a way to refine paths.

The most commonly used sampling-based method, which is the RRT, is effective in exploration tasks and offers better adaptability in large-scale environments. However, it can suffer from path irregularities and challenges in collision detection. To address these issues, several strategies have been proposed, including vectorized maps, line corner handling, and third-order Bessel curves.

Nature-inspired approaches, particularly ACO, improve search efficiency, robustness, and obstacle avoidance but require higher computational resources. To mitigate this problem, the 16-Directional 24-Neighbourhood Ant Search Approach and the Pheromone Updating Model were used.

RL-based approaches, such as grey Q-Learning, have demonstrated potential in adaptive decision-making, but their practical implementation is still limited by computational complexity and training data requirements.

Among local path planning algorithms, DWA has been identified as the most favorable. To address its limitations, the Adaptive Trajectory Evaluation Function and Fuzzy Control were employed.

Analyzing the current research, it is notable that some hybrid algorithms also demonstrated effective and optimal path planning, such as VFH-A*, Dijkstra-ACO, Dijkstra-PSO, RRT-PRM, and BFA-PSA.

Several limitations persist in analyzing the existing research studies on the selected topic, as outlined below:

- Over-reliance on simulation environments, which do not fully replicate the complexity of real underground mining conditions;
- Limited real mine experiments—only a small number of studies have been conducted in actual underground mines;
- Insufficient consideration of real-time constraints, such as sensor delays, terrain irregularities, and power limitations, which are critical for full-scale deployment in mining operations;
- High algorithmic complexity, particularly in hybrid, nature-inspired, and reinforcement learning (RL)-based approaches, which can hinder implementation and real-world adaptation;
- Lack of interpretability, as many advanced algorithms lack transparency, making them hard to evaluate or debug in applications.

Based on the analyzed research data and identified limitations, several future research directions are proposed:

- Bridging the gap between simulation and real-world implementation by testing algorithms not only in simulation environments such as Gazebo but also in modeled indoor labs and actual underground mines;
- Evaluating both common and reinforcement learning (RL)-based path planning algorithms using UGVs in underground mining environments to assess their real-time performance and adaptability;
- Investigating the effectiveness of common path optimization techniques and improving existing planning algorithms to address their current limitations in path smoothness, efficiency, and robustness.

7. Conclusions

The automation of underground mining operations has advanced significantly in recent years, with Unmanned Ground Vehicles (UGVs) playing a crucial role in improving efficiency, safety, and productivity. Unlike outdoor and structured indoor environments, underground mines present unique challenges, including complex tunnel networks, low visibility, and unstable terrain. To navigate these environments, autonomous mining vehicles rely on core navigation components, such as mapping, Simultaneous Localization and Mapping (SLAM), and path planning, all of which must be tailored to the constraints of underground operations. UGVs are deployed in exploration, monitoring structural changes, transportation, Search and Rescue Operations (SARs) tasks, and inspection, requiring robust navigation systems to ensure operational reliability.

Path planning in underground mines presents distinct challenges due to GPS-denied, unpredictable environmental changes, and the need for efficient real-time decision-making. This review comprehensively analyzed global and local path planning algorithms used in underground mining applications, covering graph-based, sampling-based, nature-inspired, and reinforcement learning (RL)-based approaches. The analysis conducted in this study addressed the research questions and produced the following key findings:

- Sensors, such as LiDAR, cameras, and radar, are critical for perceiving underground mining environments, and Robot Operating System (ROS) has emerged as the primary framework for integrating and managing sensor-based data in mining robots;
- Graph-based algorithms, including A* and Dijkstra's, work well in structured environments due to their deterministic nature, while sampling-based approaches, like Rapidly-exploring Random Tree (RRT*), offer better scalability in high-dimensional spaces. Nature-Inspired algorithms, such as Ant Colony Optimization (ACO), are useful for solving complex optimization problems. In contrast, RL-based techniques provide adaptability in dynamic and unknown environments;
- Among global path planning algorithms, A*, Dijkstra's, RRT*, and ACO are the most frequently used in underground UGV applications. For local path planning, the Dynamic Window Approach (DWA) remains the dominant method in real mining scenarios;
- Optimization strategies such as spline curves, node optimization, and potential field are commonly applied to improve path smoothness, efficiency, and safety;
- The majority of research is conducted in simulation environments, such as MATLAB, which do not fully capture real-world sensor integration challenges and environmental complexities. Only a relatively small number of studies have been validated in real underground mining environments;
- Complex and computationally intensive algorithms, especially RL-based and hybrid methods, present practical implementation challenges in real-world deployments.

These findings underline a significant research gap between theoretical development and practical deployment. Addressing this gap requires future studies to emphasize real-environment testing and the interpretability of complex algorithms in underground conditions.

This review serves as a timely and valuable resource for researchers and practitioners aiming to advance autonomous navigation systems for UGVs in underground mining environments.

Author Contributions: Conceptualization, A.A. and J.B.; methodology, A.A.; formal analysis, A.A.; investigation, A.A.; resources, J.B; data curation, A.A.; writing—original draft preparation, AA.; writing—review and editing, J.B.; visualization, A.A.; supervision, J.B.; project administration, J.B.; funding Acquisition, J.B. All authors have read and agreed to the published version of the manuscript.

Funding: This research was funded by the German Academic Scholarship Foundation, the Deutsche Forschungsgemeinschaft (DFG, German Research Foundation)—Project number 422117092 using the research infrastructure INST 267/165-1 FUGG and by "El-yurt umidi" Foundation for training specialists abroad and the dialogue with compatriots—Certificate number DR-2022-0115.

Data Availability Statement: No new data were created or analyzed in this study. Data sharing is not applicable to this article due to internal policy.

Conflicts of Interest: The authors declare no conflict of interest.

References

1. Kokkinis, A.; Frantzis, T.; Skordis, K.; Nikolakopoulos, G.; Koustoumpardis, P. Review of Automated Operations in Drilling and Mining. *Machines* **2024**, *12*, 845. [CrossRef]
2. Saleh, J.H.; Cummings, A.M. Safety in the Mining Industry and the Unfinished Legacy of Mining Accidents: Safety Levers and Defense-in-Depth for Addressing Mining Hazards. *Saf. Sci.* **2011**, *49*, 764–777. [CrossRef]
3. Li, J.; Zhan, K. Intelligent Mining Technology for an Underground Metal Mine Based on Unmanned Equipment. *Engineering* **2018**, *4*, 381–391. [CrossRef]
4. Hofmann-Wellenhof, B.; Legat, K.; Wieser, M. *Navigation: Principles of Positioning and Guidance;* Springer: Vienna, Austria, 2003. Available online: https://books.google.de/books?id=dMXcBQAAQBAJ (accessed on 26 February 2025).

5. Roberts, J.M.; Duff, E.S.; Corke, P. Reactive Navigation and Opportunistic Localization for Autonomous Underground Mining Vehicles. *Inf. Sci.* **2002**, *145*, 127–146. Available online: https://api.semanticscholar.org/CorpusID:1925516 (accessed on 8 May 2025). [CrossRef]
6. Ebadi, K.; Bernreiter, L.; Biggie, H.; Catt, G.; Chang, Y.; Chatterjee, A.; Denniston, C.E.; Deschênes, S.-P.; Harlow, K.; Khattak, S.; et al. Present and Future of SLAM in Extreme Underground Environments. *arXiv* **2022**, arXiv:2208.01787. Available online: https://api.semanticscholar.org/CorpusID:251280074 (accessed on 8 May 2025).
7. Ojeda, L.; Borenstein, J. Personal Dead-Reckoning System for GPS-Denied Environments. In Proceedings of the 2007 IEEE International Workshop on Safety, Security and Rescue Robotics, Rome, Italy, 25–27 September 2007; pp. 1–6. [CrossRef]
8. Tatsch, C.; Bredu, J.A.; Covell, D.; Tulu, I.B.; Gu, Y. Rhino: An Autonomous Robot for Mapping Underground Mine Environments. In Proceedings of the 2023 IEEE/ASME International Conference on Advanced Intelligent Mechatronics (AIM), Seattle, WA, USA, 27–30 June 2023; pp. 1166–1173. [CrossRef]
9. Mäkelä, H. Overview of LHD Navigation without Artificial Beacons. *Rob. Auton. Syst.* **2001**, *36*, 21–35. [CrossRef]
10. Li, Y.; Peng, P.; Li, H.; Xie, J.; Liu, L.; Xiao, J. Drilling Path Planning of Rock-Drilling Jumbo Using a Vehicle-Mounted 3D Scanner. *Sustainability* **2023**, *15*, 9737. [CrossRef]
11. Kim, H.; Choi, Y. Development of Autonomous Driving Patrol Robot for Improving Underground Mine Safety. *Appl. Sci.* **2023**, *13*, 3717. [CrossRef]
12. Wang, W.; Dong, W.; Su, Y.; Wu, D.; Du, Z. Development of Search-and-Rescue Robots for Underground Coal Mine Applications. *J. Field Robot.* **2014**, *31*, 386–407. [CrossRef]
13. Karimi, H. *Indoor Wayfinding and Navigation*; CRC Press: Boca Raton, FL, USA, 2015. Available online: https://books.google.de/books?id=I3oZBwAAQBAJ (accessed on 28 February 2025).
14. Galar, D.; Kumar, U.; Seneviratne, D. *Robots, Drones, UAVs and UGVs for Operation and Maintenance*; CRC Press: Boca Raton, FL, USA, 2020. [CrossRef]
15. Inostroza, F.; Parra-Tsunekawa, I.; Ruiz-del-Solar, J. Robust Localization for Underground Mining Vehicles: An Application in a Room and Pillar Mine. *Sensors* **2023**, *23*, 8059. [CrossRef]
16. Liu, L.; Wang, X.; Yang, X.; Liu, H.; Li, J.; Wang, P. Path Planning Techniques for Mobile Robots: Review and Prospect. *Expert Syst. Appl.* **2023**, *227*, 120254. [CrossRef]
17. Biggie, H.; Rush, E.R.; Riley, D.G.; Ahmad, S.; Ohradzansky, M.T.; Harlow, K.; Miles, M.J.; Torres, D.; McGuire, S.; Frew, E.W.; et al. Flexible Supervised Autonomy for Exploration in Subterranean Environments. *Field Robot.* **2023**, *3*, 125–189. [CrossRef]
18. Egerstedt, M.; Martin, C. *Control Theoretic Splines: Optimal Control, Statistics, and Path Planning*; Princeton University Press: Princeton, NJ, USA, 2009. Available online: https://books.google.de/books?id=crNiMAsPex8C (accessed on 6 March 2025).
19. Trybała, P.; Szrek, J.; Remondino, F.; Kujawa, P.; Wodecki, J.; Blachowski, J.; Zimroz, R. MIN3D Dataset: MultI-seNsor 3D Mapping with an Unmanned Ground Vehicle. *PFG–J. Photogramm. Remote Sens. Geoinf. Sci.* **2023**, *91*, 425–442. [CrossRef]
20. Peng, P.; Pan, J.; Zhao, Z.; Xi, M.; Chen, L. A Novel Obstacle Detection Method in Underground Mines Based on 3D LiDAR. *IEEE Access* **2024**, *12*, 106685–106694. [CrossRef]
21. Nielsen, K. *Robust LIDAR-Based Localization in Underground Mines*; Linköping University Electronic Press: Linköping, Sweden, 2021; Volume 1906. [CrossRef]
22. Jing, N.-B.; Ma, X.; Guo, W.; Wang, M. 3D Reconstruction of Underground Tunnel Using Depth-Camera-Based Inspection Robot. *Sensors Mater.* **2019**, *31*, 2719–2734. Available online: https://api.semanticscholar.org/CorpusID:203068830 (accessed on 8 May 2025). [CrossRef]
23. Cunha, F.; Youcef-Toumi, K. Ultra-Wideband Radar for Robust Inspection Drone in Underground Coal Mines. In Proceedings of the 2018 IEEE International Conference on Robotics and Automation (ICRA), Brisbane, Australia, 21–25 May 2018; pp. 86–92. [CrossRef]
24. Ahmad, N.; Ghazilla, A.R.; Khairi, N.M.; Kasi, V. Reviews on Various Inertial Measurement Unit (IMU) Sensor Applications. *Int. J. Signal Process. Syst.* **2013**, *1*, 256–262. [CrossRef]
25. Rodrigues, C.M. Ground Mobile Vehicle Velocity Control Using Encoders and Optical Flow Sensor Fusion. 2016. Available online: https://api.semanticscholar.org/CorpusID:202662013 (accessed on 8 May 2025).
26. Liu, K.-Q.; Zhong, S.-S.; Zhao, K.; Song, Y. Motion Control and Positioning System of Multi-Sensor Tunnel Defect Inspection Robot: From Methodology to Application. *Sci. Rep.* **2023**, *13*, 13125. [CrossRef]
27. Quigley, M. ROS: An Open-Source Robot Operating System. In Proceedings of the IEEE International Conference on Robotics and Automation, Kobe, Japan, 12–17 May 2009. Available online: https://api.semanticscholar.org/CorpusID:6324125 (accessed on 8 May 2025).
28. Aljamal, M.; Patel, S.; Mahmood, A. Comprehensive Review of Robotics Operating System-Based Reinforcement Learning in Robotics. *Appl. Sci.* **2025**, *15*, 1840. [CrossRef]
29. Al-Batati, A.S.; Koubaa, A.; Abdelkader, M. ROS 2 in a Nutshell: A Survey. *Preprints* 2024. [CrossRef]

30. Joseph, L.; Cacace, J. *Mastering ROS for Robotics Programming*; Packt Publishing: Birmingham, UK, 2018. Available online: https://books.google.de/books?id=MulODwAAQBAJ (accessed on 9 March 2025).
31. Agha, A.; Otsu, K.; Morrell, B.; Fan, D.D.; Thakker, R.; Santamaria-Navarro, A.; Kim, S.-K.; Bouman, A.; Lei, X.; Edlund, J.; et al. NeBula: Quest for Robotic Autonomy in Challenging Environments; TEAM CoSTAR at the DARPA Subterranean Challenge. *arXiv* **2021**, arXiv:2103.11470.
32. Losch, R.; Grehl, S.; Donner, M.; Buhl, C.; Jung, B. Design of an Autonomous Robot for Mapping, Navigation, and Manipulation in Underground Mines. In Proceedings of the 2018 IEEE/RSJ International Conference on Intelligent Robots and Systems (IROS), Madrid, Spain, 1–5 October 2018; pp. 1407–1412. [CrossRef]
33. Thrun, S. Learning Metric-Topological Maps for Indoor Mobile Robot Navigation. *Artif. Intell.* **1998**, *99*, 21–71. [CrossRef]
34. Gil, A.; Reinoso, Ó.; Vicente, A.; Fernández, C.; Payá, L. Monte Carlo Localization Using SIFT Features. In *Computer Aided Systems Theory—EUROCAST 2005*; Springer: Berlin/Heidelberg, Germany, 2005; pp. 623–630. [CrossRef]
35. Yang, L.; Li, P.; Qian, S.; Quan, H.; Miao, J.; Liu, M.; Hu, Y.; Memetimin, E. Path Planning Technique for Mobile Robots: A Review. *Machines* **2023**, *11*, 980. [CrossRef]
36. Javaid, M.A. Understanding Dijkstra Algorithm. *SSRN Electron. J.* 2013. [CrossRef]
37. Hart, P.E.; Nilsson, N.J.; Raphael, B. A Formal Basis for the Heuristic Determination of Minimum Cost Paths. *IEEE Trans. Syst. Sci. Cybern.* **1968**, *4*, 100–107. [CrossRef]
38. Koenig, S.; Likhachev, M. D*Lite. In Proceedings of the AAAI Conference on Artificial Intelligence (AAAI/IAAI), Edmonton, AB, Canada, 28 July–1 August 2002.
39. LaValle, S.M. Rapidly-Exploring Random Trees: A New Tool for Path Planning. *Annu. Res. Rep.* 1998. Available online: https://api.semanticscholar.org/CorpusID:14744621 (accessed on 8 May 2025).
40. Karaman, S.; Frazzoli, E. Optimal Kinodynamic Motion Planning Using Incremental Sampling-Based Methods. In Proceedings of the 49th IEEE Conference on Decision and Control (CDC), Atlanta, GA, USA, 15–17 December 2010; pp. 7681–7687. [CrossRef]
41. Kavraki, L.E.; Švestka, P.; Latombe, J.C.; Overmars, M.H. Probabilistic Roadmaps for Path Planning in High-Dimensional Configuration Spaces. *IEEE Trans. Robot. Autom.* **1996**, *12*, 566–580. [CrossRef]
42. Goldberg, D.E. *Genetic Algorithms in Search, Optimization, and Machine Learning*; Addison-Wesley: Boston, MA, USA, 1989. Available online: https://archive.org/details/geneticalgorithm0000gold (accessed on 2 February 2025).
43. Kennedy, J.; Eberhart, R. Particle Swarm Optimization. In Proceedings of the IEEE International Conference on Neural Networks, Perth, WA, Australia, 27 November–1 December 1995; pp. 1942–1948.
44. Dorigo, M.; Gambardella, L.M. Ant Colony System: A Cooperative Learning Approach to the Traveling Salesman Problem. *IEEE Trans. Evol. Comput.* **1997**, *1*, 53–66. [CrossRef]
45. Sutton, R.S.; Barto, A.G. *Reinforcement Learning: An Introduction*, 2nd ed.; MIT Press: Cambridge, MA, USA, 2018.
46. Watkins, C.J.C.H.; Dayan, P. Q-Learning. *Mach. Learn.* **1992**, *8*, 279–292. [CrossRef]
47. Mnih, V.; Kavukcuoglu, K.; Silver, D.; Rusu, A.A.; Veness, J.; Bellemare, M.G.; Graves, A.; Riedmiller, M.; Fidjeland, A.K.; Ostrovski, G.; et al. Human-Level Control through Deep Reinforcement Learning. *Nature* **2015**, *518*, 529–533. [CrossRef]
48. Al-Ansarry, S.; Al-Darraji, S. Hybrid RRT-A*: An Improved Path Planning Method for an Autonomous Mobile Robots. *J. Electr. Electron. Eng.* **2021**, *17*, 1–9. Available online: https://api.semanticscholar.org/CorpusID:236371326 (accessed on 8 May 2025). [CrossRef]
49. Lu, D.V.; Hershberger, D.; Smart, W.D. Layered Costmaps for Context-Sensitive Navigation. In Proceedings of the IEEE International Conference on Intelligent Robots and Systems (IROS), Chicago, IL, USA, 14–18 September 2014; pp. 709–715. [CrossRef]
50. Fox, D.; Burgard, W.; Thrun, S. The Dynamic Window Approach to Collision Avoidance. *IEEE Robot. Autom. Mag.* **1997**, *4*, 23–33. [CrossRef]
51. Khatib, O. Real-Time Obstacle Avoidance for Manipulators and Mobile Robots. *Int. J. Robot. Res.* **1986**, *5*, 90–98. [CrossRef]
52. Borenstein, J.; Koren, Y. The Vector Field Histogram—Fast Obstacle Avoidance for Mobile Robots. *IEEE J. Robot. Autom.* **1991**, *7*, 278–288. [CrossRef]
53. Li, J.; Sun, J.; Liu, L.; Xu, J. Model Predictive Control for the Tracking of Autonomous Mobile Robot Combined with a Local Path Planning. *Meas. Control* **2021**, *54*, 1319–1325. [CrossRef]
54. Yongzhe, Z.; Ma, B.; Wai, C.K. A Practical Study of Time-Elastic-Band Planning Method for Driverless Vehicle for Auto-Parking. In Proceedings of the 2018 International Conference on Intelligent Autonomous Systems (ICoIAS), Singapore, 1–3 March 2018; pp. 196–200. [CrossRef]
55. Chang, Z.; Wang, Y.; Cai, Y.; Li, S.; Gao, F. Fusion of Improved RRT and Ant Colony Optimization for Robot Path Planning. *Eng. Res. Express* **2024**, *6*, 045247. [CrossRef]
56. Sharma, V.D.; Lee, J.; Andrews, M.; Hadžić, I.H. Hybrid Classical/RL Local Planner for Ground Robot Navigation. *arXiv* **2024**, arXiv:2410.03066. Available online: http://arxiv.org/abs/2410.03066 (accessed on 2 February 2025).

57. Ravankar, A.; Ravankar, A.A.; Kobayashi, Y.; Hoshino, Y.; Peng, C.-C. Path Smoothing Techniques in Robot Navigation: State-of-the-Art, Current and Future Challenges. *Sensors* **2018**, *18*, 3170. [CrossRef]
58. Thrun, S.; Burgard, W.; Fox, D. *Probabilistic Robotics*; MIT Press: Cambridge, MA, USA, 2005.
59. Wang, L.; Yang, X.; Chen, Z.; Wang, B. Application of the Improved Rapidly Exploring Random Tree Algorithm to an Insect-like Mobile Robot in a Narrow Environment. *Biomimetics* **2023**, *8*, 374. [CrossRef]
60. Dorigo, M.; Maniezzo, V.; Colorni, A. The Ant System: Optimization by a colony of cooperating agents. *IEEE Trans. Syst. Man Cybern. B Cybern.* **1996**, *26*, 29–41. [CrossRef]
61. Liu, T.; Huang, Z.; Wang, C.; An, Z.; Meng, X.; Zheng, J. Path Planning of Underground Mining Area Transportation Scene Based on Improved A* Algorithm. In Proceedings of the 2023 7th International Conference on Transportation Information and Safety (ICTIS), Nanjing, China, 3–6 August 2023; pp. 2265–2270. [CrossRef]
62. Yulong, Q.; Qingyong, M.; Xu, T. Research on Navigation Path Planning for an Underground Load Haul Dump. *J. Eng. Sci. Technol. Rev.* **2015**, *8*, 102–109. Available online: https://api.semanticscholar.org/CorpusID:116378442 (accessed on 8 May 2025). [CrossRef]
63. Bao, J.-s.; Zhang, M.-y.; Ge, S.-r.; Liu, Q.; Yuan, X.-m.; Wang, M.-s.; Yin, Y.; Zhao, L. Underground Driverless Path Planning of Trackless Rubber Tyred Vehicle Based on Improved A* and Artificial Potential Field Algorithm. *J. China Coal Soc.* **2022**, *47*, 1347–1360. Available online: https://www.mtxb.com.cn/en/article/id/d7bec14d-e1ed-4d3d-9990-eb127b3e130c (accessed on 8 May 2025).
64. Zhang, C.; Yang, X.; Zhou, R.; Guo, Z. A Path Planning Method Based on Improved A* and Fuzzy Control DWA of Underground Mine Vehicles. *Appl. Sci.* **2024**, *14*, 3103. [CrossRef]
65. Zhu, D.; Zhang, Y.; Wang, J.; Ren, K.; Yang, K. Evaluation of Motion Planning Algorithms for Underground Mobile Robots. In Proceedings of the International Conference on Robotics and Applications (ICRA), Philadelphia, PA, USA, 23–27 May 2022; Springer: Singapore, 2022; pp. 368–379. [CrossRef]
66. Gu, C.; Liu, S.; Li, H.; Yuan, K.; Bao, W. Research on Hybrid Path Planning of Underground Degraded Environment Inspection Robot Based on Improved A* Algorithm and DWA Algorithm. *Robotica* **2025**, *43*, 1–22. [CrossRef]
67. Zhang, B.; Zhang, Y.; Zhang, C.; Cheng, J.; Shen, G.; Wang, X. Path Planning for Unmanned Load–Haul–Dump Machines Based on a VHF_A* Algorithm. *Proc. Inst. Mech. Eng. C J. Mech. Eng. Sci.* **2024**, *238*, 6584–6597. [CrossRef]
68. Ma, X.; Mao, R. Path Planning for Coal Mine Robot to Avoid Obstacle in Gas Distribution Area. *Int. J. Adv. Robot. Syst.* **2018**, *15*, 1–10. [CrossRef]
69. Xu, R.; Yao, S. Research on UGV Path Planning in Tunnel Based on the Dijkstra*-PSO* Algorithm. In Proceedings of the 2022 6th Asian Conference on Artificial Intelligence Technology (ACAIT), Chengdu, China, 9–11 December 2022; pp. 1–9. [CrossRef]
70. Chen, Z. Path Planning Method for Underground Survey Robot in Dangerous Scenarios. *Highl. Sci. Eng. Technol.* **2023**, *43*, 207–214. [CrossRef]
71. Ma, T.; Lv, J.; Guo, M. Downhole Robot Path Planning Based on Improved D* Algorithm. In Proceedings of the 2020 IEEE International Conference on Signal Processing, Communications and Computing (ICSPCC), Macau, China, 26–28 August 2020; pp. 1–5. [CrossRef]
72. Wang, H.; Li, G.; Hou, J.; Chen, L.; Hu, N. A Path Planning Method for Underground Intelligent Vehicles Based on an Improved RRT* Algorithm. *Electronics* **2022**, *11*, 294. [CrossRef]
73. Hopfenblatt, B.; Axel, N. A Combined RRT*-Optimal Control Approach for Kinodynamic Motion Planning for Mobile Robots. 2016. Available online: https://api.semanticscholar.org/CorpusID:203702248 (accessed on 8 May 2025).
74. Wu, J.; Zhao, L.; Liu, R. Research on Path Planning of a Mining Inspection Robot in an Unstructured Environment Based on an Improved Rapidly Exploring Random Tree Algorithm. *Appl. Sci.* **2024**, *14*, 6389. [CrossRef]
75. Steinbrink, M.; Koch, P.; Jung, B.; May, S. Rapidly-Exploring Random Graph Next-Best View Exploration for Ground Vehicles. *arxiv* **2021**, arXiv:2108.01012. Available online: http://arxiv.org/abs/2108.01012 (accessed on 28 January 2025).
76. Song, B.; Miao, H.; Xu, L. Path Planning for Coal Mine Robot via Improved Ant Colony Optimization Algorithm. *Syst. Sci. Control Eng.* **2021**, *9*, 283–289. [CrossRef]
77. Xu, J.C.; Huang, Y.R.; Xu, G.Y. A Path Optimization Algorithm for the Mobile Robot of Coal Mine Based on Ant Colony Membrane Algorithm. In Proceedings of the 2018 ACM International Conference on Computing and Sustainable Societies (COMPASS), Menlo Park and San Jose, CA, USA, 12–15 December 2018. [CrossRef]
78. Shao, X.; Wang, G.; Zheng, R.; Wang, B.; Yang, T.; Liu, S. Path Planning for Mine Rescue Robots Based on Improved Ant Colony Algorithm. In Proceedings of the 2022 8th International Conference on Control, Automation and Robotics (ICCAR), Xiamen, China, 8–10 April 2022; pp. 161–166. [CrossRef]
79. Liu, K.; Zhang, M. Path Planning Based on Simulated Annealing Ant Colony Algorithm. In Proceedings of the 2016 9th International Symposium on Computational Intelligence and Design (ISCID), Hangzhou, China, 10–11 December 2016; Volume 2, pp. 461–466. [CrossRef]

80. Fang, C.; Liu, X.; Hua, D.; Wang, H.; Li, Z.; Chauhan, S.; Vashishtha, G. Real-Time Path Planning of Drilling Robot Using Global and Local Sensor Information Fusion: Mining-Oriented Validation. *Int. J. Adv. Manuf. Technol.* **2024**, *135*, 5875–5891. [CrossRef]
81. Li, P.; Zhu, H. Immune Optimization of Path Planning for Coal Mine Rescue Robot. 2016. Available online: https://api.semanticscholar.org/CorpusID:13507361 (accessed on 8 May 2025).
82. Gao, Y.; Dai, Z.; Yuan, J. A Multiobjective Hybrid Optimization Algorithm for Path Planning of Coal Mine Patrol Robot. *Comput. Intell. Neurosci.* **2022**, *2022*, 9094572. [CrossRef]
83. Zhang, S.; Zeng, Q. Online Unmanned Ground Vehicle Path Planning Based on Multi-Attribute Intelligent Reinforcement Learning for Mine Search and Rescue. *Appl. Sci.* **2024**, *14*, 9127. [CrossRef]
84. Jiang, Y.; Peng, P.; Wang, L.; Wang, J.; Wu, J.; Liu, Y. LiDAR-Based Local Path Planning Method for Reactive Navigation in Underground Mines. *Remote Sens.* **2023**, *15*, 309. [CrossRef]
85. Tian, Z.-J.; Gao, X.-H.; Zhang, M.-X. Path Planning Based on the Improved Artificial Potential Field of Coal Mine Dynamic Target Navigation. *J. China Coal Soc.* **2016**, *41*, 589–597. [CrossRef]

mining

MDPI

Article

Fragmentation Size Distribution Measurement by GNSS-Aided Photogrammetry at Real Mine Site

Hisatoshi Toriya [1,*], Zedrick Paul L. Tungol [1], Hajime Ikeda [1], Narihiro Owada [1], Hyong Doo Jang [2], Tsuyoshi Adachi [1], Itaru Kitahara [3] and Youhei Kawamura [4,5]

[1] Faculty of International Resource Sciences, Akita University, 1-1 Tegatagakuen-Machi, Akita 010-0862, Japan; zed.tungol37@gmail.com (Z.P.L.T.); ikeda@gipc.akita-u.ac.jp (H.I.); owada@gipc.akita-u.ac.jp (N.O.); adachi.t@gipc.akita-u.ac.jp (T.A.)

[2] Western Australian School of Mines, Curtin University, Perth 6845, Australia; hyongdoo.jang@curtin.edu.au

[3] Center for Computational Sciences, University of Tsukuba, 1-1-1 Tennodai, Tsukuba 305-8577, Japan; kitahara@ccs.tsukuba.ac.jp

[4] Faculty of Engineering, Hokkaido University, Kita 8, Nishi 5, Kita-Ku, Sapporo 060-8628, Japan; kawamura@eng.hokudai.ac.jp

[5] North China Institute of Science and Technology, Xueyuan Ave, Sanhe, Langfang 065201, China

* Correspondence: toriya@gipc.akita-u.ac.jp

Abstract: In mining operations that employ explosives and mineral processing, one of the important factors for efficient and low-cost operation is the fragmentation size distribution of rock after it has been blasted. Automatic scaling is a critical component of fragmentation size distribution measurement as it will directly determine the accuracy of the size estimation. In this study, we propose a method to create a system for creating a scaled 3D CG model, without the use of ground truth data such as GCPs (Ground Control Points), for the purpose of improving fragmentation size distribution measurement using positional data such as GNSS (Global Navigation Satellite System)-aided photogrammetry. We confirmed the validation of the method through an experimental evaluation of actual muckpiles. The results showed evidence of improving the scaling aspect of 3D fragmentation measurement systems without using GCPs or manual scales, specifically in surface mines where GNSS data are available.

Keywords: point cloud scaling; fragmentation size measurement; structure from motion; global navigation satellite system

1. Introduction

A mining operation contains several factors that affect its efficient extraction of resources. One such factor is fragmentation size, a key parameter across almost all of the stages of production from mine (drill, blasting, haulage, etc.) to mill (mineral processing). Several studies have explored the effect of fragmentation size on other factors such as drilling and blasting cost [1] and the performance of crushing and grinding circuits [2,3]. Therefore, it can be considered that it is vital for companies to continuously monitor fragmentation size and make necessary changes to mine planning and execution to keep the fragmentation size that is most beneficial to the operation as a whole. Traditionally, methods such as manual sieving, boulder counting, and visual estimation have been used for fragmentation size measurement. However, due to the generally large amount of material being mined and its innately heterogeneous nature, difficulties arise from using traditional methods. In addition, limitations on sampling as well as bias make these methods relatively inefficient [4]. As such, there exists a need for a quick and accessible method of rock fragmentation size distribution determination that can surmount the limitations of physical sampling and laboratory analysis. A currently used digital solution to this problem is to employ image-based particle size analysis software. Commercial products such as

WipFrag [5,6] and Split Desktop [7,8] use images of muckpiles or orthomosaics to measure fragmentation size distribution, where the conventional studies use 2D information and the 3D size estimation accuracy is relatively low.

A 3-Dimensional Fragmentation Measurement (3DFM) system was developed that uses 3D photogrammetry to measure fragmentation size distribution at accuracies greater than that of conventional methods [9]. The 3DFM extracts the fragmentation sizes from a 3D muckpile model, which facilitates accurate fragmentation size distribution measurement without scale objects and eliminates the need for excessive manual editing. Furthermore, a representative fragmentation of an entire blasting shot can easily be analyzed by generating a 3D model of an entire blasted muckpile. A workflow for this system when applied to a mining operation, and scope of this work, is shown in Figure 1.

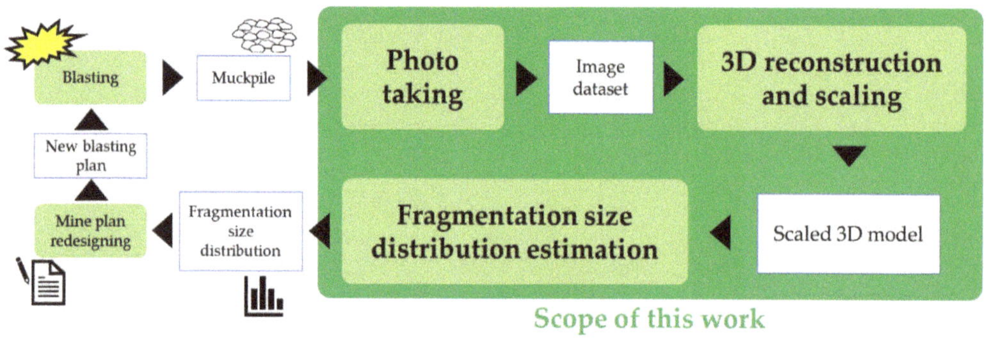

Figure 1. Application of a 3D fragmentation measurement system.

The developed system is divided into stages, utilizing multiple computational techniques in order to achieve its purpose. In a hypothetical application of the system, pictures of the muckpiles from the products of blasting are taken. The sizes of muckpiles vary considerably depending on the specifications of the hauling equipment as well as the mine plan that the operation employs. In situations where the muckpiles are too large or have parts inaccessible to photo-taking, the system can reconstruct only a representative slice of the muckpiles. The images are then processed in a high-power computer by a sequence of 3D imaging techniques that will ultimately output a scaled 3D model of the muckpiles in the form of a point cloud. The dimensional data can be used to compute the fragment size distribution of the muckpiles. Using this information, the blasting product can be judged if it is up to the expected specification. Adjustments are then made to the blasting design, such as the amount and type of explosive and blasting patterns to achieve the required distribution.

Scaling is a critical component of fragmentation size distribution measurement using photogrammetry as it will directly determine the accuracy of the size estimation. In creating a 3D model, extrinsic data such as ground truths are needed to create a properly-scaled reconstruction of the scene. Several methods are used to resolve scale in photogrammetry. Most of these methods have the same basic idea in that once the exact distance between at least two different points in a scene as a ground truth is known, a scale factor can be applied to the 3D model. One way to do this is to include an object of known length, such as scale bars, in the scene. In larger applications such as aerial mapping, GCPs (Ground Control Points) are used, marked points of known absolute or relative coordinates.

A previous work by authors [10] proposed a method using GNSS (Global Navigation Satellite System, such as GPS) to get absolute (but rough) camera positions for scaling muckpile fragmentation. The authors showed the potential of the fragmentation size distribution estimation method using rocks similar to muckpiles in the previous work, but the verification test with actual muckpiles and fragmentation size distribution estimation remained an issue.

This paper validates the method's effectiveness through practical experiments using actual muckpiles generated in a quarry site, and show that the method can be applied to fragmentation size distribution measurement.

2. Muckpile Scaling by GNSS-Aided Photogrammetry

2.1. Structure-from-Motion-Multi-View-Stereo (SfM-MVS)

Structure from Motion (SfM) has been collectively defined as the photogrammetric technique [11], the process [12], as well as the tools [13] used to generate 3D models from 2D images taken at different angles. It has been developed since the 1980s, resulting in various applications such as photogrammetric surveys, virtual reality model reconstruction, determination of camera motion [14], and odometrical scale estimation [15]. Compared to traditional photogrammetry, where the calculation is more direct, SfM uses repeating algorithms to identify matching features in a set of overlapping images, and use these matched features to calculate camera location and orientation [13]. SfM can be computed in several ways, depending on numerous factors such as camera type, image ordering, capture format, and more.

Mathematically, SfM can be described as the conversion of four coordinate systems, illustrated in Figure 2:

i. Image pixel coordinate system, which concerns the pixels on the 2D image.
ii. Imaging plane coordinate system, which lies on the same plane of the previous system, but whose origin is the plane's intersection with the camera's optical axis.
iii. Camera coordinate system, which concerns a pinhole camera's point of view of the image.
iv. World coordinate system, which is a reference system to describe the position of the camera and the objects being taken pictures of.

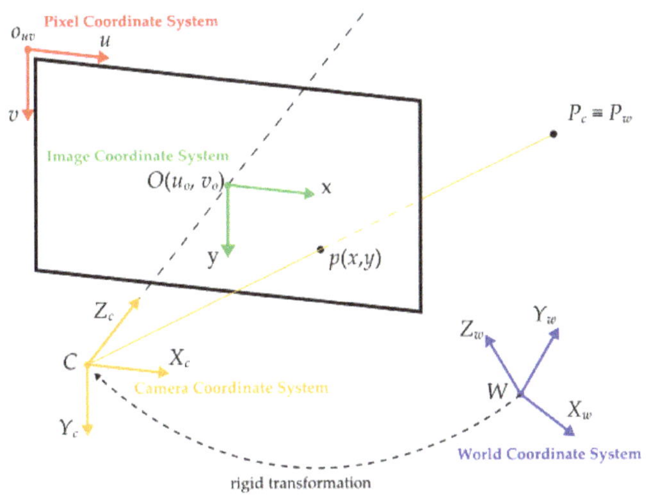

Figure 2. An illustration of the coordinate systems. Described is the conversion of world point P_w to camera point P_c, to imaging plane coordinates (x, y), and finally pixel coordinates (u, v).

The conversion of these four coordinates systems can be described by Equation (1). The u and v describe the axes in the imaging planes. The u_0 and v_0 are the coordinates of the origins of the imaging plane in the pixel coordinate system. The δ_x and δ_y represent the physical size of each pixel in the image in the imaging plane (zoom ratio). The f describe the focal length, which is the distance from the optical center of the camera to the pixel plane. The $\boldsymbol{R} \in \mathbb{R}^{3\times3}$ and $\boldsymbol{t} \in \mathbb{R}^3$ describes the rotational and translational vectors that relate the camera and the world coordinate systems. The X_w, Y_w, and Z_w are the actual coordinates of a point in the world coordinate system.

$$Z_c \begin{bmatrix} u \\ v \\ 1 \end{bmatrix} = \begin{bmatrix} \frac{1}{\delta_x} & 0 & u_o \\ 0 & \frac{1}{\delta_y} & v_o \\ 0 & 0 & 1 \end{bmatrix} \begin{bmatrix} f & 0 & 0 \\ 0 & f & 0 \\ 0 & 0 & 1 \end{bmatrix} \begin{bmatrix} \boldsymbol{R} & \boldsymbol{t} \end{bmatrix} \begin{bmatrix} X_w \\ Y_w \\ Z_w \\ 1 \end{bmatrix} = \begin{bmatrix} f_x & 0 & u_0 \\ 0 & f_y & v_0 \\ 0 & 0 & 1 \end{bmatrix} \begin{bmatrix} \boldsymbol{R} & \boldsymbol{t} \end{bmatrix} \begin{bmatrix} X_w \\ Y_w \\ Z_w \\ 1 \end{bmatrix} \quad (1)$$

Equation (1) represents the fact that in order to estimate the position of a point in the real world, the external parameter matrix of the camera (i.e., $\boldsymbol{R}$ and $\boldsymbol{t}$) needs to be measured first. Once $\boldsymbol{R}$ is known, the relative position of the object in the world coordinate system can be estimated, and once $\boldsymbol{t}$ is known, the absolute position can also be acquired. Basic SfM can relatively estimate $\boldsymbol{R}$ and $\boldsymbol{t}$.

It can be inferred that there is no single correct workflow or process in the conversion of 2D images into models. However, key processes are present in almost all applications of the method, as shown in Figure 3. Firstly, an image dataset must be taken by cameras with a GNSS receiver. As many images as possible from various viewpoints should be taken because the number of images and variation of viewpoints effect the accuracy of the result of 3D reconstruction. SfM process is composed of keypoint detection, keypoint matching, matched filtering (filtering valid corresponding points), and sparse reconstruction (bundle adjustment for 3D model with point cloud). In the bundle adjustment step, a scaled sparce 3D model is generated using the georeferenced data in the input. By combining texture information and the scaled sparse 3D model through SfM and MVS processes, the scaled and dense 3D model can be obtained.

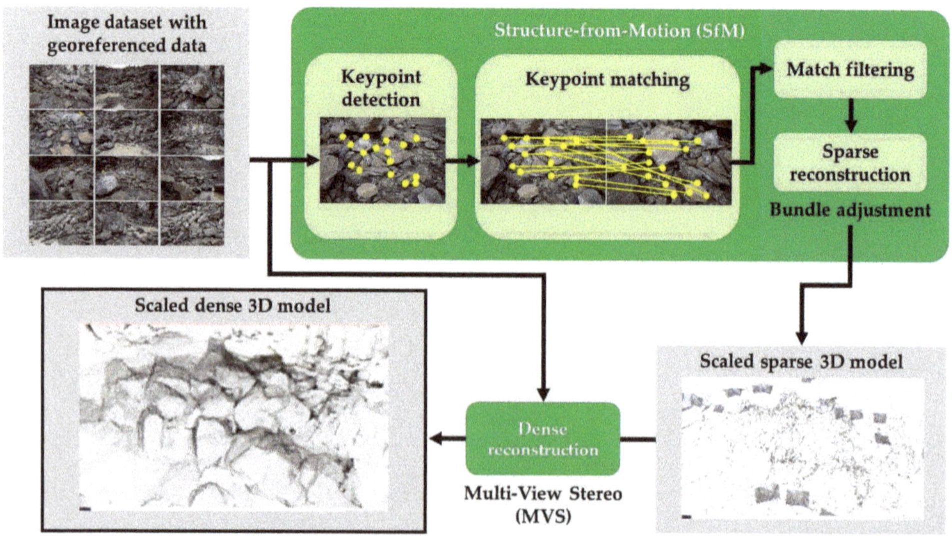

Figure 3. SfM-MVS pipeline. Through SfM and MVS processes, a scaled dense 3D model can be obtained using a georeferenced image dataset.

2.2. GNSS-Aided Scaling

The study proposes a method that makes use of GNSS (Global Navigation Satellite System) data to create scaled 3D models without the need for post-reconstruction rescaling. GNSS positional data and its sub-systems such as GPS, Beidou [16], GLONASS (GLObal NAvigation Satellite System) [17], and QZSS (Quasi-Zenith Satellite System) [18] can be utilized. A previous study was performed with regards to using GPS in reconstruction, but mostly in the context of UAV (Unmanned Aerial Vehicle) Mapping [19]. This study aimed to develop a system without needing ground truth data such as GCPs to create a properly scaled 3D model of muckpiles. This would aid greatly in the fragmentation size distribution measurement of muckpiles using photogrammetry.

It is a known fact that an inherent error exists within GNSS and its subsets, and even high-end geodetic GNSS receivers have errors in the centimeter range [20]. For this study, a smartphone is used as a GNSS receiver for the digital camera. This decision is due to the end goal of this research which is to be able use both image data and GNSS data from a smartphone, as this practicality can be important in a mining operation environment. This comes at a drawback to the GNSS accuracy, as recreational-grade GNSS chips like those found in smartphones typically have errors in the meter range [21]. To overcome this error, the study proposes to make use of an increasing number of georeferenced images to statistically decrease the scaling error of the constructed 3D model. Figure 4 shows a general overview of the proposed system for this study. Utilizing a smartphone's built-in GNSS receiver, GNSS data can be logged and sent to a camera. At the moment an image is taken, GNSS data can be embedded into the image's metadata.

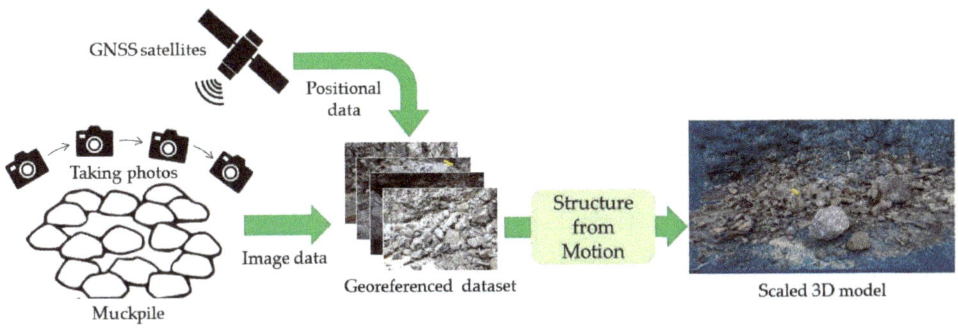

Figure 4. GNSS-aided photogrammetry workflow. Captured images with georeferenced data are used for 3D reconstruction, and finally a scaled 3D model is obtained.

Bundle adjustment is an essential process of SfM. In the bundle adjustment phase, the previous and imperfect solutions regarding camera positions and 3D features of the scene are refined [22]. More specifically, bundle adjustment is a nonlinear minimization procedure that jointly optimizes camera parameters and point position by minimizing the reprojection error between the image locations of observed and predicted image points. This minimization is done using nonlinear least squares algorithms [23]. Numerous studies have been done since its inception in the 90′s regarding bundle adjustment, with most of the research going into reducing its computational burden and accelerating the problem-solving process [24]. One such proposition is the fusion of positional data and bundle adjustment, with GNSS data being used as constraints for solving reprojection errors [25]. This concept is what this research aims to use to produce accurately-scaled 3D reconstructions of muckpiles. GNSS data are used to provide position and covariance estimates for the bundle adjustment process. The nominal form of these solutions is:

$$r^M_{GNSS}(t) = r^M_c(t) + R^M_c(t) r^c_{GNSS} + \left(b^M_{GNSS} + d^M_{GNSS}(t - t_0)\right), \quad (2)$$

where $r^M_{GNSS}(t)$ is the position of the GNSS receiver, $r^M_c(t)$ denotes the camera position, and $R^M_c(t)$ is the rotational matrix that aligns camera and mapping space axes, and r^c_{GNSS} is the difference between the GNSS receiver and camera position. The b^M_{GNSS} and d^M_{GNSS} denote bias and drift terms and are included to account for data inconsistencies and the inherent errors that exist within GNSS. A previous study [26] applied GNSS-assisted terrestrial photogrammetry to model coastal areas without the use of GCPs. With bundle adjustment being an error minimization problem with multiple factors, weights can be assigned to them, as is the case in the study's SfM workflow.

3. Evaluation Experiment in a Real Mine Site

For this experiment, actual muckpiles found in an active mining (quarry) site are recreated in 3D space. The aim of this experiment is to assess the performance of GNSS-aided photogrammetry in a practical application in an open mining environment.

The figures in this section are labeled "CGI" (Computer Generated Image) or "Photo" (real photo) in the lower left corner of each figure.

3.1. Experimental Site

The site chosen for this experiment was the Mikurahana Quarry Site of TOSEKI MATERIAL Inc. https://www.materiaru.jp/ (accessed on 12 June 2022), found in Hachirogata-town in Akita Prefecture, Japan, and the experiment was performed on 15 November 2021. The muckpiles were composed mainly of limestone and fragments that were generally large, with fragments as large as 2 m wide in its largest dimension. Satellite imagery is shown in Figure 5 to provide locational context for the quarry and the muckpiles in question.

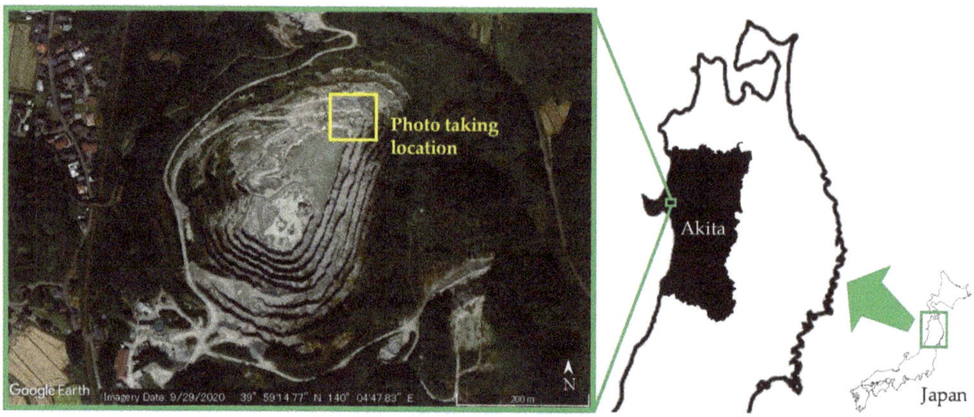

Figure 5. Satellite view of the Mikurahana quarry site in Hachirogata-town, Akita-prefecture, Japan.

3.2. Experiment with Real Muckpiles

A total of 400 images of the muckpiles were taken from varying angles facing the muckpiles using a smartphone (Xiaomi Mi 9T Pro) with a GNSS receiver on. Each image size is 4464 × 2976 pixels. Sample images are shown in Figure 6, which also feature the yellow cardboard box that was used as an absolute reference for scaling error. The box measures 15 cm × 15 cm × 60 cm, with the 60 cm-long side used for measuring scaling error.

3.3. Results of 3D Reconstruction and Scaling

Using 3DF Zephyr (version 6.010) [27], a 3D photogrammetry software, the images are used to create 3D models at different image numbers, namely 50, 100, 200, and 400 images. The camera positions, measured by GNSS and by SfM, are shown in Figure 7. The red dots in the left figure represent photo-taking positions obtained using GNSS, and green objects in the right represent actual photo-taking positions measured by SfM. The figure on the right was created by manually superimposing an orthorectified image on a satellite image. The orthorectified image was generated by a 3D model obtained by SfM. We tried to take as many images as we could from various viewpoints, as Figure 7 shows.

Figure 6. Sample images from the dataset. The yellow box is used as an absolute reference.

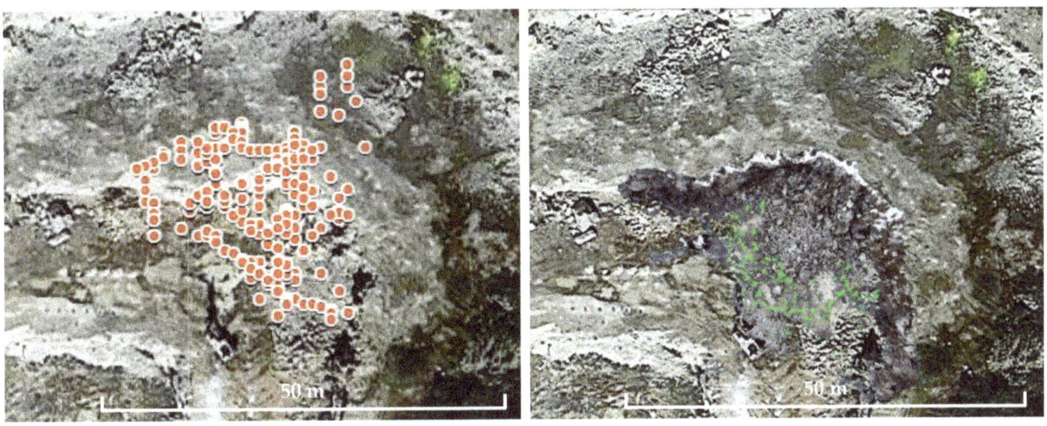

Figure 7. Camera positions measured by GNSS and by SfM. The red dots in the left figure represent photo-taking positions obtained using GNSS, and the green objects in the right figure represent actual photo-taking positions measured by SfM. The figure on the right was created by manually superimposing an orthorectified image on a satellite image. The orthorectified image was generated by a 3D model obtained by SfM.

The scaling error of each model with respect to the yellow box is measured and plotted in a graph to observe the relationship between scaling error and image number. The sparse and dense reconstruction of the muckpiles at 400 images is shown in Figure 8 at different angles for reference. A mesh reconstruction featuring measurement of the yellow box in 3DF zephyr is also shown in Figure 9.

As shown in Table 1 and Figure 10, there is a trend that at an increasing number of images used in reconstruction, the difference from the real measurement decreases. This result, as with the previous experiments, lends more credence to the hypothesis that using more images for reconstruction has the tendency to lessen scale error in 3D models, with a relatively linear relationship between image number and scaling error as shown by the trendline with an R-value of 0.93. It is noted, however, that the improvement in scaling error by increasing the number of images is lessened, as the scaling error at using only 50 images is already 36.67%, which is much higher than when using 50 images in the other experiments. The study hypothesizes that this is due to the lack of vegetation in the area, which improved the GNSS accuracy. The study achieved 11.7% scaling error at 400 images, which is close to the extrapolated 10% scaling error at 400 images in the previous experiment on monuments [10]. The error is comparable or more accurate than WipFrag [5,6].

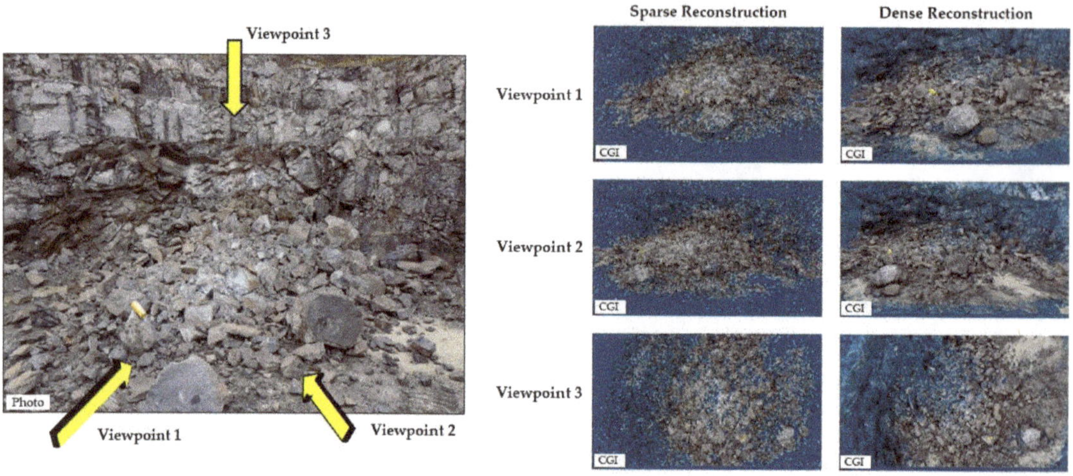

Figure 8. Sparse and dense reconstruction of limestone muckpiles using 400 images, shown at different angles.

Figure 9. Close-up of mesh reconstruction, featuring the reference yellow box and the dimension of its longest side as measured.

Table 1. Results of the experiment on the Mikurahana quarry muckpiles.

Number of Images	Measured (m)	Real Measurement (m)	Difference from Real Measurement (m)	Error (%)
50	0.82	0.60	0.22	36.7
100	0.76	0.60	0.16	26.7
200	0.74	0.60	0.14	23.3
400	0.67	0.60	0.07	11.7

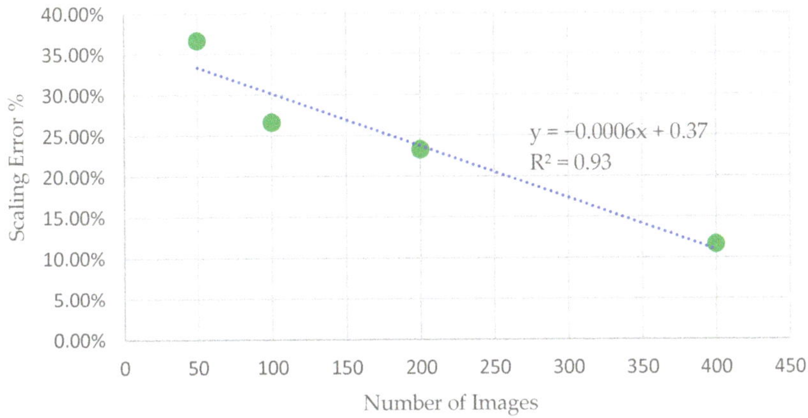

Figure 10. Graph detailing the results of the Mikurahana experiment.

3.4. Muckpile Fragmentation Size Distribution Estimation

In the previous chapters, it was confirmed that scaling by GNSS-aided 3D photogrammetry is effective for actual muckpiles. In this chapter, based on the scaling results, we measured the muckpile size distribution of actual muckpiles. The results of manually measuring the muckpile size distribution of the generated 3D model of muckpiles are shown in Figure 11. Green bars represent the percentage of each rock size measured by the 3D model, and the gray bars represent that measured by real measurements. The blue line represents the cumulative percentage of rock size measured by the 3D model, and the black dotted line represents that measured by real measurements.

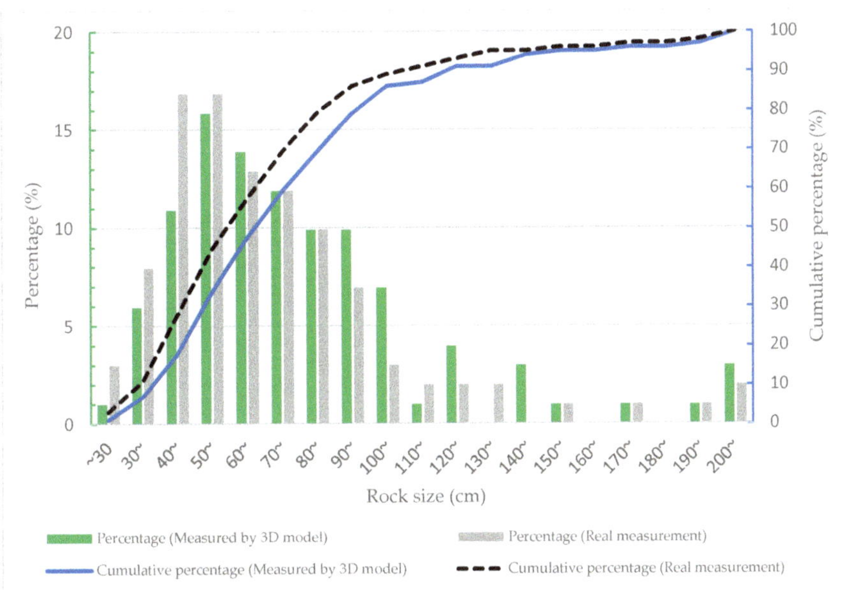

Figure 11. Measured fragmentation size distribution in the experimental site. The green bars represent the percentage of each rock size measured by the 3D model, and the gray bars represent that measured by real measurements. The blue line represents the cumulative percentage of rock size measured by the 3D model, and the black dotted line represents that measured by real measurements.

Figure 11 shows that the proposed method achieved a close result to the real measurements. Referring to the result, we concluded that it is possible to measure the muckpile fragmentation size distribution in an actual quarry.

Although it is difficult to count rocks that are hidden from view by other rocks, it is possible to count all large rocks that require heavy machines to transport. Scaling of very small rocks (a few centimeters) is also difficult due to the resolution of the camera, but this is expected to improve as camera performance improves. Many and various viewpoints of photos should be taken for improving the accuracy. It is hard to achieve by handheld cameras, but it can be easier with videos by drone. An end-to-end fragmentation size distribution estimation system using drones is one of the ultimate goals of this study.

4. Conclusions

This paper showed that the method of creating an accurately scaled 3D model by constraining camera positions through georeferenced images as input for SfM can be applied to real muckpiles and can be used to measure the fragmentation size distribution. Monitoring fragmentation size is an important procedure in optimizing mining operations that perform blasting. In recent years, a new method that involves using 3D photogrammetry to measure fragment sizes has been developed that has the potential to surpass traditional techniques. For this particular process to be accurate, a method for properly scaling 3D models with georeferenced images using GNSS was investigated. An experiment was performed on actual muckpiles in the Mikurahana quarry to test the system's accuracy in a practical application. Three-dimensional reconstructions were created at image numbers of 50, 100, 200, and 400 limestone muckpiles, and the scaling error was measured and graphed against the image number. It also showed a linear pattern with an R-squared value of 0.93. The scaling error decreases with increasing image number, albeit at a lower ratio than has been hypothesized due to the lack of interference from vegetation and buildings.

Two observations can be drawn from the experimental result: (1) increasing the number of georeferenced images in SfM will incrementally improve the scaling error of the reconstruction. These observations can help improve scale accuracy in GNSS-aided 3D fragmentation measurement; and (2) the proposed method can be applied to actual muckpiles at real mine sites. These results show examples of improving the scaling aspect of 3D fragmentation measurement systems without the use of GCPs or manual scales, specifically in surface mines where GNSS data are generally readily available. This shows that monitoring the fragmentation distribution can potentially be performed using just a camera and GNSS-enabled devices, such as smartphones or drones.

Author Contributions: Conceptualization, H.T., Z.P.L.T. and H.D.J.; methodology, H.T. and Z.P.L.T.; software, H.T. and Z.P.L.T.; validation, H.T., H.I. and Z.P.L.T.; formal analysis, N.O.; investigation, H.T. and N.O.; resources, N.O.; data curation, H.T. and Z.P.L.T.; writing—original draft preparation, H.T., Z.P.L.T. and H.I.; writing—review and editing, H.I. and Y.K.; visualization, H.T. and Z.P.L.T.; supervision, H.D.J., I.K., T.A. and Y.K.; project administration, Y.K.; funding acquisition, Y.K. All authors have read and agreed to the published version of the manuscript.

Funding: This research received no external funding.

Data Availability Statement: Not applicable.

Acknowledgments: The dataset was taken in the Mikurahana Quarry Site of TOSEKI MATERIAL Inc. https://www.materiaru.jp/ (accessed on 12 June 2022).

Conflicts of Interest: The authors declare no conflict of interest.

References

1. Afum, B.O.; Temeng, V.A. Reducing Drill and Blast Cost through Blast Optimisation—A Case Study. *Ghana Min. J.* **2015**, *15*, 50–57.
2. Kanchibotla, S.S.; Valery, W.; Morrell, S. Modelling Fines in Blast Fragmentation and Its Impact on Crushing and Grinding. In Proceedings of the Explo '99—A Conference on Rock Breaking, Kalgoorlie, Australia, 7–11 November 1999; The Australasian Institute of Mining and Metallurgy: Kalgoorlie, Australia, 1999; pp. 137–144.

3. Valery, W.; Morrell, S.; Kojovic, T.; Kanchibotla, S.S.; Thornton, D.M. Modelling and Simulation Techniques Applied for Optimisation of Mine to Mill Operations and Case Studies. In Proceedings of the VI Southern Hemisphere Meeting on Mineral Technology, Rio de Janeiro, Brazil, 27 May–1 June 2001.
4. Nefis, M.; Talhi, K. A Model Study to Measure Fragmentation by Blasting. *Min. Sci.* **2016**, *23*, 91–104.
5. Palangio, T.C.; Franklin, J.A.; Maerz, N.H. WipFrag—A Breakthrough in Fragmentation Measurement. In Proceedings of the 6th High-Tech Seminar on State of the Art Blasting Technology, Instrumentation, and Explosives Applications, Boston, MA, USA, 1995; p. 943. Available online: https://scholarsmine.mst.edu/geosci_geo_peteng_facwork/1261/ (accessed on 12 June 2022).
6. Maerz, N.H.; Palangio, T.C.; Franklin, J.A. WipFrag Image Based Granulometry System. In *Measurement of Blast Fragmentation*; Routledge: Abingdon, UK, 1996.
7. Kemeny, J.M. Practical Technique for Determining the Size Distribution of Blasted Benches, Waste Dumps and Heap Leach Sites. *Min. Eng.* **1994**, *46*, 1281–1284.
8. Kemeny, J.; Mofya, E.; Kaunda, R.; Lever, P. Improvements in Blast Fragmentation Models Using Digital Image Processing. *Fragblast* **2002**, *6*, 311–320. [CrossRef]
9. Jang, H.; Kitahara, I.; Kawamura, Y.; Endo, Y.; Topal, E.; Degawa, R.; Mazara, S. Development of 3D Rock Fragmentation Measurement System Using Photogrammetry. *Int. J. Min. Reclam. Environ.* **2020**, *34*, 294–305. [CrossRef]
10. Tungol, Z.P.L.; Toriya, H.; Owada, N.; Kitahara, I.; Inagaki, F.; Saadat, M.; Jang, H.D.; Kawamura, Y. Model Scaling in Smartphone GNSS-Aided Photogrammetry for Fragmentation Size Distribution Estimation. *Minerals* **2021**, *11*, 1301. [CrossRef]
11. Westoby, M.J.; Brasington, J.; Glasser, N.F.; Hambrey, M.J.; Reynolds, J.M. 'Structure-from-Motion'Photogrammetry: A Low-Cost, Effective Tool for Geoscience Applications. *Geomorphology* **2012**, *179*, 300–314. [CrossRef]
12. Schonberger, J.L.; Frahm, J.-M. Structure-from-Motion Revisited. In Proceedings of the IEEE Conference on Computer Vision and Pattern Recognition, Las Vegas, NV, USA, 26 June–1 July 2016; pp. 4104–4113.
13. Carrivick, J.L.; Smith, M.W.; Quincey, D.J. *Structure from Motion in the Geosciences*; John Wiley & Sons: Hoboken, NJ, USA, 2016; ISBN 9781118895832.
14. Cipolla, R.; Robertson, D.P. Chapter 13: Structure from Motion. In *Practical Image Processing and Computer Vision*; John Wiley and Sons Ltd.: New York, NY, USA, 2009.
15. Grater, J.; Schwarze, T.; Lauer, M. Robust Scale Estimation for Monocular Visual Odometry Using Structure from Motion and Vanishing Points. In Proceedings of the 2015 IEEE Intelligent Vehicles Symposium (IV), Seoul, Korea, 28 June–1 July 2015.
16. Han, C.; Yang, Y.; Cai, Z. BeiDou Navigation Satellite System and Its Time Scales. *Metrologia* **2011**, *48*, S213. [CrossRef]
17. Revnivykh, S.; Bolkunov, A.; Serdyukov, A.; Montenbruck, O. GLONASS. In *Springer Handbook of Global Navigation Satellite Systems*; Teunissen, P.J.G., Montenbruck, O., Eds.; Springer International Publishing: Cham, Switzerland, 2017; pp. 219–245. ISBN 9783319429281.
18. Hauschild, A.; Steigenberger, P.; Rodriguez-Solano, C. Signal, Orbit and Attitude Analysis of Japan's First QZSS Satellite Michibiki. *GPS Solut.* **2012**, *16*, 127–133. [CrossRef]
19. Fonstad, M.A.; Dietrich, J.T.; Courville, B.C.; Jensen, J.L.; Carbonneau, P.E. Topographic Structure from Motion: A New Development in Photogrammetric Measurement. *Earth Surf. Process. Landf.* **2013**, *38*, 421–430. [CrossRef]
20. Khomsin; Anjasmara, I.M.; Pratomo, D.G.; Ristanto, W. Accuracy Analysis of GNSS (GPS, GLONASS and BEIDOU) Obsevation For Positioning. *E3S Web Conf.* **2019**, *94*, 01019. [CrossRef]
21. Merry, K.; Bettinger, P. Smartphone GPS Accuracy Study in an Urban Environment. *PLoS ONE* **2019**, *14*, e0219890. [CrossRef] [PubMed]
22. Zhang, J.; Boutin, M.; Aliaga, D.G. Robust Bundle Adjustment for Structure from Motion. In Proceedings of the 2006 International Conference on Image Processing, Atlanta, GA, USA, 8–11 October 2006; pp. 2185–2188.
23. Lourakis, M.I.A.; Argyros, A.A. SBA: A Software Package for Generic Sparse Bundle Adjustment. *ACM Trans. Math. Softw.* **2009**, *36*, 1–30. [CrossRef]
24. Triggs, B.; McLauchlan, P.F.; Hartley, R.I.; Fitzgibbon, A.W. Bundle Adjustment—A Modern Synthesis. In *Vision Algorithms: Theory and Practice*; Lecture Notes in Computer Science; Springer: Berlin/Heidelberg, Germany, 2000; pp. 298–372. ISBN 9783540679738.
25. Kume, H.; Taketomi, T.; Sato, T.; Yokoya, N. Extrinsic Camera Parameter Estimation Using Video Images and GPS Considering GPS Positioning Accuracy. In Proceedings of the 2010 20th International Conference on Pattern Recognition, Istanbul, Turkey, 23–26 August 2010; pp. 3923–3926.
26. Jaud, M.; Bertin, S.; Beauverger, M.; Augereau, E.; Delacourt, C. RTK GNSS-Assisted Terrestrial SfM Photogrammetry without GCP: Application to Coastal Morphodynamics Monitoring. *Remote Sens.* **2020**, *12*, 1889. [CrossRef]
27. 3DF Zephyr—Photogrammetry Software—3D Models from Photos. Available online: https://www.3dflow.net/3df-zephyr-photogrammetry-software/ (accessed on 13 April 2022).

mining MDPI

Systematic Review

Slope Stability Monitoring Methods and Technologies for Open-Pit Mining: A Systematic Review

Rohan Le Roux [1,*], Mohammadali Sepehri [2], Siavash Khaksar [1] and Iain Murray [2]

[1] School of Electrical Engineering, Computing and Mathematical Sciences, Curtin University, Perth 6102, Australia; siavash.khaksar@curtin.edu.au

[2] Western Australia School of Mines: Minerals, Energy and Chemical Engineering, Curtin University, Perth 6102, Australia; mohammadali.sepehri@curtin.edu.au (M.S.); i.murray@curtin.edu.au (I.M.)

* Correspondence: rohan.leroux@curtin.edu.au

Abstract: Slope failures in open-pit mining pose significant operational and safety issues, underscoring the importance of implementing effective stability monitoring frameworks for early hazard detection to allow for timely intervention and risk mitigation. This systematic review presents a comprehensive synthesis of existing and emerging methods and technologies used for slope stability monitoring in open-pit mining, including both remote sensing and in situ methods, as well as advanced technologies, such as Artificial Intelligence (AI), the Internet of Things (IoT), and Wireless Sensor Networks (WSNs). Using the Preferred Reporting Items for Systematic Reviews and Meta-Analysis (PRISMA) 2020 guidelines, a total of 49 studies were selected from a collection of four engineering databases, and a comparative analysis was conducted to determine the underlying differences between the various methods for open-pit slope stability monitoring in terms of their performance across key attributes, such as monitoring accuracy, spatial and temporal coverage, operational complexity, and economic viability. Their juxtaposition highlighted the notion that no universally optimal slope stability monitoring system exists, due to a series of compromises that arise as a result of inherent technological limitations and site-specific constraints. Notably, remote sensing methods offer large-scale, non-intrusive monitoring, but are often limited by environmental factors and data acquisition infrequency, whereas in situ methods provide high precision, but suffer from limited spatial coverage and scalability. This review further highlights the capacity of emerging methods and technologies to address these limitations, providing suggestions for future research directions involving the integration of multiple sensing technologies for the enhancement of monitoring capabilities. This study provides a consolidated knowledge base on open-pit slope stability monitoring methods, technologies, and techniques, to guide the development of integrated, cost-effective, and scalable slope monitoring solutions that enhance mine safety and efficiency.

Keywords: mining; slope stability monitoring; remote sensing; artificial intelligence; automation; sensors

1. Introduction

1.1. Background and Context

Open-pit mining is a cornerstone of Australia's resources sector, facilitating the extraction of valuable commodities, such as iron ore, coal, and gold. Open-pit mines are typically developed as a series of descending benches excavated into the earth, with haul ramps providing access from the surface to the pit floor, as shown in Figure 1. The resulting pit walls, or slopes, are often steep and geotechnically complex, and are inherently susceptible to failure, due to geological discontinuities, such as joints, bedding planes, and faults, which serve as potential planes of weakness within the rock mass [1].

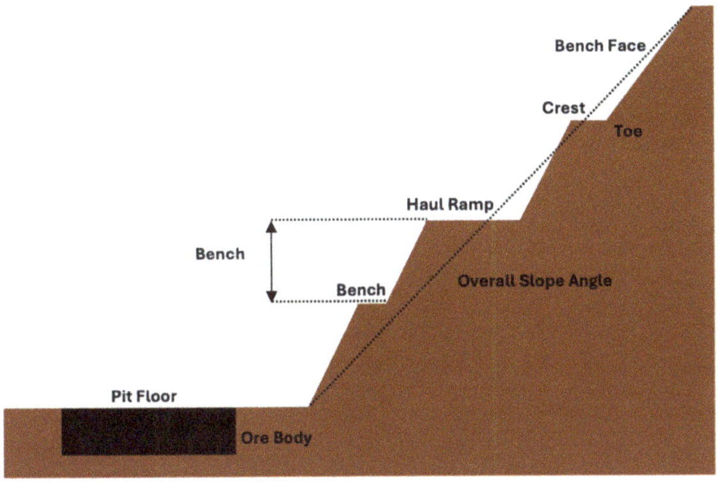

Figure 1. Open-pit wall cross-sectional schematic diagram.

Over a two-year period, spanning from 2018 to 2020, five slope failures were reported across Western Australian open-pit mines, leading to significant asset and infrastructure damage. During that same period, two fatalities occurred as a result of slope failures in open-pit mines in Queensland and the Northern Territory, underscoring the severe consequences of inadequate slope stability monitoring [2].

Slope failures can occur due to a range of internal and external factors, including geological discontinuities, weak rock masses, and weathering, which reduce the inherent strength of the slope; as well as external triggers, such as heavy rainfall infiltration and groundwater variation, which increase pore water pressure and reduce shear strength; seismic activity, which can induce vibrations that destabilize rock masses; and mining-induced disturbances, such as blasting and excavation, which can undercut or overload the slope, disrupting its equilibrium and triggering slope failure [3,4]. As shown in Figure 2, the four primary failure mechanisms for open-pit slopes, which can be initiated by the aforementioned factors, are:

1. Plane failure, wherein a large section of the pit wall slides along a planar surface, often controlled by pre-existing discontinuities, such as bedding planes, joints, or faults [5];
2. Wedge failure, the most common and hazardous mode of slope failure in open-pit mines, resulting from the intersection of two structural planes, which form a wedge-shaped rock mass, that can detach under the effects of gravity or external forces and descend the slope [6];
3. Circular failure, in which geological material is displaced vertically along a curved failure plane, most commonly observed in slopes with relatively uniform, homogenous materials, like cohesive soils or weathered rock [7];
4. Toppling failure, which occurs when steeply inclined rock columns or blocks overturn due to gravity, undercutting, or external forces, such as water infiltration or seismic activity [8].

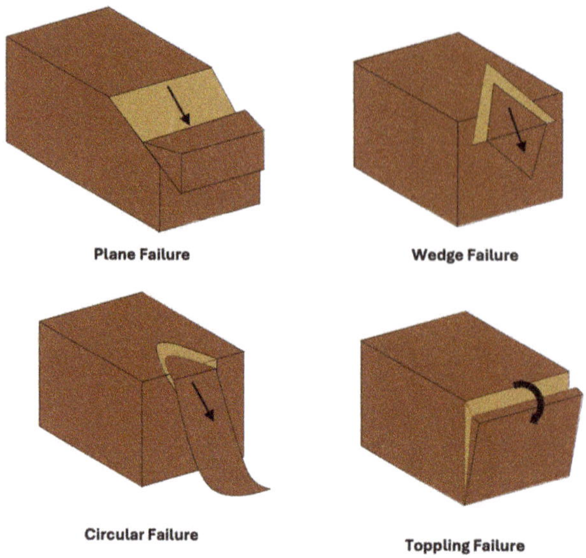

Figure 2. Slope failure mechanisms in action: plane failure (**top left**), wedge failure (**top right**), circular failure (**bottom left**), and toppling failure (**bottom right**).

Given these risks, effective slope stability monitoring is critical to ensure mine safety and operational continuity. The primary goal of slope stability monitoring is to detect and monitor the rate of slope displacement, as increasing displacement rates are typically indicative of incipient slope failure. It is widely understood that slope failures rarely occur without warning, developing instead from progressive, cumulative displacement over time and, for this reason, the early detection of slope failure indicators is essential in order to enable mine operators to intervene and mitigate risks [1].

Slope monitoring methods, broadly, fall into one of the following categories, based on the manner of data acquisition:

1. Sub-surface monitoring, which measures movements deep within the weathered material of open-pit slopes;
2. Surface monitoring, which detects visible signs of instability and quantifies surface-level deformations and displacements.

Conventional slope monitoring approaches have long relied on the use of geotechnical instruments and geodetic surveys to quantify slope displacements. While advancements in technology have enhanced these processes, there remains a fragmented understanding of the comprehensive landscape of slope stability monitoring methods available in regard to open-pit mining, attributed to the lack of review papers critically appraising existing and emerging methods [4]. For this reason, a systematic review is necessary to synthesize research in the field, to provide an objective assessment of current methods, and to identify gaps requiring further investigation, with the overarching goal of developing strategies for the future of real-time open-pit slope stability monitoring.

1.2. Objectives and Research Questions

This systematic review aims to consolidate knowledge in the field of open-pit slope stability monitoring, with the goal of examining both existing and emerging methods and trends to determine their inherent strengths and weaknesses. The PRISMA 2020 guidelines were adapted for use in this systematic review in order to follow an explicit and reproducible search methodology, ensuring a comprehensive synthesis of the most relevant research on slope stability monitoring methods. The Research Questions (RQs) guiding this systematic review were derived using the outputs from the Population, Intervention, Comparison, Outcome (PICO) mnemonic [9], as documented in Table 1.

Table 1. PICO criteria used for evaluating the effectiveness of open-pit slope stability monitoring methods.

PICO Criterion	Description	Value
Population	What is the topic?	Open-pit mines
Intervention	What is being examined?	Slope stability monitoring
Comparison	What is the control group?	N/A
Outcome	What is being measured?	Accuracy and effectiveness

Using the PICO criteria from Table 1, the following RQs were established:

1. What are the most widely used monitoring methods for open-pit slope stability, and what are their comparative advantages and disadvantages?
2. What are the emerging trends in continuous and autonomous open-pit slope stability monitoring?

The remainder of this systematic review is structured as follows: Section 2 details the study selection methodology used, while Section 3 presents the search results in the context of answering the aforementioned RQs. Section 4 provides a discussion of the implications of the key findings ascertained from the review, and offers recommendations for future studies, while Section 5 concludes by restating our findings and summarizing our primary arguments.

2. Methods

2.1. Eligibility Criteria

To ensure that the selection process captured only relevant studies, predefined study characteristics were established to determine the eligibility of studies for inclusion in the review, as presented in Table 2.

Table 2. Inclusion and exclusion criteria used in the study selection process.

Criteria	Inclusion	Exclusion
Relevance	Studies on open-pit slope stability monitoring methods and technologies	Studies focused on underground mining or civil infrastructure, such as tailings storage facilities (TSFs) or embankments
Publication Type	Journal articles, conference papers, review papers	Grey literature, textbooks, thesis papers
Publication Date	Studies published after 2005	Studies published prior to 2005
Language	Studies published in English	Studies in languages other than English

2.2. Information Sources

The primary focus of this systematic review was to explore existing and emerging methods for monitoring open-pit slope stability. For this reason, the information sources, or databases, used in this review were selected due to their comprehensive coverage of contemporary research in the field of engineering, as illustrated in Table 3.

Table 3. Database selection and rationalization.

Database	Reason for Inclusion
IEEE Xplore	Specializes in research covering advanced sensor and networking technologies, such as the IoT, WSNs, and AI
Scopus	Provides a broad, multidisciplinary repository of mining engineering research
ScienceDirect	Popular database focusing on the intersection of the applied sciences and engineering
ASCE Library	Contains a myriad of research on civil, structural, and mining engineering

2.3. *Search Strategy*

The search strategy used for this systematic review was designed to comprehensively capture relevant studies on open-pit slope stability monitoring across the selected databases. Using the PICO mnemonic criteria from Table 1, a series of keywords and synonyms were developed, as shown in Table 4, which were tailored towards the research scope.

Table 4. Keywords and synonyms used to derive the database search query.

Keyword	Synonym
Open-pit mine	Surface mine, open cast mine, open pit
Slope stability	Rock slope failure, mine slope deformation, slope failure
Monitoring	Detection, prediction

Using these keywords and synonyms, a database search query was developed using a combination of Boolean operators, as demonstrated in Table 5.

Table 5. Database search query used in the study selection process.

(“open-pit mine” OR “surface mine” OR “open cast mine” OR “open pit”) AND (“slope stability” OR “rock slope failure” OR “mine slope deformation” OR “slope failure”) AND (“monitoring” OR “detection” OR “prediction”)

The purpose of this database search query was to ensure that any relevant studies discussing slope stability monitoring, in the context of open-pit mining, would be captured, regardless of the specific wording of their titles. Filters were also applied to limit the search results to peer-reviewed journal articles, conference papers, and review papers published within the last 20 years, from March 2005 to March 2025.

2.4. *Quality Assessment*

Following the PRISMA 2020 guidelines, a quality assessment was conducted using the criteria suggested by [10], to ensure that only methodologically sound and reliable studies were included in the systematic review. Each study was evaluated based on four criteria, documented in Table 6, and those that scored below the cumulative threshold of five points were excluded.

Table 6. Quality assessment criteria and scoresheet used to evaluate selected studies.

Criteria	Yes (2 Points)	Partial (1 Point)	No (0 Points)
Study objective clarity			
Study design appropriateness			
Study results presented clearly			
Study conclusion clear and relevant			

2.5. Study Selection Process

The selection process for this systematic review followed the PRISMA 2020 guidelines using a structured, multi-stage approach to ensure that the included studies were relevant to the research scope and consistent with the inclusion criteria, with Parsifal (https://parsif.al/ accessed on 26 February 2025) serving as a central repository to facilitate study import, quality assessment, data extraction, and analysis. Study selection was conducted on 26 February 2025, with the search results compiled shortly thereafter, following these steps:

1. Searched selected databases using the Boolean search query and exported the results to Parsifal for screening;
2. Identified and removed duplicate studies through manual and automated review in Parsifal;
3. Assessed articles for relevance through title and abstract screening, and excluded studies that were misaligned with the review scope;
4. Conducted a full-text assessment of each study, evaluating them against the eligibility criteria and quality assessment guidelines;
5. Extracted selected studies into a standardized study characteristics table, and categorized each included study based on its primary focus in terms of the monitoring methods used.

3. Results

3.1. Study Selection and Classification

As shown in Figure 3, a total of 644 studies were retrieved from the databases searched using the Boolean search query outlined in Table 5. After removing the duplicate studies, of which there were 29, the remaining 615 studies underwent screening, in which their titles and abstracts were read and assessed for relevance to the research scope. This resulted in the exclusion of 421 papers, with the remaining 194 studies undergoing a full-text appraisal against the eligibility criteria and quality assessment guidelines. A total of 145 papers were excluded at this stage, due to their misalignment with the RQs (103), inaccessibility (24), poor quality (4), and ineligibility (14). This resulted in a final selection of 49 studies, which were included in the systematic review.

The number of studies that were identified and included in this systematic review, broken down by their respective databases, are shown in Figure 4, which, unsurprisingly, demonstrates a high concentration of included studies from Scopus. ScienceDirect and IEEE Xplore both contributed equal proportions of studies, while studies published across the ASCE Library were not included in the final selection, namely due to the large proportion of duplicates found across this database, which were covered by the studies included from Scopus.

The trend, evidenced by Figure 5, reflects a gradual increase in the publication of studies about open-pit slope stability from 2010 onwards, peaking in 2022. Furthermore, the results clearly demonstrate the increased topicality of emerging trends and methods for open-pit slope stability monitoring around 2019 to 2022, with all the recent studies (2023 onwards) focusing solely on remote sensing methods. Mentions of in situ methods occur more commonly in less recent studies, with all the publications in 2017 solely focused on this stability monitoring category. Despite this, 2022 saw a resurgence in studies about in situ monitoring, namely associated with their retrofitting using emerging technologies, such as the IoT and WSNs; however, this trend appears to have tapered off in the following years.

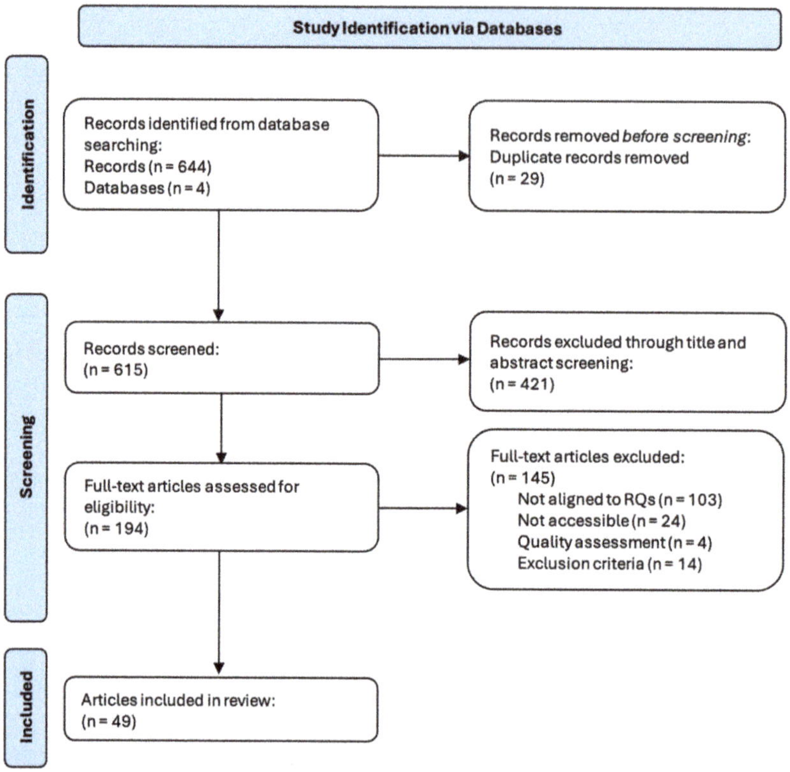

Figure 3. Flowchart of study identification, screening, and selection, using the PRISMA 2020 guidelines.

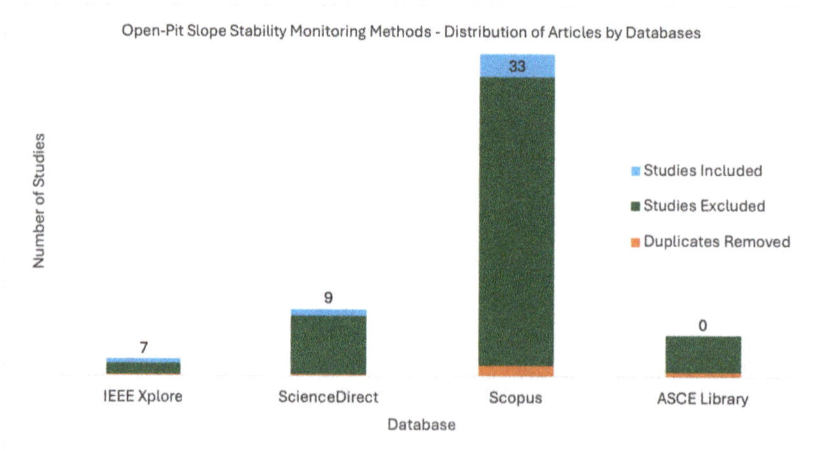

Figure 4. Breakdown of study selection by searched databases.

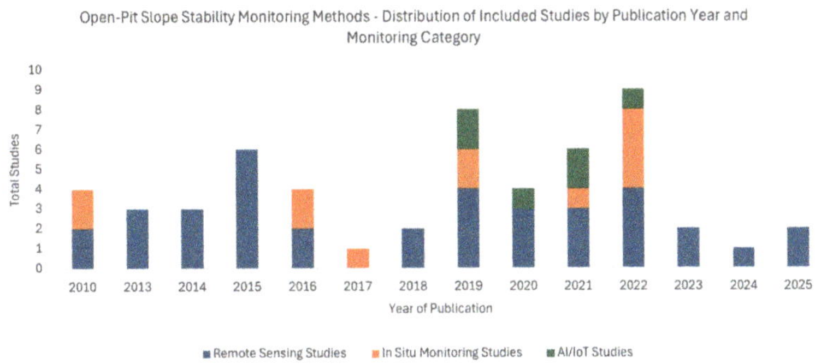

Figure 5. Distribution of included studies based on year of publication.

During the study selection process, the included studies were categorized based on the type of open-pit slope stability monitoring method they predominantly focused on. These slope stability monitoring methods, which were identified from the results of the systematic review, are shown in Figure 6.

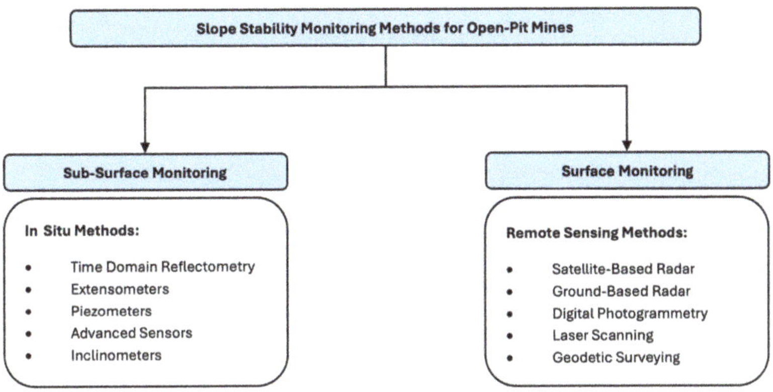

Figure 6. Categories and methods of open-pit slope stability monitoring.

With reference to Table 7, the majority of the included studies focused on remote sensing techniques (76%), while the remaining 24% examined in situ methods for slope stability monitoring.

Table 7. Categorization of included studies based on monitoring methods used.

Category	Monitoring Method	Studies (n = 49)	References
Remote Sensing	Satellite-Based Radar, Ground-Based Radar, Digital Photogrammetry, Laser Scanning, Geodetic Surveying, Thermal Imaging	37	[11–47]
In Situ Monitoring	Time Domain Reflectometry, Extensometers, Piezometers, Advanced Sensors, Inclinometers	12	[48–59]

3.2. Remote Sensing Methods

The first question in this systematic review, RQ1, focused on analyzing the methods and technologies behind some of the most pertinent forms of slope stability monitoring, in order to gain a better appreciation of their comparative advantages and disadvantages. The first category of open-pit slope stability monitoring uncovered in this systematic review was remote sensing, which uses advanced sensors and techniques to detect slope displacements from afar, by monitoring surface-level changes. Enveloped within this stability monitoring category, the most commonly discussed methods were satellite-based radar, followed by digital photogrammetry, ground-based radar, laser scanning, geodetic surveying, and, finally, thermal imaging, as presented in Table 8.

Table 8. List of remote sensing methods identified in the review.

Remote Sensing Method	References
Satellite-Based Radar	[15,20,23,25,28–30,33–36,38,42,43]
Ground-Based Radar	[11–13,18,19,21,22]
Digital Photogrammetry	[16,24,26,27,31,39,40,44–47]
Laser Scanning	[14,16,17,32,37,41]
Geodetic Surveying	[14,15,20]
Thermal Imaging	[28]

3.2.1. Satellite-Based Radar

Satellite-based radar systems employ Synthetic Aperture Radar (SAR) to generate Digital Elevation Models (DEMs) and surface velocity maps, enabling the remote detection and monitoring of slope displacements. SAR operates by transmitting microwave pulses from orbiting satellites and recording the amplitude and phase of the backscattered signals. This allows for high-resolution imaging of the Earth's surface regardless of atmospheric conditions, such as cloud cover or illumination [25].

To detect and quantify slope displacements over time, Differential Interferometric SAR (D-InSAR) is used. This technique exploits the phase component of the radar signal, which is sensitive to changes in the distance between the satellite and the ground surface. By acquiring two or more SAR images from successive satellite passes over the same area and precisely co-registering them, an interferogram is generated by computing the phase difference between corresponding pixels in the SAR images. The total phase difference contains contributions from several sources: satellite geometry, topography, surface displacement, atmospheric delay, and noise. In regard to D-InSAR, the topographic phase is removed using an external DEM, thereby isolating the displacement-related phase component. This residual phase is proportional to the change in the Line of Sight (LoS) distance between the satellite and the ground surface, allowing ground deformation to be measured with millimeter-level accuracy [20,23,34,35,38]. An example of this process is illustrated in Figure 7.

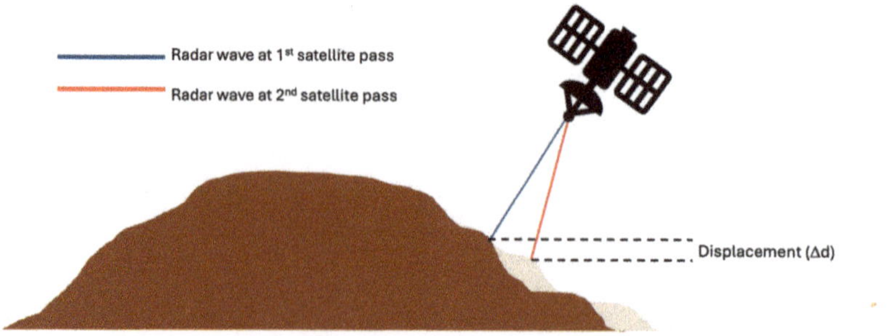

Figure 7. Data acquisition through D-InSAR, illustrating the phase difference between two SAR acquisitions at different satellite passes due to slope displacement.

Persistent Scatterer Interferometric SAR (PS-InSAR) builds upon this technique by identifying and analyzing coherent targets, known as persistent scatterers, which exhibit stable phase behavior across a long time series of SAR acquisitions. These targets, such as buildings or exposed rock faces, provide consistent radar reflections that enhance temporal coherence. By modeling and correcting for atmospheric effects and orbital errors across multiple acquisitions, PS-InSAR achieves higher precision and robustness, enabling the detection of long-term, subtle deformations that are often missed by traditional Interferometric SAR (InSAR) approaches [29,33,34,38].

Satellite-based radar systems offer broad spatial coverage due to their orbital vantage point, with the ability to monitor large geographic regions in a single pass. Their LoS-based measurement geometry enables precise displacement measurements across wide areas [20,23,28,30,33,35,36,43]. Moreover, access to freely available satellite datasets, such as those from the European Space Agency's Sentinel-1 mission, enhances the scalability and cost efficiency of SAR-based slope monitoring solutions [30].

However, several limitations affect the practical application of these techniques. The temporal resolution is constrained by satellite revisit intervals, which typically range from several days to weeks. This limits the ability of D-InSAR and PS-InSAR to capture rapid or sudden slope movements, which are characteristic of highly unstable areas [20,23,24,28,34–36,42,43]. Additionally, SAR techniques are susceptible to temporal decorrelation, especially in vegetated or rapidly changing environments, wherein the radar signal coherence between acquisitions deteriorates due to changes in the surface scattering properties. This degradation negatively impacts the quality of interferograms and reduces the accuracy of displacement measurements [25,33].

3.2.2. Digital Photogrammetry

Digital photogrammetry is a remote sensing technique that reconstructs Three-Dimensional (3D) models of objects or environments by fusing images captured from multiple angles. One of the most widely used methods in this context is Structure-from-Motion (SfM), a computer vision technique that recovers both the 3D structure of a scene and the camera's motion path by identifying and tracking distinct visual features across overlapping images taken from different perspectives, as demonstrated in Figure 8.

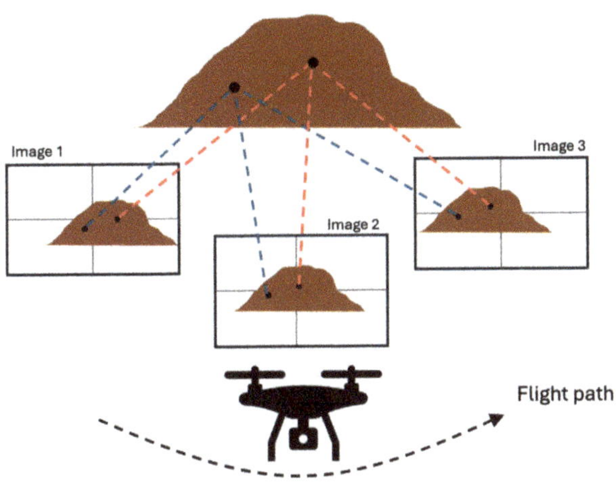

Figure 8. Digital photogrammetry data acquisition.

The SfM process begins by detecting characteristic feature points in each image, using various algorithms from software packages such as Agisoft. These feature points are then matched across multiple images, allowing the photogrammetry software to compute the relative positions and orientations of the cameras through bundle adjustment, an optimization algorithm that minimizes the re-projection error of the matched points. Once the camera poses are determined, triangulation is applied to the matched feature points to derive their 3D coordinates, resulting in the generation of a 3D point cloud, which can be converted into a 3D mesh or surface model, such as a DEM, representing the terrain geometry. Cloud-to-cloud (C2C) nearest neighbor distance computation algorithms, such as Multiscale Model-to-Model Cloud Comparison (M3C2), commonly used in software packages such as Cloud Compare, can then be used to determine the relative displacement between iterative point cloud acquisitions, thereby quantifying slope displacements [39]. These 3D models provide engineers with high-resolution spatial data, enabling the detection and quantification of surface deformations, cracks, and geological structures with centimeter-level accuracy [39,40,47]. When applied to slope stability monitoring, SfM enables the comparison of 3D models generated at different times (i.e., multi-temporal analysis), facilitating the identification of small-scale surface displacements that might indicate progressive slope failure.

Additionally, the utility of SfM is greatly enhanced when paired with Unmanned Aerial Vehicles (UAVs), which provide flexible, high-resolution data acquisition over challenging and complex terrains. UAV-based photogrammetry enables the rapid and repeatable collection of imagery from multiple altitudes and angles, offering substantial coverage and resolution advantages over fixed ground-based systems [16,26,27,31,40,44–47]. Compared to traditional slope monitoring methods, photogrammetry offers several key advantages. It provides considerable operational flexibility, since UAV-mounted cameras can access remote or hazardous areas without placing personnel at risk, supporting safer and more efficient inspections [26]. The method is also cost effective, leveraging low-cost, consumer-grade equipment, which makes it significantly more affordable than laser scanning or terrestrial radar interferometry systems [24,26,31]. In addition, the approach delivers high-resolution spatial data, with dense 3D reconstructions offering a comprehensive and detailed view of the slope surface, often with centimeter-level accuracy, depending on the flight altitude and camera quality [39,40,47]. Finally, its ability to support on-demand and frequent monitoring allows UAVs to be deployed as needed, such as after rainfall events or signs of instability, enabling timely assessments of slope behavior and a rapid response to emerging risks [26].

Despite these benefits, the method has limitations. UAV-based photogrammetry is constrained by short flight times, battery limitations, and weather dependency, including reduced performance in rainy or windy conditions [24,27]. Furthermore, low-light conditions or dense vegetation can degrade image quality, making data collection difficult at night or in shaded terrain [47]. The process also requires intensive computation, particularly during point cloud generation and model reconstruction, which limits its suitability for real-time monitoring applications [26,45,46].

3.2.3. Ground-Based Radar

Ground-based radar, much like its satellite-based counterpart, uses SAR to detect and monitor slope displacements. Unlike space-borne SAR, which uses orbital satellites to capture high-resolution spatial images, ground-based radar systems use fixed, stationary equipment, located on the ground, close to the area of interest, that move linearly along a rail to capture SAR images, as demonstrated in Figure 9 [21].

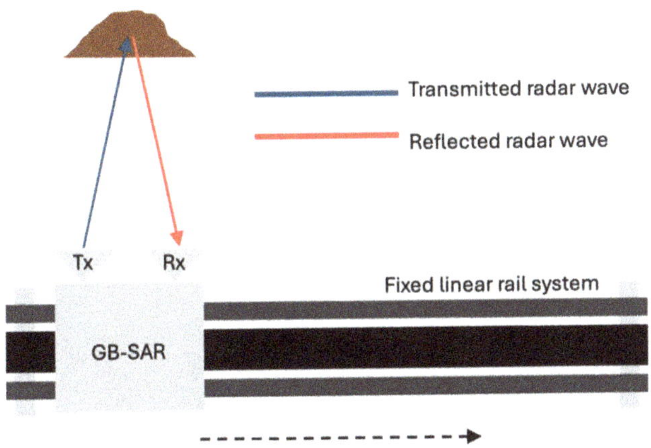

Figure 9. GB-SAR equipment.

The primary advantage of this is that ground-based radar systems can achieve high-frequency deformation monitoring in real-time [19,21]. Ground-based radar provides sub-millimeter levels of accuracy and the capacity to produce detailed DEMs through techniques such as Ground-Based InSAR (GB-InSAR), which compares multiple SAR acquisitions and measures their phase differences to quantify slope displacements, much like satellite-based D-InSAR [11–13,18,19,21,22]. However, in order to achieve this accuracy, ground-based radar necessitates stable installation and favorable weather conditions, as shadows and rain can introduce noise and impact signal propagation, thereby impacting its suitability and efficacy in regions with undesirable climates [12].

3.2.4. Laser Scanning

Light Detection and Ranging (LiDAR) is a laser-based remote sensing technology that generates high-resolution 3D point clouds of slope surfaces, facilitating the monitoring of erosion, slope failures, and subtle displacement trends through the analysis of visualizations, such as DEMs. LiDAR operates by emitting pulses of light that reflect off objects and return to the receiver. The Time of Flight (ToF) for each pulse is then measured and used to calculate the distance to the object, which allows for the derivation of precise 3D coordinates for each reflected point [11,14,16,37]. Laser scanning facilitates rapid data acquisition, yielding near real-time slope stability monitoring [37,41]. LiDAR data acquisition can be conducted using stationary Terrestrial Laser Scanners (TLSs) [11,14,16,17,28,32,37,41] or mobile UAVs [45].

A TLS provides higher accuracy and is capable of sub-millimeter precision [32], making it well-suited for detailed geotechnical assessments. Conversely, UAV-mounted LiDAR provides enhanced spatial coverage, allowing for large-scale, flexible monitoring of open-pit slopes. However, laser scanning is impacted by several factors. Its effectiveness is highly dependent on the LoS, as obstructions, such as vegetation, can reduce point cloud density [14,17,32]. Additionally, the influence of atmospheric conditions, such as rain and the reflectivity of surfaces, can negatively impact signal propagation, affecting measurement accuracy [37].

3.2.5. Geodetic Surveying

Geodetic surveying uses instruments such as Total Stations (TSs), the Global Positioning System (GPS), and Pseudolites (PLs), for high-precision deformation monitoring over vast areas [14,15,20]. These methods provide autonomous, real-time slope stability monitoring, with millimeter-level precision [57].

Among the available geodetic instruments, TSs are perhaps the most commonly used in geological surveying, operating by measuring the variations in the angles and distances between a fixed sensor and various reflective prisms located across an open-pit slope surface. Through repeated surveying, TSs can detect progressive slope movements with high precision, allowing for the tracking of long-term deformation trends. As demonstrated in Figure 10, GPS and PL technologies use differential positioning techniques to determine the relative distances between receivers deployed across an open-pit slope [14,20].

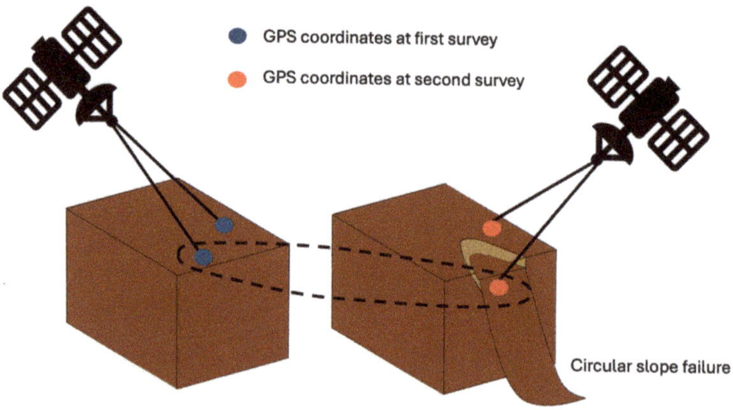

Figure 10. GPS differential positioning detecting circular slope failure.

By tracking the changes in spatial coordinates, GPS-based monitoring can detect millimeter-level slope displacements [4]. PLs act as supplementary transmitters that serve to augment GPS signal coverage in areas, such as deep open pits, where the satellite LoS might be obstructed. By integrating these technologies, monitoring networks can be established that achieve enhanced stability, negating the implications of multipath errors, which often degrade GPS measurement effectiveness [15]. Despite their real-time monitoring capabilities and strong measurement accuracy, TSs are dependent on a clear LoS and optimal weather conditions [14]. While PLs provide enhancements to GPS signal quality, their effectiveness is dependent on the positioning and configuration of orbital satellites, meaning that optimal PL positioning requires meticulous planning and may be dependent on available satellite passes, which can hinder its ability to reduce GPS multipath errors in real time [15].

3.2.6. Thermal Imaging

Thermal imaging, such as Infrared Thermography (IRT), is an emerging remote sensing technique, used to detect regions of geological instability in open-pit slopes by analyzing thermal radiation patterns. Similar to the other remote sensing methods discussed, IRT works by measuring the infrared radiation emitted by surfaces, allowing for the generation of heat maps that visualize temperature variations across terrains. These variations, which often manifest as negative temperature anomalies, can serve as indicators of precursory slope failure. While IRT offers non-contact, localized monitoring, it has several limitations, including signal attenuation due to atmospheric conditions, which can cause signal degradation over long measurement ranges [28].

3.2.7. Summary of Remote Sensing Methods

Table 9 presents a high-level summary of the remote sensing methods discussed throughout this section, highlighting their key characteristics and performance attributes.

Table 9. Summary of remote sensing methods for open-pit slope stability monitoring.

Method	Accuracy	Spatial Coverage	Temporal Resolution	Advantages	Limitations
Satellite-Based Radar	1–10 mm	Very Large	Low (Periodic)	Wide-area monitoring, effective for long-term displacement analysis, with low infrastructure investment	Low temporal resolution due to data acquisition frequency, susceptible to atmospheric interference
Digital Photogrammetry	1–10 cm	Moderate	Low	Cost effective and flexible, suitable for rapid data collection and terrain modelling	Limited measurement resolution, requires optimal ambient conditions
Ground-Based Radar	<1 mm	Large	Real time	Very high accuracy, continuous monitoring	Limited operational flexibility, high cost
Laser Scanning	<1 mm	Moderate	Near real time	High precision, operationally flexible	LoS requirements, limited coverage, high cost
Geodetic Surveying	1–10 mm	Large	Real time	High accuracy, reliable for long-term displacement tracking	Labor-intensive deployment, requires skilled operators
Thermal Imaging	N/A	Localized	Low	Effective for detecting surface anomalies to augment other methods	Low spatial resolution, influenced by surface emissivity

3.3. *In Situ Monitoring Methods*

While remote sensing techniques enable large-scale, non-contact monitoring, they are primarily used to measure surface-level displacements. In contrast, in situ methods provide direct, high-precision monitoring of both surface and sub-surface slope movements. These methods rely on traditional geotechnical instrumentation and advanced sensors to track slope displacement directly. The most commonly used in situ technologies include TDR, extensometers, piezometers, advanced sensors, and inclinometers, as shown in Table 10.

Table 10. List of in situ monitoring methods identified in this review.

In Situ Monitoring Method	References
Time Domain Reflectometry	[48,50,51,55,57–59]
Extensometers	[55–59]
Piezometers	[54,57–59]
Advanced Sensors	[49,52,53]
Inclinometers	[56–59]

3.3.1. Time Domain Reflectometry

Time Domain Reflectometry (TDR) is an in situ monitoring method used to detect slope deformations that occur at the sub-surface level. TDR uses the working principle of transmission lines to detect localized strain and deformation by measuring the reflection of electromagnetic pulses sent along a coaxial or fiber optic cable embedded within a borehole in a slope. This method provides high-resolution, real-time slope stability monitoring, enabling the detection of cracks or internal displacements along the buried wire's length [51,55,58].

Characteristic impedance varies based on a myriad of factors and is dependent on the physical parameters and dimensions of the transmission line. Shear stress and tensile strain are common by-products of sub-surface slope displacement, and these forces can affect the characteristic impedance of the wire, causing portions of transmitted pulses to reflect back to the TDR sensor [60], as demonstrated in Figure 11.

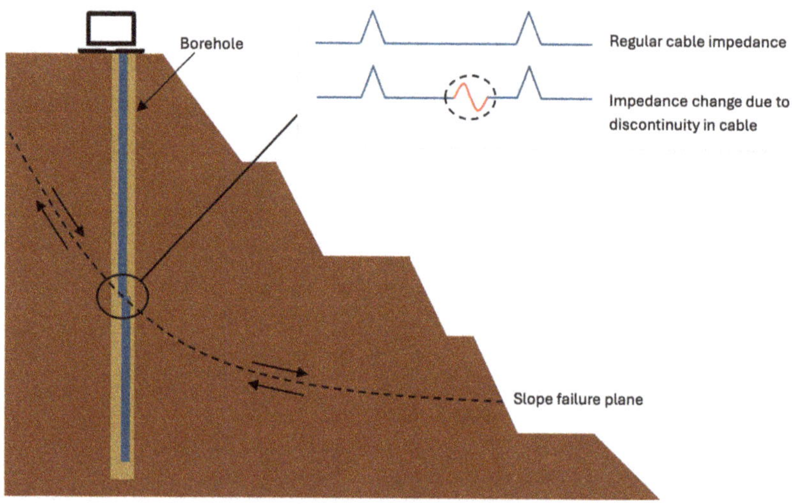

Figure 11. TDR sensor within a slope borehole, detecting the presence of a failure plane by measuring the impedance change in the reflected signal (top right).

By analyzing the travel time of the reflected pulse, the exact location of displacements can be determined with sub-millimeter precision [48], while variations in the amplitude of the signal can be used to provide an estimate of the magnitude of deformation [58]. Due to the use of durable materials, such as optic fiber cables, TDR excels in harsh environments, such as deep boreholes, making it well-suited for long-term slope stability monitoring [58]. For this reason, TDR is a fairly economical in situ monitoring technique [59].

Despite this advantage, TDR installation is often a complex procedure, requiring extensive drilling [55,58]. This discrete, point-based installation procedure also means that TDR has a limited spatial resolution, making it less effective for capturing distributed slope deformation trends across vast areas.

3.3.2. Extensometers

Visual indicators of slope failure, such as tensile cracks, are commonly monitored by using in situ techniques [3], such as wireline extensometers [55,57,58]. These geotechnical devices measure the dilation of cracks by detecting changes in the displacement of a wire that is anchored to the unstable portion of ground, as per Figure 12.

Traditionally, changes in the length of the wire were measured manually [57]; however, the advent of modern sensors, such as vibrating wire displacement transducers, has digitized this process, allowing for remote and continuous slope stability monitoring [58]. Extensometers are widely used in slope stability monitoring [55–59], due to their high precision, with some systems capable of detecting sub-millimeter deformations under optimal conditions [56]. Like the other in situ monitoring methods, extensometers require fixed, stable anchor points, impacting the ease of deployability.

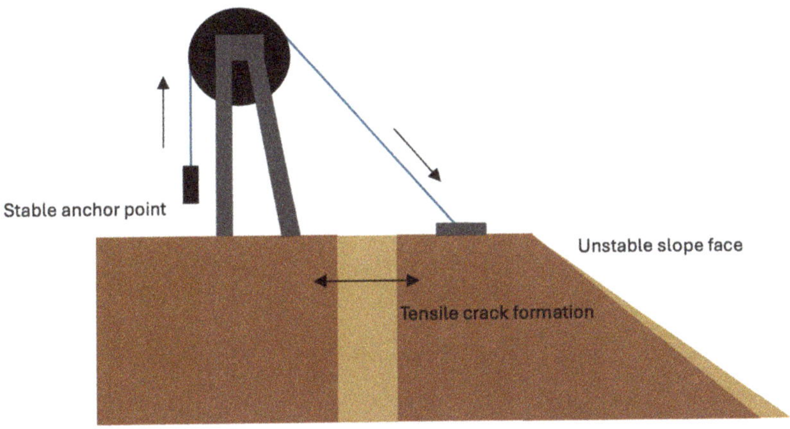

Figure 12. Traditional wireline extensometer detecting slope failure initiated by tensile crack formation.

3.3.3. Piezometers

Elevated pore water pressure reduces the effective stress in soils, weakening them and contributing to slope failure [6–8]. To combat this, piezometers are installed in open-pit slopes to quantify and monitor groundwater fluctuations. Traditional piezometers require manual, periodic data acquisition; however, modern sensors, such as automated pressure transducers, have enabled real-time, continuous monitoring of pore water pressure [56,57]. The discrete, point-based installation of piezometers means that they have a limited spatial resolution; however, this can be overcome by installing an array of sensors across vast spaces for increased coverage. Their deployability is also limited by their installation methodology, as they are often placed within boreholes, which requires extensive drilling [59].

3.3.4. Advanced Sensors

Modern sensor technologies provide alternative ways to detect physical phenomena such as strain and slope displacement. The first of these methods uses Fiber Bragg Grating (FBG) sensors, which use fiber optic technology to detect sub-surface displacements in real-time by measuring strain-induced shifts in the wavelength of reflected pulses of light. This allows FBG sensors to locate strain accumulation and detect tensile crack formation and dilation, with sub-millimeter accuracy. These sensors are highly durable and immune to electromagnetic interference (EMI), due to their use of fiber optic cables. Furthermore, they can be daisy chained to provide distributed sensing over vast distances, increasing their spatial resolution [49]. Inertial Measurement Units (IMUs) are advanced sensors that contain an accelerometer, gyroscope, and magnetometer to measure triaxial linear acceleration and angular velocity simultaneously. By applying sensor fusion algorithms, IMUs track object motion and orientation in 3D space, with high precision. IMUs are able to effectively identify the motion patterns associated with various modes of slope failure, such as toppling failure [59]. Additionally, by combining sensors capable of measuring both force and deformation, a far more comprehensive analysis of the internal dynamics of slope behavior can be attained, with sub-millimeter precision, in real time [53]. Similar to the other in situ monitoring methods, advanced sensors, despite providing precise localized measurements, lack broad spatial coverage, as they rely on discrete point measurements rather than continuous spatial data [49,52,53]. This makes them less effective for detecting widespread slope instabilities without the use of extensive sensor networks.

3.3.5. Inclinometers

Among the available geotechnical instrumentation, borehole inclinometers are widely used for slope stability monitoring [56–59], due to their ability to measure localized angular deformation and detect sub-surface displacement trends. Borehole inclinometers use Micro-Electro-Mechanical System (MEMS) sensors and probes to detect differential movements in real time, with sub-millimeter accuracy. As deformation within the slope occurs, due to shearing forces, for instance, the inclinometer measures the change in the inclination of the sensor, relative to its initial position, which allows for the quantification of the rate and magnitude of sub-surface displacement [4], as shown in Figure 13.

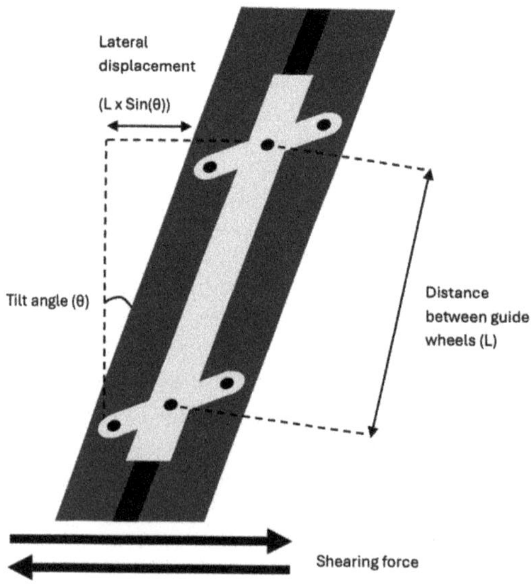

Figure 13. Borehole inclinometer schematic.

Continuing the trend seen across in situ monitoring methods, the predominant limitation of borehole inclinometers is enveloped within their complex and invasive installation procedure, which requires specialized drilling, increasing operational and maintenance costs [59].

3.3.6. Summary of In Situ Monitoring Methods

Table 11 provides an overview of the in situ monitoring methods discussed, summarizing their measurement accuracy, spatial coverage, and temporal resolution, as well as key advantages and limitations.

3.4. Comparison of Monitoring Method Categories

The previous sections outlined the most common remote sensing and in situ monitoring methods used for open-pit slope stability monitoring. While each method ultimately works towards the same common goal, their effectiveness varies widely based on a myriad of factors, such as measurement precision, spatial coverage, and temporal resolution, due to the inherent differences in the underlying technologies, sensors, and data processing techniques used. These systemic differences are presented in Table 12, which contrasts the key performance attributes associated with the slope stability monitoring methods commonly used in open-pit mining.

Table 11. Summary of in situ monitoring methods for open-pit slope stability monitoring.

Method	Accuracy	Spatial Coverage	Temporal Resolution	Advantages	Limitations
Time Domain Reflectometry	<1 mm	Localized	Real time	Very precise, cost effective for long-term monitoring	Requires borehole installation, limited to predefined zones
Extensometers	<1 mm	Localized	Real time or Periodic	Accurate for uniaxial measurements	Point-based over predefined instabilities (e.g., tensile cracks)
Piezometers	N/A	Localized	Real time	Reliable for pore water pressure monitoring	Installation challenges, affected by groundwater fluctuations
Advanced Sensors	<1 mm	Localized	Real time	Highly accurate and adaptable	Requires expert installation and data interpretation
Inclinometers	<1 mm	Localized	Real time or Periodic	Well-established method, reliable	Requires borehole installation, operationally inflexible

Table 12. Comparison of monitoring method categories for open-pit slope stability monitoring.

Aspect	Remote Sensing	In Situ Monitoring
Measurement Accuracy	Good (sub-mm to cm) [1]	Very high (<1 mm)
Spatial Coverage	Variable (localized to large) [2]	Localized
Temporal Resolution	Variable (up to several weeks) [2]	Real-time or periodic [3]
Monitoring Depth	Surface-level displacements only	Surface and sub-surface
Installation Complexity	Simple	Difficult
Real-Time Monitoring	Method dependent [4]	Yes
Primary Strength	Remote data acquisition	Reliability
Primary Limitation	Limited temporal resolution	Limited spatial coverage

[1] Remote sensing precision varies from sub-mm for GB-InSAR to cm level for UAV-borne photogrammetry (see Table 9). [2] Spatial coverage and temporal resolution are dependent on the underlying technologies used (see Table 9). [3] In situ monitoring frequency varies based on sensor type and deployment setup (see Table 11). [4] Some remote sensing methods are suitable for real-time monitoring, such as GB-InSAR, while others require extensive data post-processing, such as UAV-borne digital photogrammetry (see Table 9).

3.5. Emerging Trends in Open-Pit Slope Stability Monitoring

Traditional slope stability monitoring methods have inherent limitations and weaknesses that impact their overall effectiveness, as discussed in the prior section. However, advancements in technology have continued to drive innovations in slope stability monitoring in an effort to address these shortcomings. As shown in Table 13, of the 49 studies included in the review, 12% discussed emerging trends, such as the IoT, WSNs, and the growing role of AI and Machine Learning (ML) in detecting and analyzing indicators of slope failure. In response to RQ2, this section explores these advancements, highlighting the underlying technologies, strengths, and limitations.

Table 13. List of emerging trends identified in this review.

Emerging Trend	References
Artificial Intelligence and Machine Learning	[46,54]
The IoT and Wireless Sensor Networks	[48,50,52,59]

3.5.1. Improved Failure Detection Using Artificial Intelligence

Computer vision is increasingly being applied to slope stability monitoring to enhance predictive modeling and provide early failure detection. Computer vision leverages neural networks, a subset of ML, to enable computers to interpret and analyze visual data from the world around them. In the context of slope stability monitoring, computer vision can be used to detect tensile cracks [46] and other surface anomalies that are indicative of deformation [54].

The process starts with data acquisition using UAVs [46] or ground-based cameras [54], followed by processing techniques, such as noise reduction, data augmentation, thresholding, and edge detection. This breaks the image into smaller segments, which undergo feature extraction, wherein characteristics, such as crack patterns [46] and contours around deformed areas [54], are extracted using advanced algorithms and analyzed through Convolutional Neural Networks (CNNs) [46], which recognize, classify, and categorize these deformations. Like other remote sensing methods, a primary advantage of the use of AI-driven failure detection systems, such as the computer vision approaches outlined in [46,54], is their ability to automate manual processes and reduce operational risk. As stipulated by [46], visual inspection is still the most effective way for geotechnical engineers to identify and track tensile crack formation in open pits; however, this is an inherently hazardous process. The application of advanced ML failure detection models, as illustrated in the literature, can enhance and streamline this process.

Despite the qualitative nature of these approaches, their object detection precision is strong, with more robust algorithms, such as the Efficient Neural Network (ENet) [46] and the Improved Region Growing Segmentation Method (IRGSM) [54], achieving only a few false positives during testing. This accuracy comes at a cost to the temporal resolution, however, due to the high computational demands associated with image processing in computer vision applications [46]. Furthermore, the effectiveness of computer vision is limited when monitoring complex environments that may be heavily textured or shaded [46,54], as detection algorithms may flag false positives, affecting detection reliability. In addition to this, CNN-based image classification also requires large, high-quality datasets for training, increasing the complexity and resource demands of implementing computer vision for slope stability monitoring [46].

3.5.2. Industrial Automation for Real-Time Monitoring

The unification of the IoT and WSNs provides an avenue for industrial automation, by enabling automated data acquisition for real-time open-pit slope stability monitoring. The IoT refers to a network of ubiquitous, interconnected devices that collect, share, and interpret data remotely, while WSNs consist of spatially distributed sensor nodes that wirelessly communicate geotechnical and environmental measurements [59], such as displacement [52] and strain [48,50].

In the context of open-pit mining, Industrial IoT (IIoT) systems use smart sensors, processes, such as edge computing, and cloud-based platforms to automate and interpret slope stability data. The integration of WSNs with the IoT improves slope stability monitoring by allowing for the provision of real-time stability alerts, centralized data storage, and remote accessibility. In situ methods, such as TDR, can be enhanced through IoT-enabled sensor networks, which enable real-time, automated detection of slope deformation trends, and centralized data processing in the cloud [48]. WSNs augment the spatial coverage of sensing systems through the deployment of an array of low-power, wireless sensor nodes [50], enabling large-scale data acquisition over vast distances, while IoT infrastructure facilitates wireless data transmission and analysis. The primary advantage of this combined approach lies in its ability to provide a cost-effective and scalable alternative to in situ monitoring that minimizes infrastructure requirements, while enabling large-scale deployments in potentially hazardous or inaccessible environments [48,52,59]. Additionally, AI-driven analytics and real-time cloud computing provide these systems with enhanced failure detection, predictive analytics, and early risk mitigation capabilities [59].

Despite these transformative capabilities, two primary limitations constrain unified IoT–WSN slope stability monitoring. While recent advancements in the field of wireless communications have resulted in the development of energy-efficient techniques, such as Long Range (LoRa), power consumption remains a limiting factor, as sensor nodes require long-term energy solutions to remain operationally viable [52]. Additionally, due to the heterogeneous nature of the IoT, its lack of protocol standardization poses potential threats to data security and network vulnerability [59].

4. Discussion

4.1. Key Findings: Traditional Open-Pit Slope Stability Monitoring Methods

This systematic review examined a comprehensive collection of slope stability monitoring methods used in open-pit mining, with a focus on their accuracy, spatial coverage, and practical limitations. The findings reveal that significant trade-offs exist between key performance dimensions, as illustrated in Figure 14. Consequently, no single monitoring technique can be considered a universal or 'gold standard' solution, and, instead, the effectiveness of each method is highly context dependent, shaped by the specific geotechnical, environmental, and operational conditions of a given site.

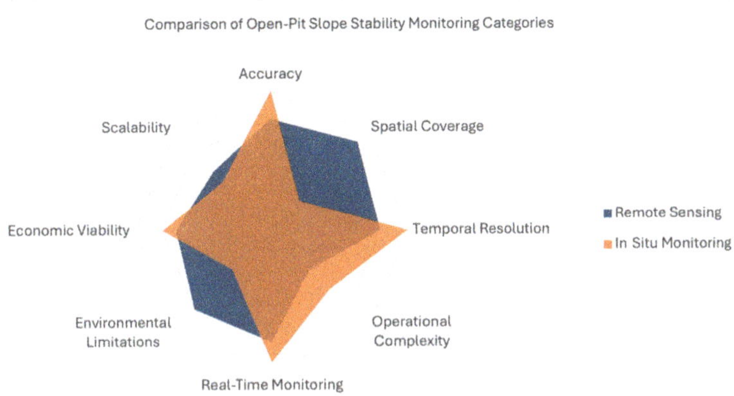

Figure 14. Radar chart comparison between remote sensing and in situ monitoring methods for assessing open-pit slope stability.

4.1.1. Measurement Accuracy and Spatial Coverage

The results indicate that the trade-off between measurement accuracy and spatial coverage is, perhaps, the most polarizing difference between remote sensing and in situ monitoring. In situ monitoring methods, such as TDR and extensometers, provide a high degree of measurement accuracy by directly quantifying local strain, stress, and displacement within slopes [4,48,49,52,53,56]. However, their reliance on point-based measurements results in limited spatial coverage, impacting their ability to provide a holistic evaluation of overall slope stability.

In contrast, remote sensing methods, such as satellite-based radar and geodetic surveying, offer enhanced spatial coverage, enabling the detection of slope instabilities across vast areas [20,23,28,30,33,35,36,43,57]. This comes at a detriment to measurement precision, however, as these methods are highly susceptible to environmental interference, due to the indirect nature of the data acquisition process used [25,33]. This trend is further demonstrated by methods using UAV-borne digital photogrammetry and LiDAR, which, despite providing good spatial coverage, are constrained due to the inherent limitations associated with data interpolation [32,39,40,47]. However, this is only broadly the case, as the review findings suggest that some remote sensing methods, such as ground-based radar and TLS, can achieve a balance between accuracy and coverage due to their fixed installation, allowing for slope stability coverage over moderately large areas [19,21,32]. For this reason, it can broadly be surmised that remote sensing covers larger areas, but may have lower precision, while in situ monitoring provides high-precision, localized open-pit slope stability monitoring.

4.1.2. Temporal Resolution and Real-Time Monitoring

The relationship between the temporal resolution of slope stability monitoring methods and their real-time monitoring capabilities are a pertinent example of the compromises that can arise due to the underlying technologies associated with each monitoring technique. The most prominent example of this is the low temporal resolution of satellite-based radar, which can vary from several days to weeks, due to the infrequency of satellite passes used to acquire SAR spatial images. For this reason, D-InSAR and PS-InSAR are unsuitable for real-time, continuous open-pit slope stability monitoring, due to the underlying limitations of the InSAR technique and its dependence on satellite revisit times [20,23,24,28,34–36,42,43].

Contrary to this, in situ monitoring methods have unanimously high temporal resolutions, with sensors capable of continuous, near real-time data acquisition, due to their high sampling rates [51,55–59]. Despite the temporal limitations of satellite-based radar, other remote sensing methods demonstrate enhanced real-time monitoring capabilities, depending on the sensing technologies used. Ground-based radar and geodetic surveying provide real-time subsidence monitoring, due to their use of high-frequency sensing instrumentation [19,21,26,28,37,41,57]. UAV-borne digital photogrammetry and LiDAR, laser scanning, and thermal imaging all provide frequent, yet periodic, slope stability monitoring, due to the need for data post-processing [26,45,46], achieving a balance between the temporal resolution and monitoring frequency. While the ability of remote sensing methods to provide continuous slope stability monitoring varies from method to method, as shown in Figure 15, it is broadly evident that in situ techniques are better equipped for real-time data acquisition and monitoring, due to the inherently high temporal capabilities of their sensors.

4.1.3. Operational Complexity and Measurement Accuracy

The results suggest that the operational complexity and installation requirements of the monitoring methods are closely related to their overall effectiveness. The primary implication of this is that more accurate techniques are typically more costly due to the specialized infrastructure and maintenance associated with their implementation and operation [55,58,59], and this increased complexity impacts their accessibility, as evidenced by the shift in the industry towards open-source stability monitoring frameworks [30,43,44]. This relationship is most distinct across the in situ monitoring methods, which require specialized sensor installation within deep boreholes, allowing for precise, localized monitoring in predetermined failure zones [55,58,59].

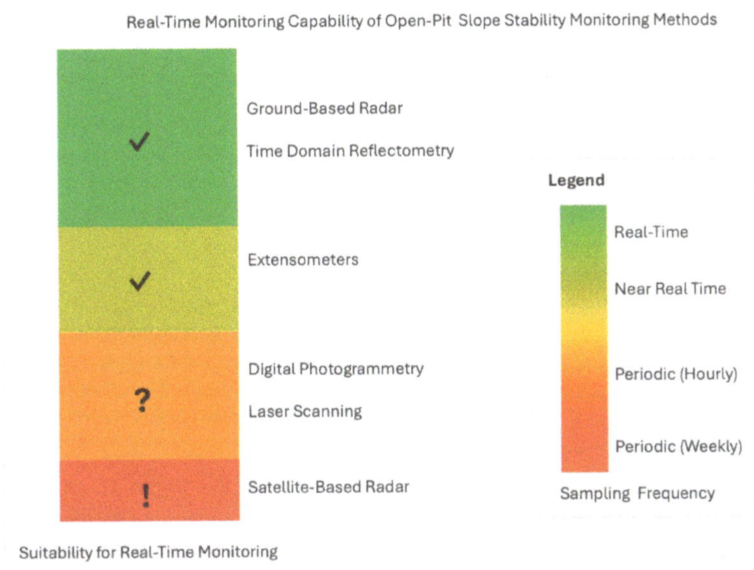

Figure 15. Heatmap showing the real-time monitoring capabilities of slope stability monitoring methods.

These methods involve time-consuming and laborious installation, maintenance, and recalibration procedures [56]; however, their measurement accuracy and ability to capture sub-surface slope deformations justify their continued use in open-pit mining applications, despite their dwindling appearance in contemporary research (see Figure 5 and Table 7). Conversely, remote sensing methods provide a compromise between operational flexibility and monitoring efficacy due to their non-intrusive and cost-effective deployability. Both satellite-based radar and UAV-borne digital photogrammetry and LiDAR provide broad spatial coverage with minimal installation effort, making them well-suited for monitoring large slopes. However, this ease of deployment comes at the cost of measurement precision, as these remote sensing methods rely on indirect measurements and data interpolation, making them more susceptible to the effects of environmental interference [20,23,24,26,28,30,31,33,35,36,43]. Alternatively, stationary or fixed sensing equipment, such as ground-based radar and TLS, offer superior measurement accuracy, albeit at a detriment to operational complexity [11–13,18,19,21,22,32].

4.1.4. Environmental Limitations

The harsh environments found across many of the world's open-pit mines can severely undermine many of the most common slope stability monitoring methods available, impacting both data quality and measurement accuracy. Remote sensing methods, such as digital photogrammetry, ground-based radar, geodetic surveying, laser scanning, and thermal imaging, all rely on clear atmospheric conditions and unobstructed visibility to accurately document surface-level slope displacements. Cloud coverage, rain, and dust can degrade image quality and impact signal propagation, introducing noise and artifacts that reduce measurement accuracy [11,12,14,27,28,37,47]. Additionally, remote sensing methods require a direct LoS to the target area, making them less effective in areas with dense vegetation, wherein obstructions can impact data acquisition and lead to unreliable results [25,33]. In contrast, in situ monitoring methods, like TDR, are less affected by environmental factors, given that they are properly installed, maintained, and recalibrated as needed. This is attributed to their use of durable materials, such as fiber optic cables, which are embedded deep within the sub-surface level of open-pit slopes [58,59].

4.1.5. Economic Considerations

The cost effectiveness of slope stability monitoring methods is broadly influenced by the balance between the initial investment costs associated with technology procurement and installation, long-term operational and maintenance expenses, and the overall risk mitigation benefits. Remote sensing methods exhibit inherent variability in their initial investment costs, due to the nature of the technology and sensors used. UAV-borne sensing using digital photogrammetry or LiDAR is considerably more economical than conventional laser scanning, geodetic surveying, and ground-based radar, due to its use of consumer-grade drones and digital cameras [24,26,31]. Satellite-based radar is similarly inexpensive due to the lack of infrastructural investment required by mine operators, who can, instead, tap into vast repositories of imaging data that are freely available via online databases [30].

The primary economic benefits associated with the use of remote sensing methodologies lie in their ability to reduce the need for manual inspections, minimizing personnel exposure to hazardous areas, while covering wide slope areas with minimal deployment efforts, yielding cost savings in the long term [26,30]. Despite the lower per unit cost of in situ monitoring methods using TDR, extensometers, piezometers, and inclinometers, their reliance on drilling boreholes and regular maintenance results in higher cumulative operational costs over time, particularly in large-scale mines where extensive monitoring is required [55,58,59]. The choice of monitoring method is, therefore, often the product of a myriad of factors, such as budget constraints, mine lifespan, and operational priorities. Remote sensing, broadly, provides a more cost-effective solution for large-scale, low-maintenance monitoring, while in situ methods remain the most accurate means for quantifying slope displacements in high-risk failure zones.

4.1.6. Suitability for Different Mine Conditions

The effectiveness of slope monitoring methods is heavily influenced by site-specific conditions, such as mine depth and visibility, slope geology, and external operational factors, such as the frequency of blasting. For deep pits with limited satellite visibility, geodetic surveying using technologies such as GPS can be impacted by the restricted LoS to satellites, degrading signal propagation and affecting measurement efficacy [14,15,20]. In such instances, ground-based radar and laser scanning provide enhanced monitoring capabilities by offering continuous, surface-level observations that are independent of any satellite constraints [21,37,41]. However, UAV-borne sensing using digital photogrammetry and LiDAR surpasses these approaches due to its unparalleled operational flexibility, allowing for remote maneuverability across vast terrains, unconstrained by elevation limitations [26].

The unique geology of each slope presents another challenge, as surface monitoring alone may not be sufficient to detect significant internal deformations, such as highly fractured rock masses. In these scenarios, in situ methods such as TDR and extensometers are more suitable, as they provide direct measurements of internal strain and shear deformation within the rock mass [55,58]. In active open-pit environments, frequent blasting is a common occurrence and, for this reason, the monitoring methods in use must account for the probability of rapid slope deformation. Ground-based radar and high-frequency geodetic surveying using TSs are particularly effective in these circumstances, due to their ability to provide real-time measurements and capture minute slope displacements, with high precision [19,26,50], with methods such as satellite-based radar being ineffective due to the limited temporal resolution [20,23,24,28,34–36,42,43]. In situ sensors also provide effective slope stability monitoring due to their real-time sensing capabilities, as per Tables 11 and 12, given that they are equipped with wireless networking technologies for remote and automated data collection.

4.1.7. Scalability and Long-Term Viability

Open-pit mines are inherently large and expansive, with the long-term viability of slope stability monitoring methods depending, in part, on their ability to scale efficiently and effectively, all while maintaining consistent accuracy and reliability. Remote sensing methods are, predominantly, well-suited for large-scale slope stability monitoring, as discussed in Section 4.1.1, with the exception of digital photogrammetry and thermal imaging.

Despite this, UAV-borne sensing, in particular, is highly scalable for periodic inspections, as it allows for rapid deployment over large areas, without the need for extensive infrastructure, due to its operational flexibility [26]. Its scalability is, however, impacted by the inherent limitations associated with contemporary drone technology, such as limited battery life and payload capacity restrictions, impacting its continuous monitoring capabilities and long-term viability [24]. Traditional in situ monitoring techniques, while offering high accuracy, can be logistically challenging to the scale of large or complex mine sites. Geotechnical instruments, such as TDR, piezometers, extensometers, and inclinometers, for instance, require extensive installation within boreholes and are limited to predetermined areas of interest, reducing their overall operational flexibility and, therefore, their scalability [55,58,59].

4.2. Key Findings: Emerging Technologies for Open-Pit Slope Stability Monitoring

The findings from this systematic review highlight the transformative implications of the use of emerging technologies, such as AI, ML, the IoT, and WSNs, in improving the accuracy, efficiency, and scalability of slope stability monitoring in open-pit mines. As discussed throughout Section 4.1, traditional monitoring methods have inherent limitations and trade-offs need to be made; however, the results indicate that emerging technologies can remedy these challenges by improving failure detection, providing real-time monitoring, and digitizing and automating manual processes, as demonstrated in Table 14.

4.2.1. The Role of Artificial Intelligence

Perhaps one of the most significant benefits of using emerging technologies for open-pit slope stability monitoring is their ability to enhance failure detection and provide predictive analytics. While visual inspection remains a widely used and reliable method for identifying early signs of slope instability [54], it exposes personnel to hazardous environments and is inherently prone to subjectivity and human error, and, significantly, is unsuitable for high-risk or rapid failure zones [56]. AI and ML techniques, particularly computer vision-based approaches, augment traditional methods by leveraging ML methods like CNNs to automate slope failure detection.

As demonstrated in [46,54], AI-driven image analysis can remotely detect surface deformations, such as tensile cracks, with high precision, serving as an effective early warning system. Unlike traditional monitoring methods, which primarily measure slope displacement or forces such as stress and strain, AI models using computer vision focus on identifying visual indicators of slope failure, such as cracking, scarping, or raveling. The benefit of this approach lies in its ability to empirically and objectively detect significant precursory indicators of slope failure [1,5–8], which may be overlooked or missed during infrequent, manual site inspections. By providing continuous, automated detection, AI-driven models can reduce the need for personnel to conduct field inspections in hazardous conditions, thereby improving operational safety.

Table 14. Comparison of traditional and emerging methods for open-pit slope stability monitoring.

Aspect	Traditional Methods	Emerging Methods
Failure Detection	Software-assisted detection of direct and indirect failure indicators (e.g., displacement, strain)	AI-driven detection of direct failure indicators (e.g., tensile cracks)
Data Collection	Ranges from periodic, infrequent measurements to real-time monitoring, depending on the method	Capacity for continuous, real-time data acquisition enabled by edge computing and cloud-based processing
Scalability	Limited scalability due to high equipment costs and fixed sensor installation	Highly scalable, as the IoT and WSNs enable flexible, cost-effective deployment over vast areas
Operational Safety	Remote sensing minimizes personnel exposure, but in situ methods require maintenance in hazardous environments	Fully remote monitoring, reducing personnel risk, WSNs offer redundancy to enhance reliability
Cost and Infrastructure	High initial investment and ongoing maintenance costs	Lower setup and operational costs due to automation and cloud-based storage
Accuracy and Resolution	High spatial resolution, but may have limited temporal resolution due to data acquisition intervals	AI improves detection speed and precision, although effectiveness is impacted by scene complexity
Reliability and Challenges	Susceptible to environmental interference, requires data post-processing and expert analysis	AI requires large, high-quality datasets for ML model training, the IoT and WSNs depend on power-efficient networking and protocol standardization

Despite these strengths, computer vision-based failure detection faces several challenges, akin to those seen across the other remote sensing methods. High computational demands, the risk of false positives due to complex terrains and environmental interference, and a dependence on large, high-quality datasets for ML model training, all impact its practical implementation [46]. Similar to digital photogrammetry and thermal imaging, AI-driven image analysis is also restricted by LoS limitations and, therefore, has a fairly localized spatial resolution. For this reason, AI-driven methods are most effective when used as complementary tools alongside traditional monitoring techniques, due to their inability to cover vast areas effectively in typical site conditions.

4.2.2. Integrated Monitoring Using the IoT and WSNs

The integration of the IoT and WSNs into geotechnical monitoring systems augments the scalability, flexibility, and responsiveness of traditional monitoring approaches. By deploying low-power wireless sensor nodes that capture and transmit physical data, such as stress, strain, or pressure, to cloud-based platforms, these systems enable real-time data acquisition, automated processing, and centralized data management [59], essentially providing iterative benefits to in situ solutions. This distributed networking architecture allows for broad spatial coverage and supports continuous monitoring across expansive and often hazardous environments, as demonstrated in [48,50,52].

Unlike conventional monitoring methods, which often face trade-offs between temporal resolution, spatial coverage, operational safety, and cost, as discussed throughout Section 4.1, the fusion of the IoT and WSNs presents a comprehensive framework, with minimal compromises. As shown in Figure 16, this approach excels across all key performance attributes in that it is scalable, cost effective, accurate, and supports continuous remote monitoring.

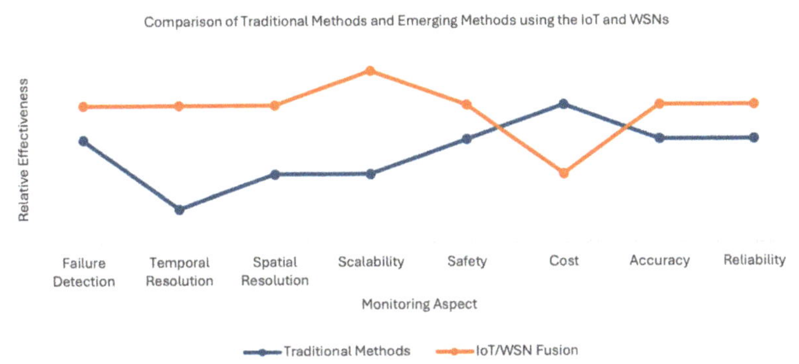

Figure 16. Parallel coordinates plot showing the differences between traditional and emerging methods for open-pit slope stability monitoring.

Furthermore, the integration of AI and edge computing within IoT frameworks enables real-time failure detection, advanced predictive analytics, and early risk mitigation strategies, enhancing both operational safety and decision making. This synergy allows for proactive slope management that can outperform traditional systems in terms of both responsiveness and precision [59].

The two primary limitations of IoT and WSN adoption within large-scale mining operations are associated with the inherent, underlying restrictions of the respective technologies. The remote nature of wireless sensor nodes presents power consumption challenges, and the lack of protocol standardization across IoT networks introduces data security and interoperability issues [52,59]. Despite these constraints, it is clear that the fusion of the IoT and WSNs serves as a transformative and complementary advancement in the field of open-pit slope stability monitoring.

4.3. Future Research Directions

In light of these findings, Table 15 presents a comprehensive evaluation of all the open-pit slope stability monitoring methods discussed in this review, collectively contrasting their strengths and weaknesses to identify potential gaps for future research. Accordingly, the following discussion focuses on these gaps, distilling them into clear, structured, future research directions.

4.3.1. Suggestion: Improve Monitoring Accuracy

Slope monitoring accuracy is a significant driver of effective operational safety in open-pit mining [1–4]. While most methods perform well in regard to capturing even the most minute slope displacements, notable limitations impact image-based techniques, such as digital photogrammetry and computer vision, particularly in challenging environmental conditions, postulating the need for further investigation.

The efficacy of digital photogrammetry is constrained by two predominant gaps, as highlighted by [26,27,47], namely: reduced effectiveness in shaded or heavily vegetated areas, and reduced DEM accuracy due to data interpolation. To address the first gap, HDR imaging technologies should be explored for open-pit mining applications, as they have demonstrated improved shadow detail preservation in other domains, including agriculture [61]. Regarding DEM accuracy, the integration of multi-sensor data extracted from in situ sources can be used to further enhance output precision, as evidenced by [62], while deep learning methods can be explored to mitigate interpolation artifacts and improve DEM fidelity [63].

Table 15. Gap analysis and future research directions.

Aspect	Identified Gap	Affected Methods	Future Research Direction
Accuracy	Reduced accuracy in shaded or vegetated terrain	Digital Photogrammetry	Investigate High Dynamic Range (HDR) technology for the capturing of enhanced shadow details
	Reduced accuracy due to data interpolation		Enhance DEM capabilities by augmenting with in situ sensor data
	Reduced accuracy in regard to complex scenes	Artificial Intelligence (Computer Vision)	Investigate other ML models and sensor fusion
	Requires solid datasets for ML model training		Investigate the efficacy of synthetic datasets using Generative AI (GenAI)
Spatial Coverage	Limited slope coverage	In Situ Methods	Investigate retrofitting in situ frameworks with WSN capabilities and distributed sensor nodes
		Digital Photogrammetry, Artificial Intelligence (Computer Vision)	
Temporal Resolution	Periodic or infrequent data collection	Satellite-Based Radar	Investigate integrated sensing approaches using ground control elements (e.g., Trigger Action Response Plan (TARP) and Rock Mass Classification (RMC))
	Delays due to data processing	Digital Photogrammetry, Artificial Intelligence (Computer Vision)	Investigate edge computing and edge AI for real-time data processing, and more efficient models, such as You Only Look Once (YOLO)
Operational Complexity	Specialist data interpretation	Both Remote Sensing and In Situ Monitoring	Leverage AI-powered data interpretation tools, such as computer vision
Real-Time Monitoring	Lack of continuous monitoring capabilities	Satellite-Based Radar	Investigate integrated sensing to provide real-time monitoring during satellite 'downtime'
		Digital Photogrammetry, Artificial Intelligence (Computer Vision)	Investigate integrated sensing using UAVs, in combination with WSNs and edge computing, to provide real-time hazard detection using ML models like YOLO
Environmental Limitations	Signal attenuation due to weather conditions	Remote Sensing Methods	Investigate monitoring frameworks with redundancy to provide auxiliary monitoring during poor weather conditions
	Sensor degradation due to extreme conditions	Extensometers, Inclinometers, Piezometers	Investigate rugged sensors
Economic Viability	High equipment and maintenance costs	Ground-Based Radar, Thermal Imaging, Laser Scanning, TDR, Extensometers, Piezometers, Inclinometers	Investigate the efficacy of low-cost, modular sensors, such as the advanced sensors discussed in Section 3.3.4
Scalability	High cost and complexity limit large-scale deployment	Ground-Based Radar, TDR, Extensometers, Piezometers, Inclinometers	Investigate the applicability of IoT-enabled WSNs using low-power sensor nodes and networking technologies, such as LoRa

Similarly, computer vision-based image analysis is also impacted by two principal limitations, in that its hazard detection accuracy is reduced in complex environments, and its object recognition capabilities are dependent on rigorously trained ML models that use high-quality datasets [46,54]. To address these issues, other ML models or algorithms should be explored, as analogous studies have demonstrated the effectiveness of various model optimization techniques in enhancing object detection accuracy, particularly in domains such as complex industrial construction environments [64]. Furthermore, the difficulty associated with obtaining high-quality datasets for ML model training may be circumvented by exploring the applicability of GenAI for producing synthetic datasets [65], particularly in the context of hazardous open-pit mines, where accessibility might be limited.

4.3.2. Suggestion: Enhance Spatial Coverage

Comprehensive spatial coverage is essential for providing holistic insights into overall slope stability, given the particularly complex geomechanics of vast open pits [54].

In situ methods, fundamentally, face limitations in this regard due to their point-based, localized nature [4,48,49,52,53,56]. In contrast, remote sensing techniques typically excel in this regard as they offer a superior LoS [19,21,32], with the exception of image-based approaches, such as digital photogrammetry and computer vision [26,46], which are tempered due to their underlying data collection methodologies. To overcome the spatial limitations inherent to in situ techniques, such as TDR, extensometers, inclinometers, and piezometers, future research efforts should investigate the retrofitting of traditional in situ monitoring frameworks with WSN capabilities and distributed sensor nodes, as described by [48,50,59], to achieve increased spatial granularity, while also facilitating real-time, autonomous data collection.

The spatial performance of image-based remote sensing methods can be enhanced by leveraging context-aware sensing frameworks, guided by key geotechnical parameters. Specifically, incorporating ground control elements, such as TARPs and RMC criteria, into deployment and data interpretation strategies [66] for autonomous UAVs could optimize monitoring frequency, thereby improving spatial coverage in high-risk areas. These context-aware sensing frameworks, which have been shown to greatly enhance the efficacy of computer vision-based image analysis in other domains [67,68], would allow monitoring systems to adapt dynamically to geotechnical risk profiles, enhancing overall slope coverage and maximizing monitoring efficiency.

4.3.3. Suggestion: Improve Temporal Resolution and Real-Time Monitoring

The effectiveness of any slope stability monitoring method is largely determined by its ability to provide advanced notice of incipient slope failure. Accordingly, temporal resolution, defined by the frequency of data acquisition, is a key performance metric, as systems capable of continuous or near real-time monitoring enable more proactive risk mitigation and timely intervention [1,4,54]. Despite this, popular remote sensing methods, such as satellite-based radar and digital photogrammetry [20,23,24,26,28,34–36,42,43], as well as emerging technologies, such as computer vision-based image analysis [46], provide unsatisfactory performance across this criterion, necessitating the requirement for further research.

The periodic and infrequent nature of SAR acquisitions is a limiting factor for both D-InSAR and PS-InSAR, restricting its ability to function effectively in volatile and evolving landscapes, such as open-pit mines [33,36]. For this reason, future research should explore the development of context-aware monitoring frameworks that integrate geotechnical parameters, such as slope angle, excavation rates, and ground control elements, to dynamically prioritize monitoring frequency and sensor deployment. These frameworks could enhance the temporal resolution through coordinated, integrated sensing strategies that complement satellite observations with ground-based sensors or UAV-based monitoring during satellite pass intervals.

The temporal resolution of digital photogrammetry is principally constrained by the high computational demands of data processing, particularly for 3D models and point cloud generation [26,45]. This issue is further compounded when coupled with ML-based computer vision techniques [46,54], which typically require significant Graphics Processing Unit (GPU) resources. However, advancements in edge computing have demonstrated the feasibility of performing such computationally intensive tasks in near real time, as shown in analogous domains like powerline inspection [69]. Similarly, recent studies in autonomous UAV navigation and mapping [70] have highlighted the benefits of offloading processing tasks to edge nodes to significantly improve responsiveness and data throughput.

To overcome the temporal limitations associated with conventional computer vision models, future research should investigate more efficient algorithms, such as YOLO, or lightweight, bespoke models designed for resource-constrained environments. When deployed in conjunction with edge AI on UAV platforms, these solutions could facilitate real-time hazard detection and adaptive monitoring in critical areas. This integrated approach has shown promise in comparable applications, such as autonomous road object detection on mine haul ramps, and presents a viable path toward improving the temporal resolution in open-pit slope stability monitoring systems [71].

4.3.4. Suggestion: Address Environmental Limitations

Environmental conditions, such as adverse weather, can significantly impact the reliability and effectiveness of remote sensing and in situ slope monitoring technologies.

This trend is most evident across remote sensing methods, particularly those reliant on optical and radar imaging, wherein the results indicated that the impacts of dust, rain, and cloud cover could lead to signal attenuation and degraded image quality [11,12,14,27,28,37,47]. For this reason, future research should investigate the suitability of redundant monitoring frameworks that leverage complementary sensor technologies capable of operating under varying environmental conditions. Such redundancy, in tandem with context-aware sensing, would ensure continuity in data acquisition during weather-induced disruptions, by, potentially, combining radar-based systems with rugged in situ sensors, thereby offering auxiliary monitoring to augment the overall reliability of the mine's monitoring efforts.

While in situ monitoring systems, such as extensometers, inclinometers, and piezometers are inherently designed to operate under extreme conditions, they are, nonetheless, susceptible to degradation over time, leading to issues such as sensor drift and damage [72]. For this reason, to improve system durability, future research should investigate fiber optic technologies, which are resistant to EMI and environmental degradation, as demonstrated in analogous contexts, such as landslide monitoring [73].

4.3.5. Suggestion: Improve Economic Viability and Scalability

High equipment and ongoing maintenance costs pose significant limitations for various slope stability monitoring methods. Remote sensing technologies, including ground-based radar, thermal imaging, and laser scanning, along with in situ tools, such as TDR, extensometers, piezometers, and inclinometers, typically require substantial initial capital investment, are labor intensive to install and operate, and demand long-term maintenance, especially at scale [24,26,55,58,59]. To address these financial barriers, future research should explore the viability of low-cost, modular sensors, such as those discussed in Section 3.3.4. Despite the preliminary nature of these studies, advanced sensors like IMUs have been shown to provide a promising, cost-effective alternative to traditional in situ methods, due to their capacity to maintain acceptable accuracy and reliability [59]. For this reason, similar technologies should be examined for their applicability to open-pit slope stability monitoring.

Aside from equipment costs, the challenge of scaling these methods across large, geotechnically complex mine sites remains a significant gap in the research. This limitation is most evident in regard to in situ methods, which are inherently localized and require substantial infrastructure [55,58,59]. When implemented over extensive areas, the cumulative costs of installation and maintenance can rise quickly, hindering their feasibility for broad deployment across vast slope regions. To address this issue, future research should explore the use of IoT-enabled WSNs, as mentioned in Section 3.5.2, utilizing low-power sensor nodes and long-range, efficient networking protocols like LoRa. Their successful implementation in fields such as landslide monitoring and environmental hazard detection [74,75] showcases their potential to facilitate cost-effective, large-scale deployments in open-pit mining.

4.4. Limitations of This Review

While this systematic review followed a structured research methodology and presented a comprehensive analysis of prominent open-pit slope stability monitoring methods, it is evident that there is a noticeable bias toward certain technologies, particularly remote sensing methods, such as satellite-based radar and UAV-borne sensing. This trend likely reflects a topical bias within the available body of the literature, which may overrepresent novel approaches, potentially skewing the review's results. As a result, more traditional methods using in situ sensors, such as piezometers, inclinometers, and extensometers, may appear elusive and antiquated. Although this does not constitute a methodological flaw in the review itself, it introduces a research focus bias that may influence comparative assessments between traditional and emerging methods. More broadly, this limitation clearly delineates the lack of contemporary research covering in situ methods, once more supporting this review's recommendations regarding the integration of the IoT and WSN technologies into traditional slope monitoring frameworks.

We also acknowledge that the inaccessibility of several studies during the full-text screening stage may have introduced a risk of selection bias, despite extensive efforts to obtain these articles. However, we believe that the thematic analysis of the included studies reflects consistent patterns and key findings across a diverse range of sources, suggesting that the overall trends identified in this review are robust. While we cannot entirely rule out the potential for bias, we do not expect that the exclusion of these inaccessible studies would significantly alter the conclusions drawn in this review.

Another limitation of this review lies in its exclusive focus on slope stability monitoring methods, deliberately excluding studies related to slope design, modeling, and stability analysis. In addition to this, our review does not consider health and safety aspects associated with asset tracking and monitoring. While this scoping decision allowed for analytical depth, it also constrained the breadth of the findings, particularly regarding integrated risk management practices. This limitation is especially relevant in the context of Sections 4.3.2 and 4.3.3, which explore opportunities for context-aware sensing and real-time hazard detection, both of which benefit from holistic integration with geotechnical design and modeling data.

In addition, this review concentrated solely on monitoring applications within the context of open-pit mining, excluding analogous domains such as TSFs, embankments, and landslides. Although this focus was intended to ensure contextual relevance, it may have led to the exclusion of valuable insights, given that these applications often employ similar instrumentation, sensors, and monitoring approaches.

Lastly, several of the emerging methods identified, such as AI-based computer vision and IoT-enabled WSNs, are still in the nascent stages of application to open-pit mining. Consequently, their validation is often limited to controlled or simulated laboratory environments, which creates a significant data gap concerning their large-scale deployment, reliability, and long-term effectiveness.

5. Conclusions

This systematic review has examined, analyzed, and summarized 49 studies on slope stability monitoring methods in open-pit mining, comparing remote sensing and in situ technologies, while identifying emerging trends, such as AI-driven computer vision and IoT-enabled WSNs.

The findings reveal that remote sensing technologies constitute the bulk of contemporary research, focusing on advanced modalities like InSAR and LiDAR, while exploring the use of UAVs to generate DEMs, using digital photogrammetry techniques. These methods are broadly accurate, precise, and effective at detecting and monitoring indicators of incipient slope failure, namely, geological displacement and visual signals like scarping, raveling, and cracking. In situ methods, conversely, are far less common in state-of-the-art literature. Despite this, the review results indicate that reliable geotechnical instruments, like TDR, extensometers, piezometers, and inclinometers, are unanimously praised for their effectiveness in tracking and quantifying sub-surface slope displacement trends, despite their rudimentary operating principles and methodologies. Emerging technologies, such as AI, ML, the IoT, and WSNs, act as a mediator between the old and the new, providing enhanced predictive monitoring capabilities and continuous real-time stability monitoring.

Through comparative analyses, the findings suggest that no single method is universally optimal, as each exhibits trade-offs between key performance attributes, such as measurement accuracy, spatial coverage, temporal resolution, operational complexity, and cost. While traditional methods, such as remote sensing and in situ monitoring, remain highly effective in certain domains, their limitations, particularly in regard to scalability, automation, and adaptability, underscore the value of integrating emerging technologies.

Further to this, the review also identified several prominent research gaps, particularly in the areas of real-time monitoring, economic viability, and adaptability to complex geotechnical and environmental conditions. These limitations served as the basis for the stipulation of a detailed set of future research directions, focused on improving accuracy through sensor fusion, enhancing spatial and temporal coverage via distributed sensor networks built on ground control elements, and reducing operational constraints through the use of modular, low-cost equipment.

Overall, the findings of this review highlight that the future of open-pit slope stability monitoring lies in the implementation of hybrid frameworks, leveraging the strengths of both traditional and emerging technologies, such as AI. By pursuing interdisciplinary research across geotechnical and electrical engineering and data science, the mining industry can push towards smarter, safer, and more efficient slope stability monitoring solutions, to ensure that every mine worker goes home safely after their shift.

Author Contributions: Conceptualization, R.L.R., M.S., S.K., and I.M.; methodology, R.L.R., M.S., S.K., and I.M.; writing—original draft preparation, R.L.R.; writing—review and editing, R.L.R., M.S., S.K., and I.M.; visualization, R.L.R.; supervision, M.S., S.K. and I.M. All authors have read and agreed to the published version of the manuscript.

Funding: This research received no external funding.

Data Availability Statement: The review protocol and quality assessment documentation is available upon request from the corresponding author.

Conflicts of Interest: The authors declare that there are no conflicts of interest.

References

1. Geotechnical Considerations in Open Pit Mines. Available online: https://www.worksafe.wa.gov.au/publications/geotechnical-considerations-open-pit-mines-guideline (accessed on 26 February 2025).
2. Mines Safety Bulletin, No. 171 Pit Wall Failure Management. Available online: https://www.worksafe.wa.gov.au/publications/mines-safety-bulletin-no-171-pit-wall-failure-management (accessed on 26 February 2025).
3. Narendranathan, S.; Nikraz, H. Optimal System Design for Instrumented Slope Monitoring in Open Pit Mines. In Proceedings of the International Conference on Advances in Geotechnical Engineering (ICAGE 2011), Perth, Australia, 7–9 November 2011.
4. Mohammed, M.M. A Review on Slope Monitoring and Application Methods in Open Pit Mining Activities. *Int. J. Sci. Technol. Res.* **2021**, *10*, 181–186.
5. Raghuvanshi, T.K. Governing factors influence on rock slope stability—Statistical analysis for plane mode of failure. *J. King Saud Univ. Sci.* **2019**, *31*, 1254–1263. [CrossRef]
6. Singh, B.; Goel, R.K. Chapter 17–Types of Failures of Rock and Soil Slopes. In *Engineering Rock Mass Classification*, 1st ed.; Singh, B., Goel, R.K., Eds.; Elsevier: Boston, MA, USA, 2011; pp. 211–230.
7. Mohtarami, E.; Jafari, A.; Amini, M. Stability analysis of slopes against combined circular–toppling failure. *Int. J. Rock Mech. Min. Sci.* **2014**, *67*, 43–56. [CrossRef]
8. Bobet, A. Analytical solutions for toppling failure. *Int. J. Rock Mech. Min. Sci.* **1999**, *36*, 971–980. [CrossRef]
9. Carrera-Rivera, A.; Ochoa, W.; Larrinaga, F.; Lasa, G. How-to conduct a systematic literature review: A quick guide for computer science research. *MethodsX* **2022**, *9*, 101895. [CrossRef]
10. Kmet, L.; Lee, R. Standard Quality Assessment Criteria for Evaluating Primary Research Papers from a Variety of Fields. *HTA Initiat.* **2024**. [CrossRef]
11. Agliardi, F.; Crosta, G.B.; Meloni, F.; Valle, C.; Rivolta, C. Structurally-controlled instability, damage and slope failure in a porphyry rock mass. *Tectonophysics* **2013**, *605*, 34–47. [CrossRef]
12. Monserrat, O.; Crosetto, M.; Luzi, G. A review of ground-based SAR interferometry for deformation measurement. *ISPRS J. Photogramm. Remote Sens.* **2014**, *93*, 40–48. [CrossRef]
13. Mecatti, D.; Macaluso, G.; Barucci, A.; Noferini, L.; Pieraccini, M.; Atzeni, C. Monitoring open-pit quarries by interferometric radar for safety purposes. In Proceedings of the 7th European Radar Conference (EURAD 2010), Paris, France, 30 September 2010.
14. Francioni, M.; Salvini, R.; Stead, D.; Giovannini, R.; Riccucci, S.; Vanneschi, C.; Gullì, D. An integrated remote sensing-GIS approach for the analysis of an open pit in the Carrara marble district, Italy: Slope stability assessment through kinematic and numerical methods. *Comput. Geotech.* **2015**, *67*, 46–63. [CrossRef]
15. Wang, J.; Gao, J.; Liu, C.; Wang, J. High precision slope deformation monitoring model based on the GPS/Pseudolites technology in open-pit mine. *Min. Sci. Technol.* **2010**, *20*, 126–132. [CrossRef]
16. Francioni, M.; Salvini, R.; Stead, D.; Litrico, S. A case study integrating remote sensing and distinct element analysis to quarry slope stability assessment in the Monte Altissimo area, Italy. *Eng. Geol.* **2014**, *183*, 290–302. [CrossRef]
17. Kromer, R.A.; Hutchinson, D.J.; Lato, M.J.; Gauthier, D.; Edwards, T. Identifying rock slope failure precursors using LiDAR for transportation corridor hazard management. *Eng. Geol.* **2015**, *195*, 93–103. [CrossRef]
18. Liu, B.; Ge, D.; Li, M.; Zhang, L.; Wang, Y.; Zhang, X. Using GB-SAR technique to monitor displacement of open pit slope. In Proceedings of the 2016 IEEE International Geoscience and Remote Sensing Symposium (IGARSS), Beijing, China, 10–15 July 2016; pp. 5986–5989. [CrossRef]
19. Qi, Y.; Zhang, M.; Qi, L.; Tan, W.; Sun, J.; Huang, P.; Xu, W. Application of Ground-based Deformation Monitoring Radar in Mine Slope Monitoring-Taking a mine in Inner Mongolia as an example. In Proceedings of the 2019 IEEE International Conference on Signal, Information and Data Processing (ICSIDP), Chongqing, China, 11–13 December 2019; pp. 1–4. [CrossRef]
20. Wei, L.; Liu, S.; Mao, Y.; Wang, R.; Wu, L.; Xu, J. Joint application of DInSAR and ground measurements: A case study on dump stability analysis. In Proceedings of the 2016 IEEE International Geoscience and Remote Sensing Symposium (IGARSS), Beijing, China, 10–15 July 2016; pp. 6002–6005. [CrossRef]
21. Dick, G.J.; Eberhardt, E.; Cabrejo-Liévano, A.G.; Stead, D.; Rose, N.D. Development of an early-warning time-of-failure analysis methodology for open-pit mine slopes utilizing ground-based slope stability radar monitoring data. *Can. Geotech. J.* **2014**, *52*, 515–529. [CrossRef]
22. Farina, P.; Coli, N.; Yön, R.; Eken, G.; Ketizmen, H. Efficient real time stability monitoring of mine walls: The çöllolar mine case study. *Int. Mining Congr. Exhib. Turkey* **2013**, *1*, 111–117.
23. Tripolitsiotis, A.; Steiakakis, C.; Papadaki, E.; Agioutantis, Z.; Mertikas, S.; Partsinevelos, P. Complementing geotechnical slope stability and land movement analysis using satellite DInSAR. *Open Geosci.* **2014**, *6*, 56–66. [CrossRef]
24. Deguchi, T.; Sugiyama, T.; Kishimoto, M. R&D of Drone-Borne Sar System. *ISPRS Int. Arch. Photogramm. Remote. Sens. Spat. Inf. Sci.* **2019**, *XLII-2/W13*, 263–267. [CrossRef]

25. Intrieri, E.; Carlà, T.; Farina, P.; Bardi, F.; Ketizmen, H.; Casagli, N. Satellite Interferometry as a Tool for Early Warning and Aiding Decision Making in an Open-Pit Mine. *IEEE J. Sel. Top. Appl. Earth Obs. Remote Sens.* **2019**, *12*, 5248–5258. [CrossRef]
26. Bar, N.; Kostadinovski, M.; Tucker, M.; Byng, G.; Rachmatullah, R.; Maldonado, A.; Pötsch, M.; Gaich, A.; McQuillan, A.; Yacoub, T. Pit Slope Failure Evaluation in Near Real Time using UAV Photogrammetry and 3D Limit Equilibrium Analysis. *Aust. Geomech. J.* **2020**, *55*, 33–47.
27. Li, J.; Yang, T.; Deng, W.; Du, S.; Zhang, Z.; Ma, H.; Wang, H.; Guo, J.; Liu, L. Evaluating failure mechanisms of excavation-induced large-scale landslides in Xinjing open-pit coal mine through integrated UAV imagery and 3D simulation. *Landslides* **2025**, *22*, 2021–2036. [CrossRef]
28. Lu, Y.; Jin, C.; Wang, Q.; Han, T.; Li, G.; Zhong, X.; Chen, G. Combining InSAR and infrared thermography with numerical simulation to identify the unstable slope of open-pit: Qidashan case study, China. *Landslides* **2023**, *20*, 1961–1974. [CrossRef]
29. Lee, H.; Moon, J.; Lee, H. Activity of Okgye Limestone Mine in South Korea Observed by InSAR Coherence and PSInSAR Techniques. *Remote Sens.* **2022**, *14*, 6261. [CrossRef]
30. Haupt, S.; Engelbrecht, J.; Sibolla, B.; Mdakane, L.W. Time Series Insar Analysis for Slope Stability Monitoring Using Sentinel-1 in Open Pit Mining. *ISPRS Int. Arch. Photogramm. Remote. Sens. Spat. Inf. Sci.* **2023**, *XLVIII-1/W*, 945–951. [CrossRef]
31. Chand, K.; Mankar, A.K.; Koner, R.; Naresh, A.R.V.S. Dump slope change detection and displacement monitoring using UAV close-range photogrammetry. *Sādhanā* **2024**, *49*, 277. [CrossRef]
32. Hu, H.; Fernandez-Steeger, T.M.; Dong, M.; Azzam, R. Deformation Monitoring and Recognition of Surface Mine Slope Using LiDAR. In *Engineering Geology for Society and Territory—Volume 2*; Lollino, G., Giordan, D., Crosta, G.B., Corominas, J., Azzam, R., Wasowski, J., Sciarra, N., Eds.; Springer International Publishing: Berlin/Heidelberg, Germany, 2015; pp. 451–454.
33. Hartwig, M.E.; Paradella, W.R.; Mura, J.C. Detection and Monitoring of Surface Motions in Active Open Pit Iron Mine in the Amazon Region, Using Persistent Scatterer Interferometry with TerraSAR-X Satellite Data. *Remote Sens.* **2013**, *5*, 4719–4734. [CrossRef]
34. Paradella, W.R.; Mura, J.C.; Gama, F.F.; Santos, A.R.; Silva, G.G.; Galo, M.; Camargo, P.O.; Silva, A.Q. Complementary Use of Information from Space-Based Dinsar and Field Measuring Systems for Operational Monitoring Purposes in Open Pit Iron Mines of Carajás Mining Complex (Brazilian Amazon Region). *ISPRS Int. Arch. Photogramm. Remote. Sens. Spat. Inf. Sci.* **2015**, *XL-7/W3*, 905–911. [CrossRef]
35. Hartwig, M.E. Detection of mine slope motions in Brazil as revealed by satellite radar interferograms. *Bull. Eng. Geol. Environ.* **2016**, *75*, 605–621. [CrossRef]
36. Carlà, T.; Intrieri, E.; Raspini, F.; Bardi, F.; Farina, P.; Ferretti, A.; Colombo, D.; Novali, F.; Casagli, N. Perspectives on the prediction of catastrophic slope failures from satellite InSAR. *Sci. Rep.* **2019**, *9*, 14137. [CrossRef]
37. Bazarnik, M. Slope stability monitoring in open pit mines using 3D terrestrial laser scanning. *E3S Web Conf.* **2018**, *66*, 01020. [CrossRef]
38. Fornós, J.R.I.; Sánchez, C. The use of InSAR for monitoring slope stability of rock masses. In Proceedings of the ISRM 1st International Conference on Advances in Rock Mechanics—TuniRock 2018, Hammamet, Tunisia, 29 March 2018.
39. Preston, R.P. Implementation of a Pit Slope-Monitoring Network for Inactive Open-Pit Mines Using Photogrammetry. *Rock Mech. Rock Eng.* **2021**, *54*, 3377–3385. [CrossRef]
40. He, Y.; Chen, S.; Zhang, B.; Chen, K. Multimedia Tilt Photography-Assisted Remote Sensing Technology in Mine Ecological Restoration. *Comput. Intell. Neurosci.* **2022**, *1*, 1442738. [CrossRef]
41. Zhao, Y.; Wang, S.; Liu, S.; Yang, J. Real time light-independent slope-failure monitoring using LiDAR and 2D-3D semantic segmentation. In Proceedings of the 2022 IEEE Sensors, Dallas, TX, USA, 30 October–2 November 2022; pp. 1–4. [CrossRef]
42. Tang, W.; Motagh, M.; Zhan, W. Monitoring active open-pit mine stability in the Rhenish coalfields of Germany using a coherence-based SBAS method. *Int. J. Appl. Earth Obs. Geoinf.* **2020**, *93*, 102217. [CrossRef]
43. Bar, N.; Dixon, R. Unveiling unknowns: Practical application of InSAR for slope performance monitoring and risk management across multiple surface mines. *Eng. Geol.* **2021**, *293*, 106326. [CrossRef]
44. Bar, N.; Kostadinovski, M.; Tucker, M.; Byng, G.; Rachmatullah, R.; Maldonado, A.; Pötsch, M.; Gaich, A.; McQuillan, A.; Yacoub, T. Rapid and robust slope failure appraisal using aerial photogrammetry and 3D slope stability models. *Int. J. Min. Sci. Technol.* **2020**, *30*, 651–658. [CrossRef]
45. Coccia, S.; Marwan, A.; Klein, E. Combination of UAV-Borne Lidar and UAV-Borne Photogrammetry to Assess Slope Stability. In Proceedings of the 7th World Congress on Civil, Structural, and Environmental Engineering (CSEE'22), Lisbon, Portugal, 10–12 April 2022.
46. Winkelmaier, G.; Battulwar, R.; Khoshdeli, M.; Valencia, J.; Sattarvand, J.; Parvin, B. Topographically Guided UAV for Identifying Tension Cracks Using Image-Based Analytics in Open-Pit Mines. *IEEE Trans. Ind. Electron.* **2021**, *68*, 5415–5424. [CrossRef]

47. Mankar, A.K.; Koner, R. Drone Technology and 3D Reconstruction Approach for Efficient Management of Mine Dump Slope. In *Sustainable and Innovative Mining Practices*; Gorai, A.K., Ram, S., Bishwal, R.M., Bhowmik, S., Eds.; Springer Nature: Cham, Switzerland, 2025; pp. 586–595.
48. Yadav, D.K.; Karthik, G.; Jayanthu, S.; Das, S.K. Design of Real-Time Slope Monitoring System Using Time-Domain Reflectometry with Wireless Sensor Network. *IEEE Sensors Lett.* **2019**, *3*, 2500304. [CrossRef]
49. Yang, B.; Aminossadati, S.M.; Chen, Z.; Kizil, M.S. Fibre Optic Sensing Based Slope Crest Tension Crack Monitoring System for Surface Mines. In Proceedings of the 2nd International Conference for Fibre-optic and Photonic Sensors for Industrial and Safety Applications (OFSIS), Brisbane, QLD, Australia, 8–10 January 2017; pp. 27–32. [CrossRef]
50. Yadav, D.K.; Mishra, P.; Jayanthu, S.; Das, S.K. Fog-IoT-Based Slope Monitoring (FIoTSM) System with LoRa Communication in Open-Cast Mine. *IEEE Trans. Instrum. Meas.* **2021**, *70*, 9514611. [CrossRef]
51. Yadav, D.K.; Mishra, P.; Jayanthu, S.; Das, S.K. On the Application of IoT: Slope Monitoring System for Open-cast Mines Based on LoRa Wireless Communication. *Arab. J. Sci. Eng.* **2022**, *47*, 1387–1398. [CrossRef]
52. Giri, P.; Ng, K.; Phillips, W. Monitoring toppling failures using wireless inertial measurement unit sensors. *Geotech. Res.* **2022**, *9*, 83–94. [CrossRef]
53. Zhou, Y.; Li, Y.; Chen, Y. Remote Real-Time Monitoring System for Mine Slope Based on Cloud Computing. *Water* **2022**, *14*, 714. [CrossRef]
54. Li, Q.; Song, D.; Yuan, C.; Nie, W. An image recognition method for the deformation area of open-pit rock slopes under variable rainfall. *Measurement* **2022**, *188*, 110544. [CrossRef]
55. Kumar, A.; Villuri, V.G.K. Role of Mining Radar in Mine Slope Stability Monitoring at Open Cast Mines. *Procedia Earth Planet. Sci.* **2015**, *11*, 76–83. [CrossRef]
56. Vaziri, A.; Moore, L.; Ali, H. Monitoring systems for warning impending failures in slopes and open pit mines. *Nat. Hazard.* **2010**, *55*, 501–512. [CrossRef]
57. Nunoo, S.; Tannant, D.D.; Newcomen, H.W. Slope monitoring practices at open pit porphyry mines in British Columbia, Canada. *Int. J. Min. Reclam. Environ.* **2016**, *30*, 245–256. [CrossRef]
58. Osasan, K.S.; Afeni, T.B. Review of surface mine slope monitoring techniques. *J. Min. Sci.* **2010**, *46*, 177–186. [CrossRef]
59. Yadav, D.K.; Jayanthu, S.; Das, S.K.; Chinara, S.; Mishra, P. Critical review on slope monitoring systems with a vision of unifying WSN and IoT. *IET Wirel. Sens. Syst.* **2019**, *9*, 167–180. [CrossRef]
60. Beranek, L.L.; Mellow, T.J. Chapter 1—Introduction and terminology. In *Acoustics: Sound Fields and Transducers*; Beranek, L.L., Mellow, T.J., Eds.; Academic Press: Cambridge, MA, USA, 2012; pp. 1–19.
61. Suh, H.K.; Hofstee, J.W.; van Henten, E.J. Improved vegetation segmentation with ground shadow removal using an HDR camera. *Precis. Agric.* **2018**, *19*, 218–237. [CrossRef]
62. Schiewe, J. Integration of multi-sensor data for landscape modeling using a region-based approach. *ISPRS J. Photogramm. Remote Sens.* **2003**, *57*, 371–379. [CrossRef]
63. Ruiz-Lendínez, J.J.; Ariza-López, F.J.; Reinoso-Gordo, J.F.; Ureña-Cámara, M.A.; Quesada-Real, F.J. Deep learning methods applied to digital elevation models: State of the art. *Geocarto Int.* **2023**, *38*, 2252389. [CrossRef]
64. Cho, J.; Kim, K. Detection of moving objects in multi-complex environments using selective attention networks (SANet). *Autom. Constr.* **2023**, *155*, 105066. [CrossRef]
65. Sengar, S.S.; Hasan, A.B.; Kumar, S.; Carroll, F. Generative artificial intelligence: A systematic review and applications. *Multimed. Tools Appl.* **2024**, *67*, 1–40. [CrossRef]
66. Elmouttie, M.; Dean, P. Systems Engineering Approach to Slope Stability Monitoring in the Digital Mine. *Resources* **2020**, *9*, 42. [CrossRef]
67. Wang, X.; Zhu, Z. Context understanding in computer vision: A survey. *Comput. Vis. Image Underst.* **2023**, *229*, 103646. [CrossRef]
68. Mastouri, M.; Bouyahia, Z.; Haddad, H.; Horchani, L.; Jabeur, N. A Context-Aware, Computer-Vision-Based Approach for the Detection of Taxi Street-Hailing Scenes from Video Streams. *Sensors* **2023**, *23*, 4796. [CrossRef] [PubMed]
69. Pu, S.; Xie, L.; Ji, M.; Zhao, Y.; Liu, W.; Wang, L.; Yang, F.; Qiu, D. Real-Time Powerline Corridor Inspection by Edge Computing of Uav Lidar Data. *ISPRS Int. Arch. Photogramm. Remote. Sens. Spat. Inf. Sci.* **2019**, *XLII-2/W13*, 547–551. [CrossRef]
70. Messous, M.A.; Hellwagner, H.; Senouci, S.M.; Emini, D.; Schnieders, D. Edge Computing for Visual Navigation and Mapping in a UAV Network. In Proceedings of the ICC 2020–2020 IEEE International Conference on Communications (ICC), Dublin, Ireland, 7–11 June 2020; pp. 1–6. [CrossRef]
71. Xu, X.; Zhao, S.; Xu, C.; Wang, Z.; Zheng, Y.; Qian, X.; Bao, H. Intelligent Mining Road Object Detection Based on Multiscale Feature Fusion in Multi-UAV Networks. *Drones* **2023**, *7*, 250. [CrossRef]
72. Saha, A.; Mi, Y.; Glassmaker, N.; Shakouri, A.; Alam, M.A. In Situ Drift Monitoring and Calibration of Field-Deployed Potentiometric Sensors Using Temperature Supervision. *ACS Sens.* **2023**, *8*, 2799–2808. [CrossRef] [PubMed]
73. Zheng, Y.; Zhu, Z.; Li, W.; Gu, D.; Xiao, W. Experimental research on a novel optic fiber sensor based on OTDR for landslide monitoring. *Measurement* **2019**, *148*, 106926. [CrossRef]

74. Wang, C.; Guo, W.; Yang, K.; Wang, X.; Meng, Q. Real-Time Monitoring System of Landslide Based on LoRa Architecture. *Front. Earth Sci.* **2022**, *10*, 899509. [CrossRef]
75. Inbavalli, A.; Praveena, R.; Kaviya, M. Assessing Metrics using LoRa Technology for Real-Time Landslide Monitoring within an IoT Framework. In Proceedings of the 4th International Conference on Pervasive Computing and Social Networking (ICPCSN), Salem, India, 3–4 May 2024; pp. 656–662. [CrossRef]

mining

MDPI

Article

GIS-Based Subsurface Analysis and 3D Geological Modeling as a Tool for Combined Conventional Mining and In-Situ Coal Conversion: The Case of Kardia Lignite Mine, Western Greece

Pavlos Krassakis [1,*], Konstantina Pyrgaki [1], Vasiliki Gemeni [1], Christos Roumpos [2], Georgios Louloudis [2] and Nikolaos Koukouzas [1]

[1] Centre for Research and Technology, Hellas (CERTH), 15125 Athens, Greece; pyrgaki@certh.gr (K.P.); gemeni@certh.gr (V.G.); koukouzas@certh.gr (N.K.)

[2] Department of Mining Engineering, Public Power Corporation of Greece, 10432 Athens, Greece; c.roumpos@dei.gr (C.R.); g.louloudis@dei.gr (G.L.)

* Correspondence: krassakis@certh.gr

Abstract: The development of three-dimensional geological models has proven to be critical for conceptualizing complex subsurface environments. This is crucial for mining areas due to their various hazards and unstable conditions. Furthermore, three-dimensional (3D) models can be the initial step for the development of numerical models in order to support critical decisions and sustainable mining planning. This paper illustrates the results and the development phases of a 3D geological model within the boundaries of the Kardia lignite deposit in western Macedonia, Greece. It also highlights the usefulness of a Geographic Information System (GIS) methodology in the subsurface geological and hydrogeological analysis regarding the Underground Coal Gasification (UCG) methodology. In addition, the work focuses on the integrated geospatial framework that was developed to support the Coal-to-Liquids Supply Chain (CLSC) integration in unfavorable geological settings. A 3D subsurface geological model of the study area was developed to identify a suitable area for in situ coal conversion and UCG considering criteria related to specific coal thickness and depth. In this context, the suggested integrated geomodelling workflow can positively contribute to the implementation of conventional and innovative mining, saving time and reducing the cost to improve the quality of information needed to support decisions related to UCG implementation.

Keywords: underground coal gasification; GIS; rockworks; 3D geological model; lignite; western Macedonia

1. Introduction

Energy costs and demand have risen significantly during the recent COVID-19 and Ukraine crises (Figure 1) [1]. According to the International Energy Agency (IEA) [2], in 2021, high-level increases in natural gas prices have prompted a significant switch to coal rather than natural gas to produce electricity in the energy markets of Europe, Asia, and the United States of America. Underground Coal Gasification (UCG) is a promising technology that converts coal into product gas, being more environmentally friendly than conventional mining techniques in subsurface environments.

Correctly managed UCG is an advanced clean coal technology that offers a potential solution for these challenges and broad application prospects. UCG opportunity arises in the energy transition period, where there is a need to reduce emissions significantly to comply with climate change commitments [3]. UCG is not a new technology; it involves the gasification of unmined coal and the production of synthetic gas (syngas) for use in power generation or as chemical feedstock [4,5]. UCG is appealing to expand recoverable coal reserves. It provides access to coal deposits that would otherwise remain unused due to complex and challenging mining and geological conditions (i.e., too deep/steep, low grade, thin seams) using conventional mining technologies [6,7].

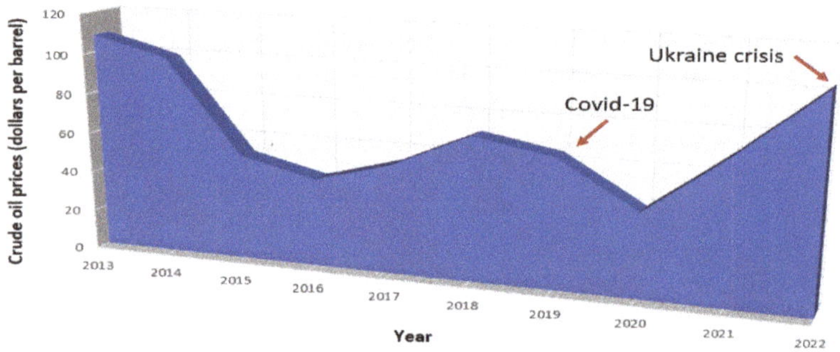

Figure 1. Crude oil prices from 2013 to 2022, based on data from US Energy Information Administration—EIA [1]. COVID-19 and recent Ukraine crisis-derived price increases are shown (red arrows).

The main challenge is the development of competitive technologies for the production of synthesis gas and the production of electricity, heat, and synthetic liquid fuels on their basis [8]. By combining UCG with Gas to Liquids (GTL) technologies, high-quality synthetic liquid transportation fuels can produce environmental and economic advantages over alternate conventional and unconventional processes [9].

Despite the extensive research, there are many areas of improvement of the UCG process and adaptation to local conditions that must be addressed through additional research. These improvements need to advance the selection of the best target area and the effectiveness of the gasification process, making sure it does not impact the environment negatively.

Many countries have already researched implementing UCG technologies, including the long-term commercial operation of several UCG plants in the former Soviet Union. However, there is limited information on applying this technology in Greece. In the ODYSSEUS project (Coal-to-liquids supply chain integration because of operational, economic, and environmental risk assessments under unfavorable geological settings), we have identified and reviewed the criteria for selecting potential UCG locations.

Studying a subsurface environment within the boundaries of a large area is a complex task. This procedure consists of many parameters, which can be examined separately under conventional methods but needs added workload and resources. On the other hand, a developed relational database can manipulate and manage various data related to GIS approaches. Furthermore, recent publications have shown that integrating a geospatial framework is an efficient tool for analyzing and mapping subsurface geological and hydrogeological environments [10,11].

In addition, geomodelling requires GIS databases and tools [12], due to their ability to interact with the data spatially, to understand the subsurface and its mechanisms better, and visualize it in 2D and 3D dimensions [11,13]. Even though much work has been done in the last century concerning geomodelling and developed GIS databases, there is no standard widely accepted workflow. Due to the high cost of various software packages and lack of data and resources, 3D geological modeling of the subsurface has not been applied frequently in publications and various studies [14,15].

Within such a framework, the present study aims to develop an efficient methodology of integrating borehole and various data to map and research the spatial distribution of lignite seam thickness in the Kardia mine. The developed 3D geological model was integrated with specific geospatial analysis tools for the lignite deposit and was applied to locate and identify potentially suitable sites for UCG implementation employing combined conventional mining and in situ coal conversion.

2. Geological and Geomorphological Settings

The lignite mine is located in the Ptolemais Lignite Basin, dominated by E-W trending normal Quaternary faults [16]. The Ptolemais Basin covers a surface area of approximately 600 km^2. The longitudinal axis of the lignite segment of the Ptolemais Basin, with an NW-SE direction, exceeds 20 km in length, while the maximum width reaches about 20 km. The basin (Figure 2) is filled with Late Miocene to Pleistocene lake sediments, including intercalated lignites and alluvial deposits with a total thickness of up to 600 m.

The sediments consist of sandy clays, sandy marls, clayey marls, calcareous sands, and conglomerates. Above lay Quaternary conglomerates of terrestrial and fluvioterrestrial origin. The exposed stratigraphic sequence in the Kardia lignite mine belongs to the Early Pliocene and Ptolemais Formation. Based on the borehole data, the stratigraphic sequence in the lignite mine has an average thickness of about 300 m and consists of an alternation of lignite and marls, with intercalated sands and silts [17–19]. The Ptolemais Basin is part of a tectonic trench over 250 km extending from northern Greece into North Macedonia [20].

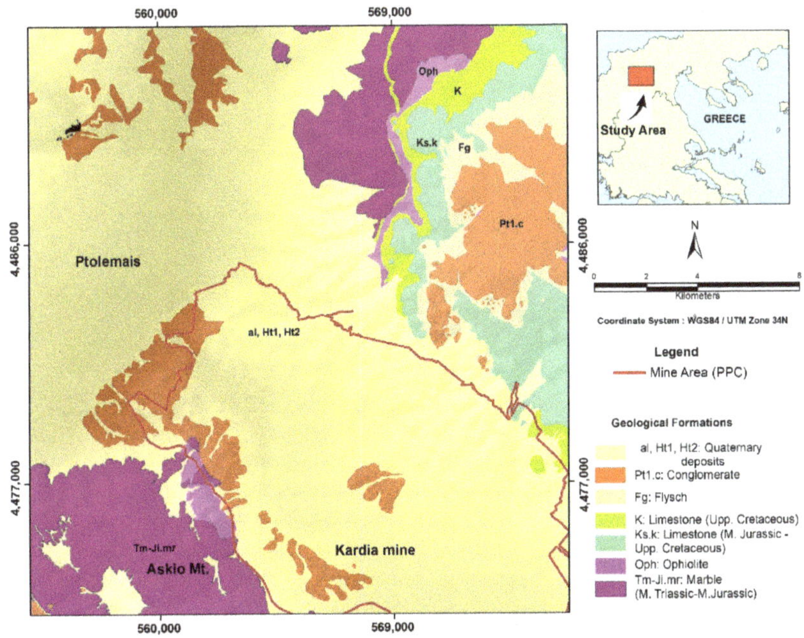

Figure 2. Simplified geological map of the Ptolemais area, located NW and SE of the city of Ptolemais (based and modified on Institute of Geology and Mineral Exploration (IGME) geological sheet, Ptolemais 1:50.000, [21]).

According to previous literature, the thickness of the lignite-bearing layers (including intercalations) in the adjacent mining area ranges between 80 m and 140 m at the western boundary of the mine near the "Askio" (Siniatsiko) mountain. After that, it increases to the SW to more than 210 m, where the overlying strata have a thickness of about 150 m. In the central and north-western parts of the mine, the thickness of the overlying strata is about 20 m–60 m, and the thickness of the lignite seams varies [22].

Regarding the Digital Elevation Model (DEM), the Kardia mine is an open and excavated area with surface elevations ranging between +622 m and +812 m above sea level (a.s.l.) from E to W direction, respectively. DEM is the most common digital 3D representation of a topographic surface concerning any reference datum. Therefore, DEM is frequently used to determine terrain attributes that include elevation at any point, slope, and aspect [23]. In our study area, DEM was imported from satellite data using the Advanced Spaceborne Thermal Emission and Reflection Radiometer (ASTER), with a nominal horizontal accuracy of 15 m and 20 m, respectively [24].

In addition, the surface elevation value for each borehole was obtained from the DEM dataset. The mining area covers approximately 18 km^2 near the Pontokomi village (Figure 3). The lignite basin of the deposit progressively broadens from W to E with a gradational decrease in surface altitudes.

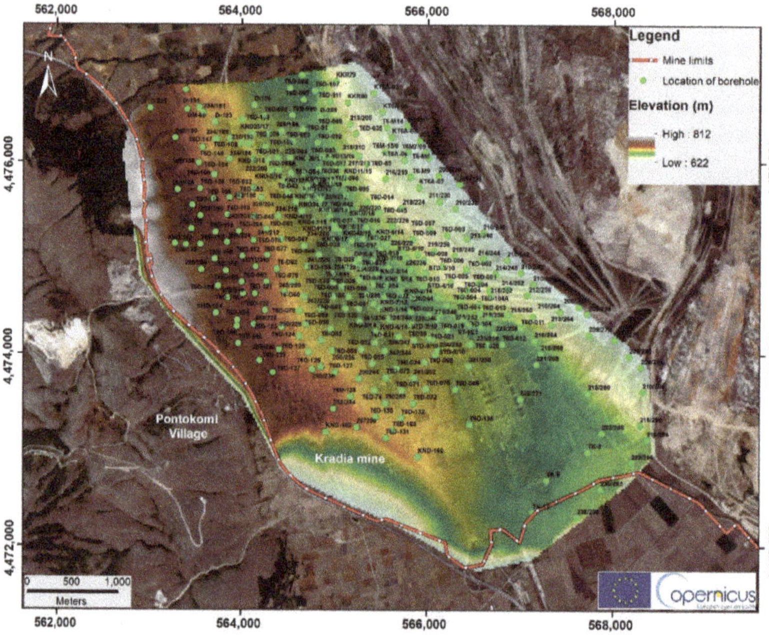

Figure 3. The map represents the spatial location of the analyzed boreholes (304) combined with the Digital Elevation Model in the study area using ArcGIS 10.4 version. The base map is provided by Sentinel 2b satellite European Union—Copernicus program) with sensing date: 31 August 2021.

3. Materials and Methods

Geospatial Analysis Framework and Resolution

Rockware's Rockworks17 and ESRI ArcGIS 10.4 software were used for geomodelling and spatial analysis processing tasks. The Rockworks software is a geomodeling software able to create block models and cross-sections in an area, providing the option to convert the exported data into ArcGIS shape easily or grid files and vice versa as input data. Thus, for the 3D geological modeling in the Rockworks environment, the lateral blending technique or Lithoblending algorithm was implemented for interpolating geological units' spatial distribution regarding horizontal and vertical (values) axes [25].

In this study, two spatial data categories have been utilized: (i) discrete representation of objects and (ii) representation with continuous fields. The type of discrete representation consists of well-defined objects in terms of shape, color, and form. Under this context, a geographical area can be described as a region with lakes, rivers, faults, land cover, or geological formations representing polylines, points, or polygons.

In order to produce valuable and comparable outcomes, preparation and filtering of the original geospatial data were applied before the thorough spatial analysis. According to the following schematic diagram (Figure 4), a specific workflow was implemented hierarchically using spatial analysis and specific geoprocessing tools within the environment of ArcGIS 10.4 version (left part of Figure 4) [26] and Rockworks (right part of Figure 4).

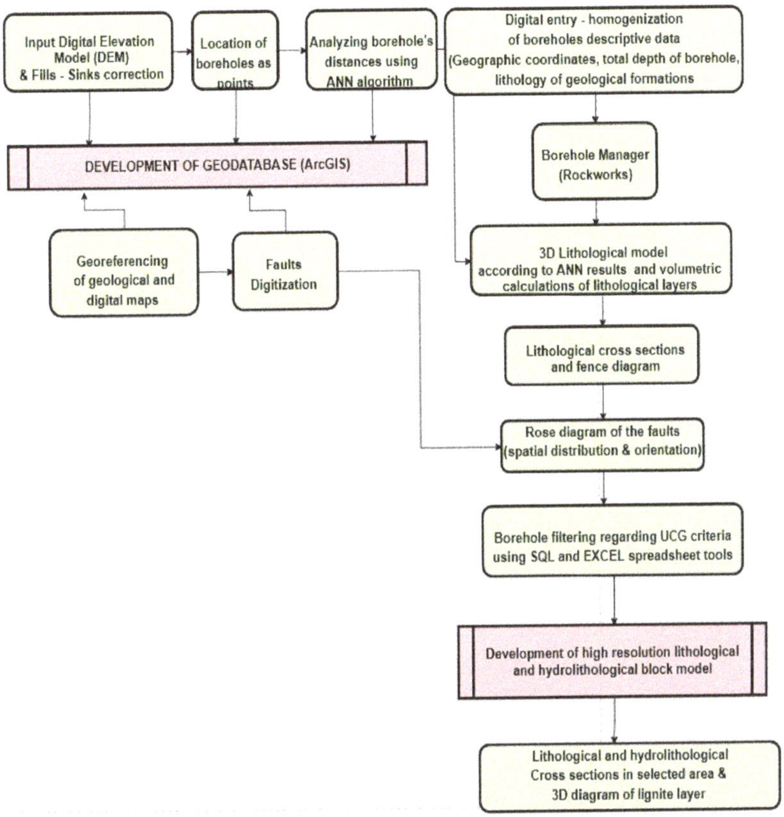

Figure 4. Workflow of the data preparation for the geospatial analysis and creation of the database using various datasets.

According to the upper part of the workflow, the first step was the selection and correction of the Digital Elevation Model of the study area. Then, the DEM correction was applied in the ArcGIS environment using the Fill geoprocessing tool. According to ESRI, peaks and sinks are often errors due to the resolution of the data or rounding of elevations to the nearest integer value. The fill tool uses the equivalents of several tools, such as Focal Flow, Flow Direction, Sink, Watershed, and Zonal Fill, to locate and fill the sinks [27–30]. Thus, the tool iterates until all peaks or sinks within the specified z limit are filled. Subsequently, after implementing the sink tool, more than 15 sink areas were identified with a total area of more than 15,000 m^2 and removed from the initial Digital Elevation Model dataset. After the correction with the Fill algorithm of the DEM, the elevation values were extracted to add the information of each borehole within the boundaries of the study area. According to the presented workflow, georeferencing and digitization of faults according to previous studies and available references were applied to be analyzed in Rockworks software as linear lineaments to produce rose diagrams (lower part of the flow chart). After the integration of borehole data in the geodatabase, followed the visualization and spatial analysis of the 304 boreholes dataset under the process of the study area in order to compare the observed point patterns to the ones generated by an independent random process (IRP), also known as complete spatial randomness (CSR). This technique was implemented using the Average Nearest Neighbour (ANN) tool from ArcGIS 10.4 version. According to the reference ANN algorithm, which is included in

Arcmap software [31,32], checks whether or not the observed first-order nearest neighbor is consistent with a distribution of points if this process was completely random. The average nearest neighbor (ANN) tool computes the first nearest neighbor mean distance for all points [33,34]. The following equation (1) is used to calculate the ANN_{ratio}:

$$ANN_{ratio} = \frac{OMD}{EMD} \tag{1}$$

where *OMD* is the observed mean distance between each feature and its nearest neighbor (2):

$$OMD = \frac{\sum_{i=1}^{n} d_i}{n} \tag{2}$$

and *EMD* (3) is the Expected Mean Distance for the features given in random pattern:

$$EMD = \frac{0.5}{\sqrt{n/A}} \tag{3}$$

In the above equations, d_i equals the distance between feature i and its nearest neighboring feature, n corresponds to the total number of features, and A is the area of the minimum enclosing rectangle around all features or it can be specified.

The ANN z-score is calculated as (4):

$$z = \frac{OMD - EMD}{SE} \tag{4}$$

where SE is calculated as follows (5):

$$SE = \frac{0.26136}{\sqrt{n^2/A}} \tag{5}$$

ANN results showed that the Observed Mean Distance was approximately 134.4 m, and the Expected Mean Distance (EMD) was calculated at approximately 100 m. Also, the ANN ratio or Nearest Neighbor Ratio (NNR) was calculated as 1,355, and regarding the z-score (Figure 5), the spatial distribution pattern was categorized as dispersed. EMD is a key aspect of calculating the distance at which locations between boreholes (points) are no longer related where the spatial dependence ceases.

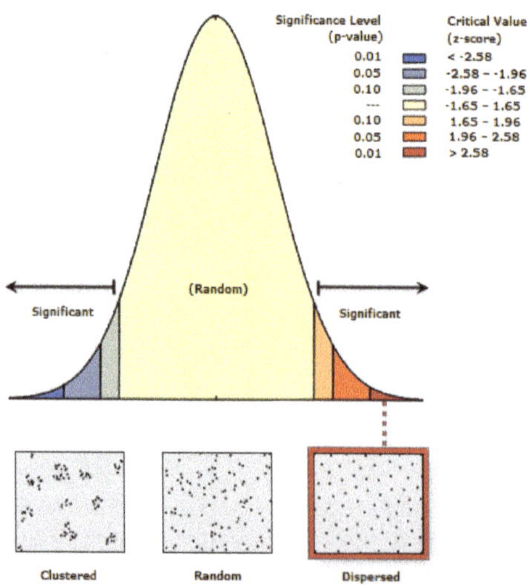

Figure 5. Distribution pattern of boreholes regarding Average Nearest Neighbor (ANN) report.

The EMD as 100 m was considered to evaluate the horizontal and vertical distance spacing in Rockworks software and, therefore, develop the 3D geological model to have an efficient, accurate, but minimum computational effort discretization. According to the upper part of the workflow, a critical step was developing and homogenizing a catalog with 304 boreholes, containing their exact location, elevation above sea level, total depth, and the lithological unit at each depth.

The description of the lithological units that were used and their codes are presented in Table 1. Nine lithological units were coded and visualized (Table 1). In addition, the Rockworks Borehole Manager tool was utilized to create the grid models for the subsurface of each lithological unit in order to visualize them in 2D cross-sections and 3D diagrams (lowest part of Figure 4).

Table 1. Lithological units that correspond in the borehole database.

Lithological Units	Code
Soil	SO
Clay	AL
Sand	SN
Silt	SI
Marl	MR
Conglomerate	KC
Coal	CO
Limestone	LI
Breccia	BC

For the creation of the block model, in terms of the upper surface and lower subsurface filters, Inverse Distance Gridding (IDG) was implemented, which is suggested in Rockworks software and is based on Newton's Gravity Equation (6). According to this method, each value assigned to a grid node is a weighted average of all data points or several directionally distributed neighboring control points. Specifically, the value of each data point is weighted according to its inverse distance from the grid node, taken to a user-selected power. The greater the exponent specified value, the more localized the gridding since distant points will have less influence on the value assigned to each grid node.

$$Z - node = \frac{\Sigma(Z - \frac{point}{dn})}{\Sigma\left(\frac{1}{dn}\right)} \tag{6}$$

Z-node is the value of the assigned node values, where the value of each data point is weighted according to the inverse of its distance *d* from the grid node, taken to the selected power (6). When the distance factor *d* is greater than one, the values of distant points have less influence than nearby points. For the upper and lower surface of the developed block model, the automatic methodology was selected based on borehole collar elevations, where the software automatically interpolated a surface representing the elevations at the top of the boreholes (using current project dimensions and the inverse-distance gridding method).

The advantage of using the IDG approach is that it can produce a smooth and continuous grid without exaggeration regarding extrapolations beyond the given data points. On the other hand, IDG can create a "bulls-eye" effect in some areas of the produced dataset. According to Rockware, the range of grid values will be smaller than the data point range. In other words, the highest value of the grid will be less than the maximum data point, and the lowest grid value will be greater than the minimum data point [25].

Regarding the workflow (Figure 4), statistical analysis has been applied to filter the suitable boreholes according to the UCG criteria. The primary criteria that have been used refer to the coal seam thickness, where single seam thickness should be at least 1 m [35,36] and coal seam depth greater than 200 m to minimize potential gases [37,38]. After filtering the 304 boreholes, 10 boreholes in a specific region were selected, where a new high-resolution lithological model was produced to illustrate coal seams better and visualize the suggested subsurface environment in greater detail.

The implemented methodology provides a distinct and integrated workflow regarding how the lithological model has been developed in a specific area. The approach followed regarding Greek open-pit lignite regions needs specific steps to be implemented to analyze the geological subsurface in the context of the UCG technique. In addition, this workflow attempted to present a specific flow in terms of how geomodelling combines GIS databases and tools to understand better the subsurface and its mechanisms, illustrating it in 2D and 3D dimensions according to UCG criteria. Previous works have been published in lithological modeling concerning geomodelling and developed GIS databases, but there is no standard widely accepted workflow [14].

Thus, due to the high cost of various software packages and lack of data and resources, 3D geological modeling of the subsurface has not been widely utilized in publications and various studies [15].

4. Results and Discussion

4.1. Geospatial Analysis Results

An essential part of the work was the investigation and the georeferencing process of the faults' traces (red polylines) on the surface of the mine area, based on the literature and geological studies. Most of the tectonic lines within the boundaries of the studied area were digitized and imported into the database. More specifically, more than 300 faults were mapped (Figure 6) in the broader area, where 31 of them were inside the polygon of the mine.

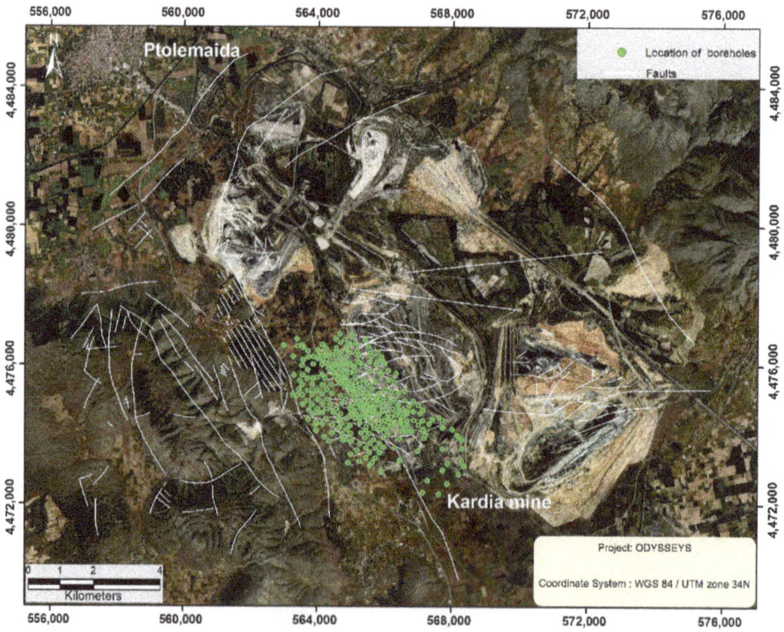

Figure 6. The map represents the spatial location of the digitized faults in the broader area of the Kardia lignite mine. The base map is provided by Sentinel 2b satellite European Union—Copernicus program) with sensing date: 31 August 2021.

According to the georeferenced faults' traces (Figure 6) and the following rose diagram (Figure 7), faults dissect the mining field into several fault blocks.

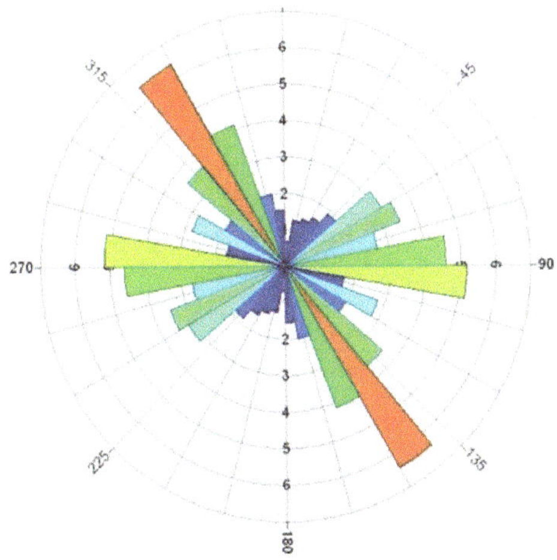

Figure 7. Rose diagram showing the distribution of fault trends at the broader area of Kardia mine.

The highest percentage of all faults' lineation (>150 faults segments) by length is following an NW–SE direction in the broader area of the mine. The total length of the faults is 224 km, the average bin population is 51.81, and the standard deviation of the bin population is 30.22.

4.2. Three-Dimensional Geological Modelling

Currently, 3D geological modeling is a rapidly expanding discipline as demand for knowledge about the structure and composition of the subsurface is increasing. In order to create the 3D model, existing borehole data were grouped in lithological units to be imported into the RockWorks Borehole Manager.

After grouping the lithological layers, the study area boundaries were calculated with the Minimum/Maximum Eastern and Northern limits. Then, based on the borehole data and the boreholes' coordinates, a grid model was created with specific dimensions and nodes. The Spatial Reference System utilized was the Universal Transverse Mercator (UTM) 34 (Northern Hemisphere)/WGS 84. It is worth mentioning that lithological units are repeated with depth in each borehole (sand, clay, coal, or other formations). Therefore, the following procedure coded the lithological types like integer or real number values, based on the "G" value for each rock/soil type. For instance, "CO = coal" was coded with a "2" and "sand" with a "4.5". The "G" value is just a numeric value that Rockworks uses to represent the material in lithology models. A sample 3D profile of the T6D-070 borehole illustrates the lithological layers using a RockPlot3-D window in the following image (Figure 8).

The three-dimensional lithological model (Figure 9) was generated taking into account three primary characteristics of the solids: extent, base, and top-level of each lithological layer, depth, thickness, and descriptive data related to lithological type. For the better interpolation of the model, a gradational method was applied to standardize and group similar lithological types. The 3D block model represents the different lithology types, creating a 3D dimensional voxel diagram that illustrates these lithological types. Lithology models are displayed as all-voxel diagrams, with each color-coded voxel based on its assigned material type. For example, in our project, the black color corresponds to a coal formation and the yellow color to a marl formation.

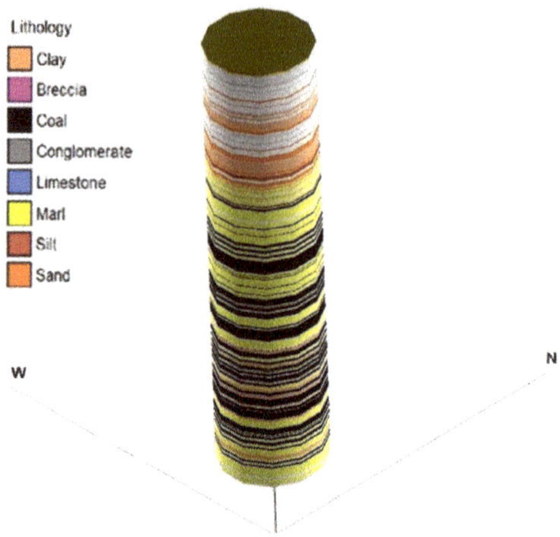

Figure 8. Sample profile of the T6D-070 borehole (Vertical Exaggeration =3x, 1350/320 compass bearing/angle from the horizon). The development of the 3-D model of the study area was created with RockWorks version 17 (standard license).

The development of the model was calculated in three dimensions as a 3D voxel/solid diagram. A 3D model of lithological variations within the basin was developed by extrapolating 304 boreholes using a 3-dimensional gridding process. Subsurface lithology was defined through the identification of distinctive lithological packages. A log correlation was performed using the lithoblending algorithm to create the model. This unique algorithm exports lithology types outward from the borehole data and trenches into a solid block model. This approach produces "lithozones" around each borehole which can end suddenly when a lithological zone from a near borehole is encountered [39]. More specifically, this algorithm acts as a nearest neighbor geospatial methodology, exclusively designed for lithological modeling in the Rockworks environment. The lithoblending algorithm detects the horizontal plane around each log. Then, it iteratively assigns the corresponding numeric value, which has been set in the lithology table, to the next node unless it bumps into a node that has already been interpolated, following a growing circle around each log. It should be noted that the horizontal weighting capacity of the lithoblending algorithm activated the illustration of the horizontal lenses from the different formations as frequently observed in environments or basins with fluvial sedimentation [40].

Available subsurface data provided sufficient detail to distinguish major lithological types confidently and with sufficient detail within these units to develop a subsurface geologic model. In our case (Figure 9), 9 primary lithological layers were depicted across the study area. In particular, breccia, conglomerate, sand/sandstone, clay, marl, coal, limestone, soil, and silt/siltstone formations were displayed in the 3D block model. In the study area, 10 vertical and horizontal cross-sections were produced (Figure 9) for the 3D model displaying the results of the numerical interpolation as a fence diagram that highlights lithological variations within the basin fill.

The cross sections' colors correspond to the lithological units in the fence diagram (Figure 10). The upper surfaces of the cross-sections were then visualized according to the elevation of each borehole (highest 732.96 m and lowest 634.13 m). Finally, the model base was visualized according to the borehole data's lowest elevations (the difference between the depth and elevation). According to the derived model, the marly, sandy, clayey formations and coal layers seem to dominate the basin's subsurface, according to the cross-sections.

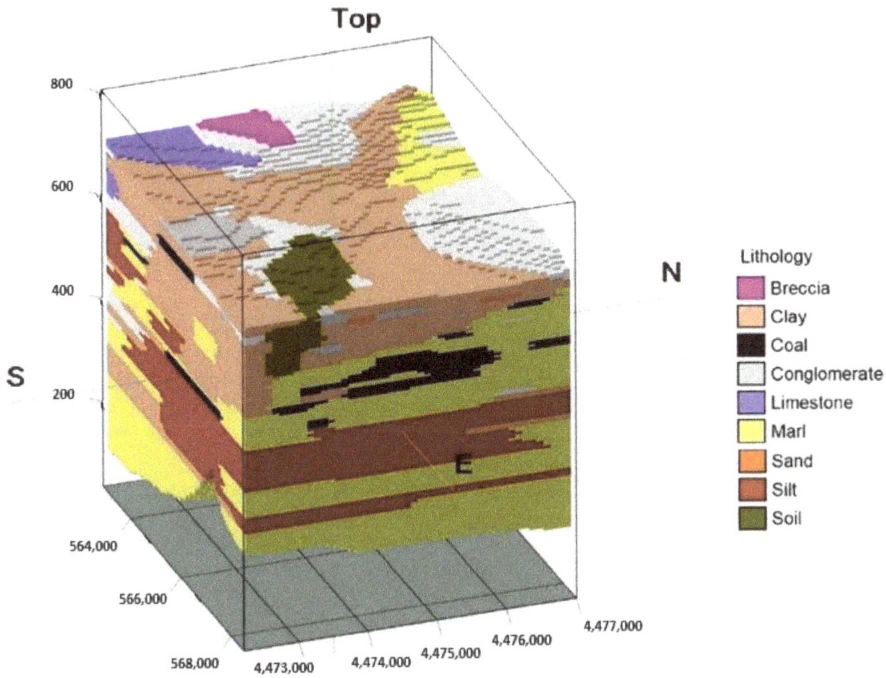

Figure 9. Interpolation of the 3D block model (Vertical Exaggeration = 8x, 1100/260 compass bearing/angle from the horizon). The development of the 3-D model of the study area was created with RockWorks version 17 [41]. Vertical axis represent a.s.l values.

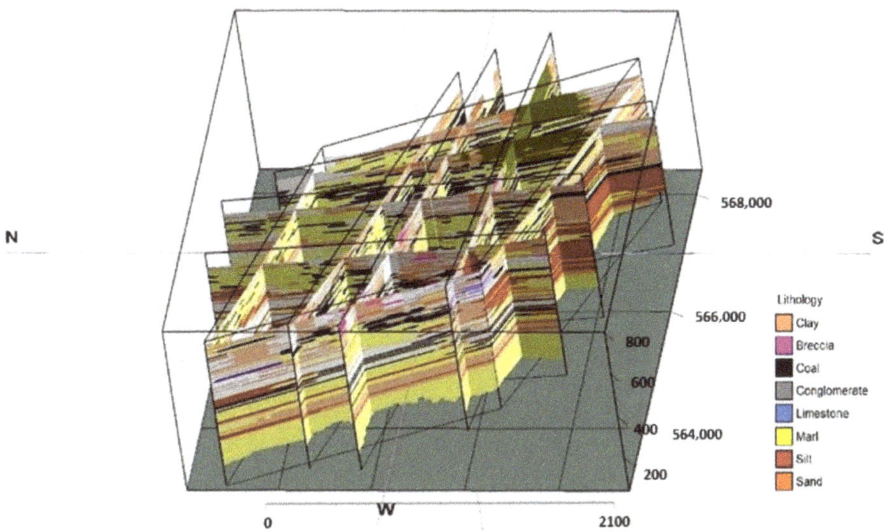

Figure 10. Grid lithology fence panels (Horizontally and vertically; 262°/400) are displayed in 3D using the background colors based on the lithology patterns.

According to the previous sections, specific tools and methodologies have been used to create the 3D geological model and manage the geospatial data in the study area. The result of this integration was the geological interpretation and modeling of the study area's subsurface environment within the boundaries of the Kardia lignite mine regarding the 304 borehole data and its spatial distribution. The relational database, which has been developed, contains the digitized faults traces, borehole data, 2D and 3D maps, and datasets for future analysis to enhance the geological and hydrogeological understanding of the area under examination.

4.3. UCG Site Identification

The next step was to identify and delineate a proposed subsurface suitable area for combined conventional mining and in situ coal conversion within the boundaries of the created block lithological model, consisting of 304 borehole data. According to reference and UCG technology, two significant criteria were considered (i) borehole location with a total depth greater than 200 m and (ii) coal thickness greater than 1 m. A series of Structured Query Language (SQL) queries combined with statistical results were implemented into Rockworks in order to filter the 304 boreholes to find suitable areas fitting the UCG above-mentioned criteria, considering the needed depth and coal continuity. After the integrated analysis and the spatial distribution of the different lithologies in compliance with the primary criteria, a particular region was identified consisting of only 10 boreholes (Table 2). The developed lithological model was filtered and cut off in order to surround these borehole locations and their lithological distribution. Specifically, the developed block model was isolated with spatial filtering to contain the 10 boreholes and reconstructed in higher resolution, considering horizontal and vertical distances for better visualization and interpretation.

Table 2. Coordinates (UTM WGS84,34N), elevation, and depth of selected boreholes.

Borehole	X (UTM WGS84)	Y (UTM WGS84)	Elevation (m a.s.l.)	Total Depth (m)
KND-5/14	565,555.3	4,474,715	670.86	354
T6D-051	565,422.6	4,474,529	669.44	384
KND-3/14	565,454.5	4,474,827	666.89	381
T6D-040	565,346.6	4,474,617	669.49	367
T6D-134	565,125.3	4,474,595	674.54	394
T6D-050	565,079.4	4,474,711	671.98	421
242/228	564,988.3	4,474,596	672.54	386
T6D-059	564,859.4	4,474,473	676.86	430
T6-D60	564,680.1	4,474,598	680.60	343
T6D-079	564,631.2	4,474,449	685.93	386

In this framework, the data of the 10 boreholes (Table 2): KND-5/14, KND-3/14, T6D-040, T6D-134, T6D-050, 242/228, T6-D60, T6D-059, T6D-051, and T6D-079 (Table 3), were considered, incorporating the interpolated lithological sequence in the suggested area, and the alternation of lignite and marls, with intercalated sands and silts. The results validated and highlighted more clearly the sequence corresponding to the Late Miocene to Pleistocene formations, which fill the basin, including intercalated lignite and recent deposits with a total thickness of up to 600 m. Two distinct lignite horizons have been identified (Figure 11), altering with marly and clayey formations. The developed new high-resolution solid model and the obtained cross-section from W to E (Figure 12) were generated with 1 m vertical and horizontal distances as long as 10 m horizontal spacing.

According to the following cross-section from W to E direction (Figure 12), a specific area with lignite lithological units (black dashed line) was delineated and selected with an extent of approximately 135,000 m^2. Top and bottom elevation heights for the volumetric of the coal were set from 500 m to 350 m a.s.l. (185 m to 325 m depth), respectively. In addition, spatial filtering was applied in X-Y coordinates from 564,600 m to 565,500 m and from 4,474,450 m to 4,474,600 m (UTM 34 N).

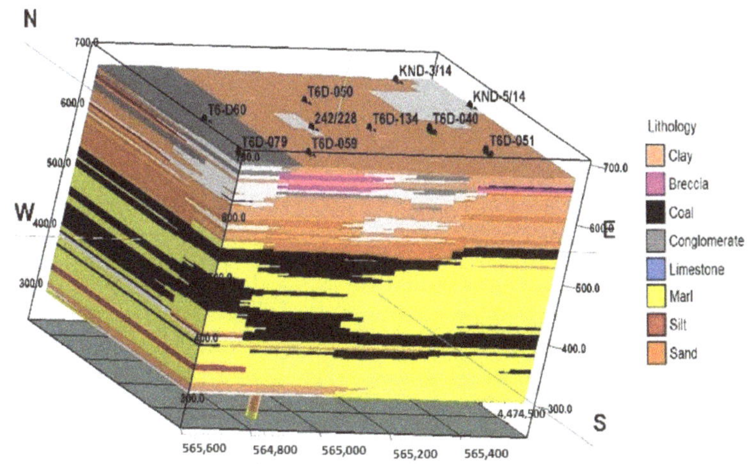

Figure 11. High-resolution 3D lithological block model in the suggested area according to UCG criteria. Vertical axis represent a.s.l values.

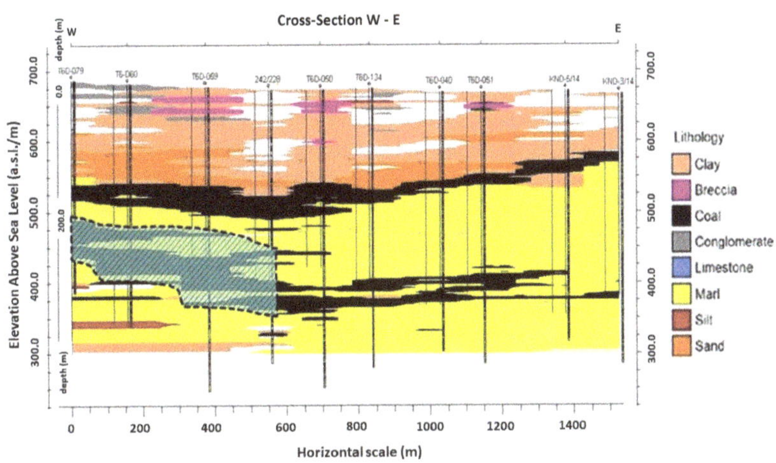

Figure 12. W-E cross-section on selected boreholes with Vertical Exaggeration x1.7. The black dashed line represents the delineated lignite area for the UCG approach. Vertical axis represent elevation above sea level (a.s.l) and depth values.

Based on the created cross-section of the solid model (Figure 12), the total volume of the coal in the suggested area (black dashed outline) was calculated as more than $4 * 10^6$ m^3. In addition, Figure 13 illustrates a 3D diagram in the same area (red rectangle), visualizing only the isolated and selected lignite layers of the developed high-resolution block model with vertical exaggeration x6.

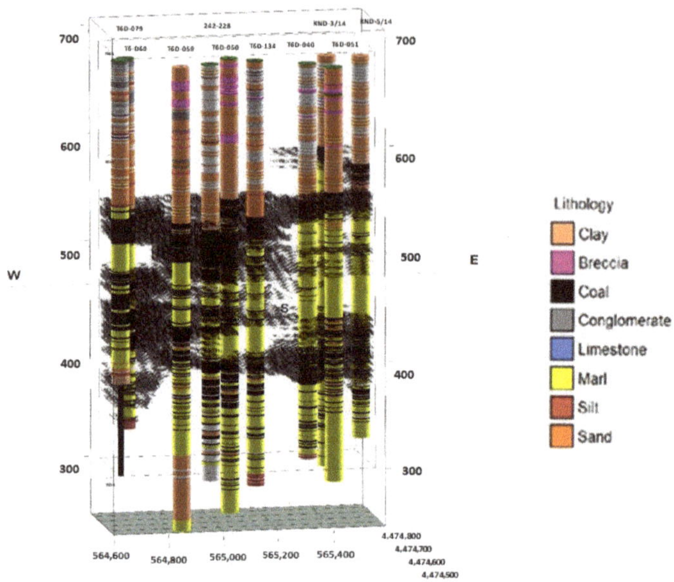

Figure 13. 3D diagram of interpolated lignite layer (black color) correlated with lithogical boreholes regarding lithological model with Vertical Exaggeration x4. Vertical axis represent elevation above sea level (a.s.l) values.

4.4. Three-Dimensional Hydrolithological Model

The main aquifer system is developed in the sand, conglomerate, and clay alternations consisting of numerous confining layers that are delimited by impermeable clay with the piezometric level ranging from 640 m to 660 m a.s.l. [42].

According to pumping test results, in the broader area of Ptolemais mines, the hydraulic conductivity of clay and sand alternations ranges from 1.03 to 25.92 m/day, and marl and lignite alternations range from 0.05 to 0.32 m/day (Table 3) [42,43].

Table 3. Hydrolithological units based on the lithological model.

Lithology	Hydrolithological Unit	Hydraulic Conductivity Range (m/d)	Average Value (m/d)
Sand, Clay, Breccia, and Conglomerate	Pleistocene–Pliocene formation: Sandy Clay & Sand alternations	1.03 to 25.92	13.48
Marl and Lignite	Lignite sequence: Marl and Lignite alternations	0.05 to 0.32	0.191

The main hydrolithological units, which are the Pleistocene–Pliocene formation and the lignite sequence, were identified based on the 3D lithological block models in the targeted site. Within such a framework, the 3D model of the main hydrolithological units was visualized based on the 3D lithological model by matching each hydrolithological unit with its respective average hydraulic conductivity (Figure 14).

According to the following hydrolithological cross-section (Figure 15), the maximum depth of the main aquifer developed in Pleistocene–Pliocene formations is approximately 150 m from the surface.

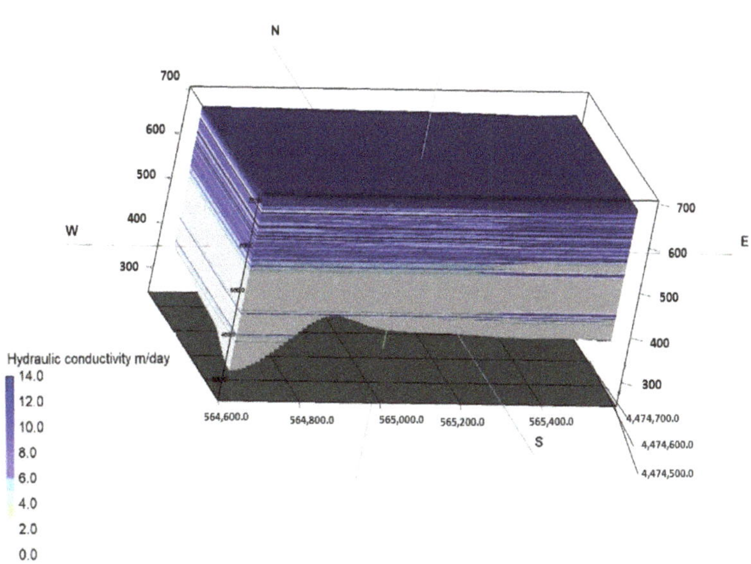

Figure 14. 3D model of the two hydrolithological units in the targeted area, regarding hydraulic conductivity m/day. Vertical axis represent elevation above sea level (a.s.l) values.

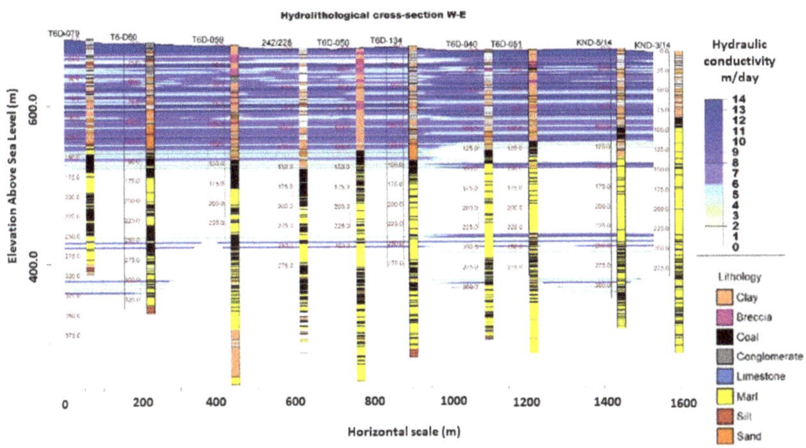

Figure 15. W-E Hydrolithological cross-section on selected boreholes with Vertical Exaggeration x2. Vertical axis represent elevation above sea level (a.s.l) values.

5. Conclusions

It is well known that climate policies are becoming increasingly stringent, and the closure phase of lignite mines has already started in Europe. In addition, regions adjacent to lignite mines face an unpredictable future. Therefore, the transition phase must be based on proper planning that considers the existing infrastructure and social awareness. The main objective of this study is to highlight the geospatial capabilities for geological and hydrogeological conceptualization within the boundaries of the Greek lignite mine of Kardia regarding the UCG approach.

The development of a geological model plays a key role in better understanding the subsurface regarding the study area. In addition, the created database can offer helpful information for future investigation and technical studies utilizing the existing spatial database and the produced datasets. Thus, 3D geological models illustrate the importance of 3D subsurface modeling in a complex geological environment underlain by multi-layer lignite horizons. Spatial results of the developed lithological model revealed the most suitable location for combined conventional mining and in situ coal conversion in the southwest region of the mine.

This work also proves that developing a 3D modeling approach is an efficient way to capture the subsurface complexity of unfavorable geological settings in the context of an integrated approach and better knowledge of the subsurface regime. Hence, identifying the most suitable region for UCG methodology according to the two major criteria, as long as creating a relational database with useful datasets, can be utilized in future studies and other research projects. A limitation of this work is that UCG methodology is more complex regarding its criteria and needs further study and geotechnical analysis within the boundaries of the suggested region. Although the developed high-resolution model, which filtered down the initial block diagram, seems to be an efficient approach to map the various lithological units and their spatial distribution. In this way, they can support sustainable practices and policies within and around the boundaries of lignite mines. To summarize, the methodology that was applied in the study area consists of a fivefold process:

- The creation of a geoinventory and the development of a GIS-managed database;
- The generation of 2D maps and 3D models as long as datasets;
- The creation of 3D lithological and hydrolithological models through a solid modeling methodology and regarding spatial analysis and SQL queries;
- The identification of a suitable area regarding borehole information, UCG major criteria, and spatial distribution of lithology,
- The development of a new high-resolution model to better interpret and understand the subsurface conditions.

Further geological and geotechnical studies can be implemented using the existing relational database and the developed 3D lithological and hydrolithological models regarding UCG methodology. Furthermore, this work showed that the iterated combination of GIS and Rockworks algorithms could be used efficiently to understand better and interpret complex subsurface parameters and mechanisms, providing a key role for UCG technology and local decision-making. From this point of view, 3D modeling seems to be the most satisfying way to interpret the subsurface complexity of the geological environment. In addition, the generated block models can be used in future studies of thorough groundwater flow and contaminant transport, where the developed databases could be used as required input data.

In conclusion, modern approaches and technologies can be combined to be transformed into diversified and sustainable resources that provide a better and more sustainable future regarding the environmental, economic, and public aspects. Therefore, the presented workflow can be applied to develop conceptual and initial models in sedimentary basins and complex geological environments, illustrating a meaningful and helpful approach combining results from conceptual models and GIS datasets.

Author Contributions: Conceptualization, P.K., V.G., K.P. and C.R.; methodology, P.K., K.P, V.G., C.R., G.L. and N.K.; software, P.K. and K.P.; validation, P.K., K.P. and G.L.; formal analysis, P.K., K.P, V.G., C.R., G.L. and N.K.; investigation, P.K., K.P., V.G., C.R., G.L. and N.K.; resources, P.K., K.P., V.G., C.R. and G.L.; data curation, P.K., K.P., V.G., C.R., G.L. and N.K.; writing—original draft preparation, P.K.; writing—review and editing P.K., K.P., V.G., C.R., G.L. and N.K; visualization, P.K. and K.P.; supervision P.K., K.P., V.G., C.R., G.L. and N.K.; project administration, P.K., K.P., V.G. and C.R. All authors have read and agreed to the published version of the manuscript.

Funding: The present study received funding from the Research Fund for Coal and Steel under grant agreement No 847333.

Data Availability Statement: The data is confidential.

Acknowledgments: Maps and diagrams throughout this work were created using ArcGIS® software by Esri. ArcGIS® and ArcMap™ are the intellectual property of Esri and are used herein under license. Copyright © Esri. All rights reserved. For more information about Esri® software, please visit www.esri.com. Block modeling, cross-sections, and lithological diagrams throughout this work were created using Rockworks software by Rockware. Rockworks is the intellectual property of Rockware and is used herein under license. Copyright Rockware. All rights reserved. For more information about Rockware software, please visit https://www.rockware.com/product/rockworks/. Many thanks are also given to colleagues from PPC, Eleni Mertiri and Katerina Servou, for their participation in the interpretation and pre-processing of the provided data, as well as Evangelia Zygouri for her help with editing.

Conflicts of Interest: The authors declare no conflict of interest.

References

1. IEA. Crude Oil Prices Rise Above $100 Per Barrel after Russia's Further Invasion into Ukraine. Available online: https://www.eia.gov/todayinenergy/detail.php?id=51498 (accessed on 10 March 2022).
2. IEA. Coal 2021, IEA, Paris. 2021. Available online: https://www.iea.org/reports/coal-2021 (accessed on 17 January 2022).
3. Pershad, S.; van der Riet, M.; Brand, J.; van Dyk, J.; Love, D.; Feris, J.; Strydom, C.C.; Kauchali, S. SAUCGA: The potential, role, and development of underground coal gasification in South Africa. *J. South. Afr. Inst. Min. Metall.* **2018**, *118*, 1009–1019. [CrossRef]
4. Burton, E.; Friedmann, J.; Upadhye, R. *Best Practises in Underground Coal Gasification;* Lawrence Livermore National Laboratory: Livermore, CA, USA, 2006.
5. Koukouzas, N.; Katsimpardi, I.; Merachev, D. Simulation of Underground Coal Gasification process in a Bulgarian Coal field. Bulletin of the Geological Society of Greece, vol. XLVII No 3. In Proceedings of the 13th International Congress, Chania, Greece, 21 January 2013; pp. 2090–2100. [CrossRef]
6. Sarhosis, V.; Lavis, S.; Mostade, M.; Thomas, H.R. Towards commercialising underground coal gasification in the EU. *Environ. Geotech.* **2016**, *4*, 113–122. [CrossRef]
7. Yang, D.; Koukouzas, N.; Green, M.; Sheng, Y. Recent Development on Underground Coal Gasification and Subsequent CO2 Storage. *J. Energy Inst.* **2016**, *89*, 469–484. [CrossRef]
8. Tyurina, E.A.; Elsukov, P.Y.; Mednikov, A.S. Synthetic liquid fuels: Prospects for innovative technologies based on underground coal gasification. In Proceedings of the E3S Web of Conferences, Irkutsk, Russia, 23 November 2020; Volume 209, p. 3026, EDP Sciences. [CrossRef]
9. Perkins, G.; du Toit, E.; Koning, B.; Ulbrich, A. Unconventional Oil Production from Underground Coal Gasification and Gas to Liquids Technologies. In Proceedings of the Society of Petroleum Engineers (SPE) Unconventional Resources Conference and Exhibition-Asia Pacific, Brisbane, Australia, 11 November 2013. [CrossRef]
10. Kadam, A.K.; Kale, S.S.; Pande, N.N.; Pawar, N.J.; Sankhua, R.N. Identifying potential rainwater harvesting sites of a semi-arid, basaltic region of Western India, using SCS-CN method. *Water Resour. Manag.* **2012**, *26*, 2537–2554. [CrossRef]
11. Trabelsi, F.; Tarhouni, J.; Mammou, A.B.; Ranieri, G. GIS-based subsurface databases and 3-D geological modeling as a tool for the set up of hydrogeological framework: Nabeul–Hammamet coastal aquifer case study (Northeast Tunisia). *Environ. Earth Sci.* **2013**, *70*, 2087–2105. [CrossRef]
12. Anderson, M.P.; Woessner, W.W.; Hunt, R.J. *Applied Groundwater Modeling: Simulation of Flow and Advective Transport;* Academic Press: London, UK, 2015; ISBN 978-0-12-058103-0.
13. Kohn, K.E. Conceptualization and characterization of ground-water systems using Geographic Information Systems. *Eng. Geol.* **1996**, *42*, 111–118. [CrossRef]
14. Brassington, F.C.; Younger, P.L. A proposed framework for hydrogeological conceptual modelling. *Water Environ. J.* **2010**, *24*, 261–273. [CrossRef]
15. Hassen, I.; Gibson, H.; Hamzaoui-Azaza, F.; Negro, F.; Rachid, K.; Bouhlila, R. 3D geological modeling of the Kasserine Aquifer System, Central Tunisia: New insights into aquifer-geometry and interconnections for a better assessment of groundwater resources. *J. Hydrol.* **2016**, *539*, 223–236. [CrossRef]
16. Delogkos, E.; Childs, C.; Manzocchi, T.; Walsh, J.J. The nature and origin of bed-parallel slip in Kardia Mine, Ptolemais Basin, Greece. *J. Struct. Geol.* **2018**, *113*, 115–133. [CrossRef]
17. Koukouzas, N. Mineralogy and geochemistry of diatomite associated with lignite seams in the Komnina Lignite Basin, Ptolemais, Northern Greece. *Int. J. Coal Geol.* **2007**, *71*, 276–286. [CrossRef]
18. Van Vugt, N.; Steenbrink, J.; Langereis, C.G.; Hilgen, F.J.; Meulenkamp, J.E. Magnetostratigraphy-based astronomical tuning of the early Pliocene lacustrine sediments of Ptolemais (NW Greece) and bed-to-bed correlation with the marine record. *Earth Planet. Sci. Lett.* **1998**, *164*, 535–551. [CrossRef]
19. Steenbrink, J.; Van Vugt, N.; Hilgen, F.J.; Wijbrans, J.R.; Meulenkamp, J.E. Sedimentary cycles and volcanic ash beds in the lower Pliocene lacustrine succession of Ptolemais (NW Greece): Discrepancy between 40Ar/39Ar and astronomical ages. *Palaeogeogr. Palaeoclimatol. Palaeoecol.* **1999**, *152*, 283–303. [CrossRef]

20. Koukouzas, N.; Ward, C.; Li, Z. Mineralogy of lignites and associated strata in the Mavropigi field of the Ptolemais Basin, northern Greece. *Int. J. Coal Geol.* **2009**, *81*, 182–190. [CrossRef]
21. Koukouzas, C. *Geological Map of Greece, Ptolemais Sheet, Scale: 1:50.000*; IGME: Athens, Greece, 1997.
22. Roumpos, C.P.; Paraskevis, N.I.; Galetakis, M.J.; Michalakopoulos, T.N. Mineable Lignite Reserves Estimation in Continuous Surface Mining. In *Proceedings of the 12th International Symposium Continuous Surface Mining-Aachen 2014*; Niemann-Delius, C., Ed.; Lecture Notes in Production Engineering; Springer: Cham, Switzerland, 2015; pp. 177–194. [CrossRef]
23. Koukouzas, N.; Krassakis, P.; Koutsovitis, P.; Karkalis, C. An integrated approach to the coal deposits in the Mesohellenic Trough, Greece. *Bull. Geol. Soc. Greece* **2019**, *54*, 34–59. [CrossRef]
24. Jacobsen, K. DEM generation from satellite data. In Remote Sensing in Transition. In Proceedings of the 23rd EARSeL Symposium, Ghent, Belgium, 2–5 June 2003; pp. 513–525.
25. Rockware 2010, RockWorks 15. User's Manual, CO, USA, 312p. Available online: https://jaketa.hu/files/szoftverek/rockworks/RockWorks_15_Manual.pdf (accessed on 3 May 2022).
26. Environmental Systems Research Institute (ESRI). *ArcGIS Release*; 10.1.ESRI: Redlands, CA, USA, 2012.
27. Planchon, O.; Darboux, F. A fast, simple and versatile algorithm to fill the depressions of digital elevation models. *Catena* **2002**, *46*, 159–176. [CrossRef]
28. Tarboton, D.G.; Bras, R.L.; Rodriguez-Iturbe, I. On the extraction of channel networks from digital elevation data. *Hydrol. Processes* **1991**, *5*, 81–100. [CrossRef]
29. Krassakis, P.; Kazana, S.; Chen, F.; Koukouzas, N.; Parcharidis, I.; Lekkas, E. Detecting subsidence spatial risk distribution of ground deformation induced by urban hidden streams. *Geocarto Int.* **2019**, *36*, 622–639. [CrossRef]
30. Nomikou, P.; Krassakis, P.; Kazana, S.; Papanikolaou, D.; Koukouzas, N. The Volcanic Relief Within the Kos-Nisyros-Tilos Tectonic Graben at the Eastern Edge of the Aegean Volcanic Arc, Greece and Geohazard Implications. *Geosciences* **2021**, *11*, 231. [CrossRef]
31. Ebdon, D.; Blackwell, B. *Statistics in Geography*, 2nd ed.; Oxford: New York, NY, USA, 1985.
32. Mitchell, A. *The ESRI Guide to GIS Analysis*; ESRI Press: Redlands, CA, USA, 2005; Volume 2, ISBN 978-1589481169.
33. Gimond, M. GitHub Repository. 2019. Available online: https://mgimond.github.io/Spatial/index.html (accessed on 29 April 2022).
34. Moore, D.A.; Carpenter, T.E. Spatial analytical methods and geographic information systems: Use in health research and epidemiology. *Epidemiol. Rev.* **1999**, *21*, 143–161. [CrossRef]
35. Williams, B. *Assessment of Underground Coal Gasification in Bituminous Coals*; Final Report; Williams Bros. Engineering Co.: Tulsa, OK, USA, 1982.
36. Mastalerz, M.; Drobniak, A.; Parke, M.; Rupp, J.A. Site Evaluation of Subsidence Risk, Hydrology, and Characterization of Indiana Coals for Underground Coal Gasification (UCG), Indiana Geological Survey Report of Progress 45, 2013, Bloomington, Indiana. Available online: https://igws.indiana.edu/bookstore/details.cfm?Pub_Num=RP45# (accessed on 3 May 2022).
37. Shafirovich, E.; Mastalerz, M.; Rupp, J.; Varma, A. *The Potential for Underground Coal Gasification in Indiana; Phase 1 Report to CCTR. Slide Presentation*; Discovery Park Energy Centre, Purdue University: Indianapolis, IN, USA, 2008; Available online: https://www.purdue.edu/discoverypark/energy/assets/pdfs/cctr/researchReports/UCG-Phase1-08-31-08.pdf (accessed on 3 May 2022).
38. Beath, A. *Underground Coal Gasification Resource Utilisation Efficiency*; Technical Report; Elsevier: Amsterdam, The Netherlands, 2006.
39. Afzal, P.; Ahari, H.D.; Omran, N.R.; Aliyari, F. Delineation of gold mineralized zones using concentration–volume fractal model in Qolqoleh gold deposit, NW Iran. *Ore Geol. Rev.* **2013**, *55*, 125–133. [CrossRef]
40. Jerbi, H.; Massuel, S.; Leduc, C.; Tarhouni, J. Assessing groundwater storage in the Kairouan plain aquifer using a 3D lithology model (Central Tunisia). *Arab. J. Geosci.* **2018**, *11*, 236. [CrossRef]
41. Rockware. *Rockworks17 Software Help*; RockWare: Boulder, CO, USA, 2016.
42. Louloudis, G.; Stathopoulos, N.; Kasfikis, G. *Hydrogeological Report for Ptolemais Mines*; Technical Report; The Public Power Corporation: Athens, Greece, 2020. (In Greek)
43. Intergeo. *Execution of 9 Drillings, Permeability Tests and Chemical Analyzes of Soil and Water in the Area of the Mavropigi Mine in Ptolemaida*; Technical Report; Intergeo: Essen, Germany, 2009. (In Greek)

mining

MDPI

Review

Real Time Mining—A Review of Developments Within the Last Decade

Keyumars Anvari and Jörg Benndorf *

Faculty of Geosciences, Geoengineering and Mining, Institute of Mine Surveying and Geodesy, TU Bergakademie Freiberg, 09599 Freiberg, Germany; keyumars.anvari@doktorand.tu-freiberg.de
* Correspondence: joerg.benndorf@mabb.tu-freiberg.de

Abstract: Real-time mining (RTM) has become increasingly significant in response to the growing need for sustainable mineral resource extraction, driven by global population growth and technological progress. This innovative approach addresses critical challenges, such as declining ore grades, deeper and less accessible deposits, and rising energy costs, by integrating advanced online grade monitoring, data analysis, and process optimization. By employing real-time grade control, dynamic mine planning, and production optimization, it enhances the efficiency of resource extraction while minimizing environmental and social impacts. Originally proposed about a decade ago, RTM has gained attention for its potential to revolutionize the industry. This review examines recent advancements in closed-loop concepts, emphasizing the integration of advanced sensors and data analytics to enable continuous monitoring and adaptive decision making across the mining value chain. It highlights the role of online sensor technologies in providing high-resolution data for process optimization and evaluates various mining optimization techniques. The paper also explores data assimilation methods, such as Kalman filters and artificial intelligence (AI), showcasing their ability to continuously update models and reduce operational uncertainties. Ultimately, it proposes a comprehensive framework for adaptive, data-driven mining operations that promote sustainable development, enhance profitability, and improve decision-making capabilities.

Keywords: sustainable resource extraction; artificial intelligence; Kalman filters; advanced real-time sensors; production optimization; adaptive decision making; machine learning

1. Introduction

The global demand for mined resources has accelerated over the past decade, driven by population growth, technological advancements, and economic expansion By 2050, the demand for metals is projected to increase significantly, with aluminum rising by 215%, copper and nickel each by 140%, iron by 86%, zinc by 81%, and lead by 46% [1]. Metals such as copper, gold, zinc, and aluminum are essential for the renewable energy transition and achieving net-zero targets due to their critical role in decarbonization technologies, such as wind turbines, solar panels, and electric vehicles [2,3].

This escalating demand intensifies pressure on the mining sector, which faces challenges such as declining ore grades, deeper and more inaccessible deposits, and increasing energy costs. Moreover, mining activities often raise social and environmental concerns, leading to what is known as the "mineral resource dilemma". This dilemma highlights the tension between the necessity for mineral resources to support technological and societal advancement and the reluctance to endorse mining due to its adverse impacts [4,5]. Balancing these competing interests necessitates greater transparency and sustainable practices in mining operations. The industry must optimize resource extraction while minimizing environmental footprints to meet global demands without exacerbating ecological degradation.

Traditionally, the mining industry has followed a linear value chain, comprising sequential steps from exploration to mine closure. The process begins with exploration, where data is collected to define the spatial extent, geometry, and geotechnical and hydrological conditions of potential sites using methods such as remote sensing, geological mapping, geophysics, and drilling [6]. Resource modeling then creates a 3D orebody model, classifying ore, waste, and rock types to minimize ore loss and dilution [7–9]. Mine planning utilizes this data to design optimal extraction sequences, involving production control and planning to ensure consistent product quality through effective blending [10–12].

Despite its widespread use, this linear approach has significant drawbacks. It is often subjective, time consuming, and prone to inconsistencies, especially in complex geological settings. Uncertainties in mineral deposits can lead to production deviations from expectations, as resource modeling relies on uncertain data, causing planning inaccuracies [13–15]. Furthermore, the traditional fragmented approach involves specialized experts working independently with minimal communication, resulting in delayed feedback and issues that surface only during operations. This lack of integration and real-time communication hinders prompt adaptation and leads to suboptimal decisions. Integrating advanced technologies and real-time data can enhance data integration and cross-phase communication, improving modeling and planning efficiency while reducing operational uncertainty [6].

To address these challenges, the concept of real-time mining (RTM) emerged about a decade ago [16]. RTM aims to optimize resource extraction and minimize environmental impacts through continuous monitoring and immediate data analysis. By employing real-time data from advanced sensors and analytics, RTM enhances process efficiency, maximizes mineral recovery, and facilitates immediate adjustments based on the latest information [17,18]. The core idea is to create a closed-loop management system where operational data feeds back into decision-making processes in real time, enabling dynamic optimization of extraction, processing, and overall efficiency. In general, key constituents of the RTM approach are advanced real-time sensors, optimization algorithms, and data assimilation. Continuous grade control becomes possible, allowing for accurate ore characterization and immediate responses to changes in ore quality [6].

Recent years have seen the emergence of megatrends that have the potential to transform the mining industry. Digital disruption, characterized by the integration of digital technologies into all aspects of operations, is reshaping traditional practices. Advanced online sensors, Internet of Things (IoT) devices, and data analytics methods, such as artificial intelligence (AI) and geostatistics, enable the collection and analysis of vast amounts of real-time data [19–22]. These innovations improve predictive maintenance, process optimization, and decision making, enhancing efficiency, safety, and sustainability. AI-driven tools also support intelligent mining systems, enabling precise ore deposit modeling, real-time monitoring, geo-metallurgical modeling, and continuous grade control [7,23–26]. Collectively, these advancements improve resource utilization, reduce environmental impacts, and promote transparency in addressing the mineral resource dilemma.

Significant advances have been made in key constituents of the RTM approach over the past decade, particularly in sensor development and data analytics. Cutting-edge sensors provide near-continuous monitoring of critical indicators, such as ore grade and environmental contamination, enabling early risk detection and timely interventions [6,21,27–30]. Additionally, data analytics techniques, including AI, geostatistics, and Kalman filters, have enhanced data interpretation, facilitating real-time updates to grade control models, optimizing extraction processes, and reducing ore dilution [16,31–33].

Despite its potential, the full RTM concept remains largely theoretical, with limited practical implementation due to technological and operational barriers. Factors such as inadequate sensor designs for field applications, high initial costs, and the need to demonstrate practical utility have hindered widespread adoption. Integrating advanced sensors into the harsh and variable conditions of mining environments poses significant technical challenges that demand robust and durable designs [21]. The technology readiness level (TRL) framework, which ranges from basic principles (TRL 1) to fully operational systems (TRL 9), highlights the substantial gap between research and application. To date, no operational mine has fully integrated RTM approaches, underscoring the low TRL status of these technologies. This status emphasizes the considerable development and testing required to advance key constituents of the RTM approach toward practical deployment.

Additionally, developing real-time data processing algorithms capable of handling the volume and complexity of mining data in operational settings is still underway [7,34]. The absence of pilot projects and field demonstrations means that mining companies are hesitant to invest in key constituents of the RTM approach without clear evidence of their practical benefits and return on investment. Elevating key constituents of the RTM approach to higher TRL levels through collaborative efforts between researchers, technology developers, and industry stakeholders is essential for bridging the gap between concept and practical application.

The primary objective of this review paper is to evaluate the advancements in key constituents of the RTM approach over the past decade, identify the barriers preventing their practical implementation, and propose strategies to bridge the gap between research and industry application. By analyzing existing literature and technological developments, this paper aims to provide insights into transitioning RTM from theoretical frameworks to practical, large-scale operations that meet global demands and optimize ore extraction processes.

2. The RTM Concept

Over the past decade, the concept of RTM has emerged as a transformative approach in the mining industry, shifting from traditional discontinuous and intermittent monitoring systems to continuous process and quality management frameworks. RTM integrates a real-time feedback control loop that immediately links live data collected at the mining face, throughout material handling and processing, with an updatable sequential resource model (see Figure 1) [6]. This integration facilitates near-real-time optimization of long-term planning, short-term sequencing, and production control, providing critical insights to improve efficiency and reduce deviations from production targets.

A closed-loop mineral resource management (CLRM) system in RTM builds a model-based representation of the real system that is continuously updated as new data become available. It builds upon several core elements, as illustrated in Figure 2. These are sensor networks for production monitoring, system models, optimization algorithms for decision support, and data assimilation for real-time model update. Together, these interconnected elements provide a comprehensive, data-driven approach to optimizing mining operations in real time.

The "real system" in this context embodies the physical mining environment, which includes activities such as drilling, blasting, hauling, and processing that interact directly with the orebody. The effectiveness of these activities is influenced by control variables, such as equipment schedules and mine plans. The system is also influenced by inherent variability and uncertainty, which is depicted as "noise" in the figure.

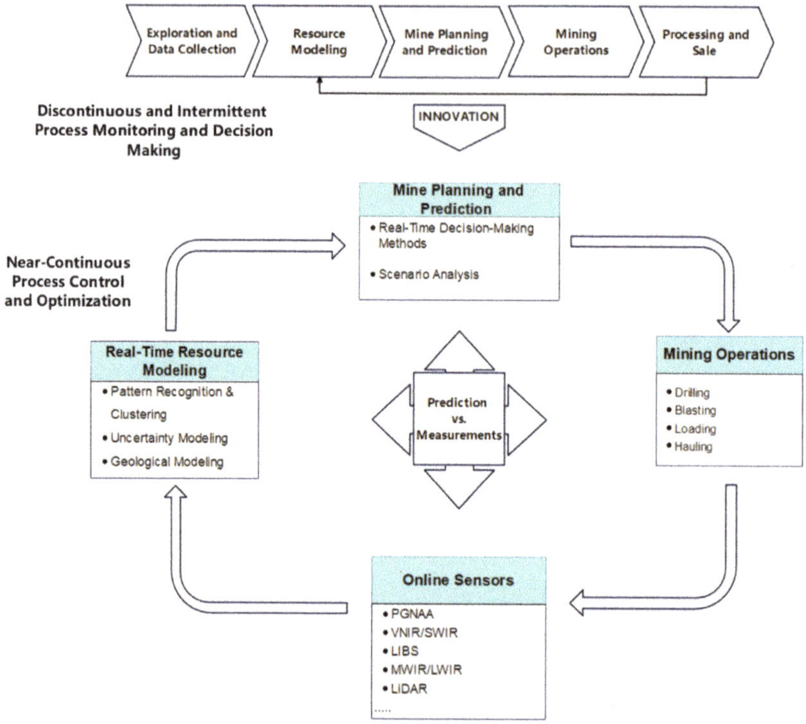

Figure 1. The concept of closed-loop management in RTM.

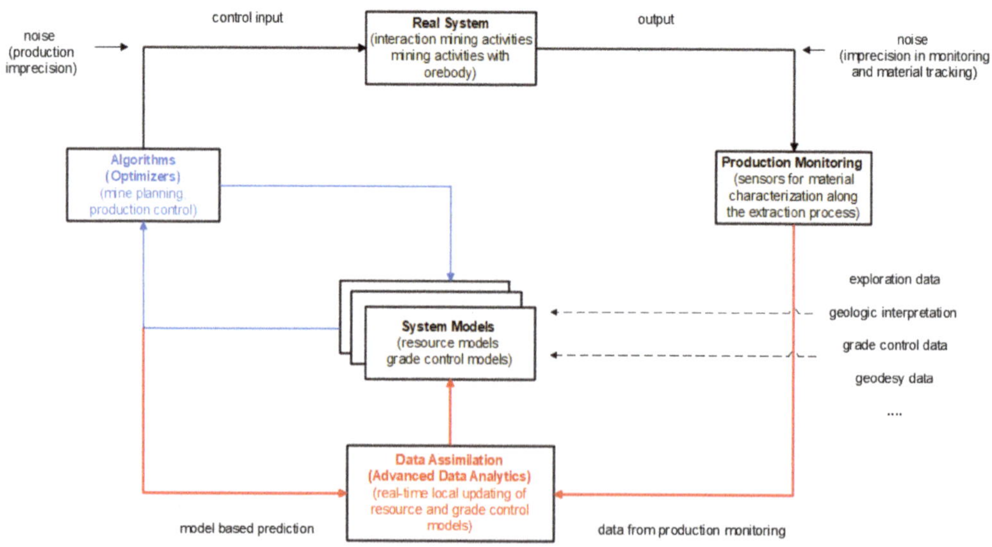

Figure 2. Key parts of closed-loop management system (adapted from after Jansen, et al. [35]).

Production monitoring involves sensors that collect real-time data during production. These sensors might measure ore grades, production rates, or equipment performance. Such monitoring can also include higher-level tasks, such as interpreting geological information or assessing grade control sampling. This process provides essential feedback about ore characteristics such as grade and hardness. The concept of "noise" here refers to measurement uncertainties and discrepancies in material tracking. These models are developed and continuously updated using data such as exploration data, geological interpretation, geodesy data, and grade control data.

The system models, including resource and grade control models, provide a mathematical or computational representation of the mining system. Resource models estimate the quantity and quality of mineral resources, while grade control models are focused on the quality of the ore being extracted. Due to the uncertainty inherent in subsurface geology, multiple models are typically created to provide estimates of mineral distribution and orebody characteristics and its uncertainty, using data, such as exploration results, geological assessments, and grade control measurements.

Within the closed-loop management system, optimization algorithms play a critical role in decision making. These algorithms generate efficient mine plans and production control strategies, optimizing parameters, such as equipment schedules and production targets. Optimizers work on different planning horizons, from short-term tasks, such as daily equipment scheduling, to long-term plans, such as designing pushbacks. The "blue loop" in the figure represents the continuous optimization process, where real-time data from production monitoring is used to refine mine planning and operations.

Data assimilation is the process of updating the system models to accurately reflect the performance of the real system. Through advanced data analytics, such as AI and geostatistics techniques, data assimilation reconciles discrepancies between model predictions and observed outcomes, thus enhancing model reliability. Whenever measured outputs deviate from model predictions, the model parameters are updated to reduce these discrepancies, thereby improving the predictive accuracy of the models. This process also helps manage the noise present in both control inputs and system outputs. The red loop in the figure represents this iterative updating of system models, which is informed by real-time production data.

The integration of these elements into a closed-loop mineral resource management system is designed to address uncertainties and improve operational outcomes. By combining real-time monitoring, system models, optimization algorithms, and data assimilation, the system ensures that physical operations are continuously observed, and the data gathered is used to enhance planning and operations. This closed-loop structure facilitates adaptive, data-driven decision making, which leads to increased resource recovery, reduced waste, and minimized operational risks.

3. Transformation in Mining Operations Through Sensors

Sensor technologies have transformed the mining industry by offering deep insights into material properties, providing rich qualitative, quantitative, and semi-quantitative data across various dimensions, such as geochemistry, mineralogy, and texture [21]. By operating within specific electromagnetic spectrums, sensors offer unique chemical information that is critical for improving mining operations. Additionally, integrating technologies such as belt scales and geometrical sensors further optimizes process efficiency and ensures operational accuracy.

Sensors enhance efficiency and accuracy across different stages of mining. During exploration, sensors reduce costs by minimizing the need for extensive physical sampling and costly laboratory tests. With real-time data, decisions can be made more swiftly, expediting the exploration process. The high-resolution data from sensors helps precisely determine ore grades and other crucial factors, thereby defining resources with greater accuracy.

In the production and grade control stage, sensors monitor ore extraction in real time, ensuring adherence to the production plan, minimizing deviations, and distinguishing accurately between ore and waste. This precision prevents costly errors, such as processing waste or discarding valuable ore, leading to improved operational efficiency and reduced equipment downtime.

In pre-sorting or pre-upgrading of materials along the production process, sensors play a key role in eliminating waste and low-grade material before reaching the processing plant, thereby simplifying the subsequent processes and making them more energy and cost efficient. In the post-stockpile and pre-processing stages, sensors further refine the material by optimizing feed quality, which helps operators increase recovery rates and control costs.

During post-processing and quality control, sensors provide real-time feedback on product quality, allowing immediate adjustments to meet customer specifications. This enhances product value, reduces penalties for off-spec materials, and ensures customer satisfaction.

By integrating sensors at each stage, mining operations are optimized for precision, efficiency, and cost effectiveness, ultimately improving both financial performance and product quality [20]. Table 1 highlights the different impacts of sensors on mining operations.

Table 1. Effects of sensors in mining operations [20].

Mine Stage	Sensor Effects	Advantages of Sensors
Exploration	Sensors replace physical sampling and provide real-time data for resource definition (e.g., grade, geometry).	• Reduce costs and time for sampling. • Provide immediate data for decision making. • Eliminate extensive need for offline laboratory analysis.
Production & Grade Control	Sensors monitor ore grade and material properties at the face to influence real-time extraction decisions.	• Improve compliance with mining plans. • Reduce misclassification of ore and waste. • Optimize production sequencing and equipment allocation.
Pre-sorting/ Pre-upgrading	Sensors sort material before crushing, controlling dispatch to specific destinations (e.g., waste or stockpile).	• Reduce material variability. • Increase product homogeneity. • Improve efficiency by removing waste early.
Post-stockpile & Pre-processing	Sensors further refine material by removing residual variability in grade and chemistry.	• Increase feed quality for processing plants. • Enhance process control. • Improve precision of material characteristics.
Post-processing & Quality Control	Sensors ensure final product quality by analyzing the material stream in real time.	• Maintain product specifications. • Provide immediate feedback for process adjustments. • Eliminate off-line quality control analysis.

In contemporary mining operations, the adoption of advanced sensor technologies has resulted in a rich data ecosystem, enabling continuous process monitoring and optimization. One example is measurement while drilling (MWD), an online data recording technology used to optimize blast patterns based on geomechanical properties of rock masses [36,37]. MWD data facilitates the classification of rock benches by blastability, allowing targeted blast optimization that enhances mine-to-mill (M2M) efficiency [38,39]. This M2M optimization is a holistic, data-driven approach that integrates drilling, blasting, and downstream processing operations to maximize overall mining efficiency and productivity [40–42].

Another advanced technology is prompt in situ neutron activation analysis (PGNAA), which serves as an elemental analyzer for real-time composition measurement of primary crushed material, significantly enhancing ore evaluation [43]. Fast neutrons from a source, such as Californium-252, interact with atomic nuclei in the mining medium, producing thermal neutrons that emit prompt in situ rays. The intensity and energy of these in situ rays are analyzed to determine the specific elements present, providing rapid, in situ analysis that eliminates the need for physical sampling and reduces errors [44].

Additionally, several emerging sensor technologies, including hyperspectral imaging, infrared technologies, Raman spectroscopy, laser-induced breakdown spectroscopy (LIBS), and light detection and ranging (LiDAR), are transforming raw material characterization across broad spatial and temporal scales [45].

Raman spectroscopy is particularly effective for identifying mineral compositions by detecting vibrational modes of molecules. For instance, it reveals microstructural features of carbonaceous materials through identifying the D (disorder) and G (graphitic) bands, which relate to the degree of crystallinity. Raman spectroscopy is advantageous as it requires minimal sample preparation and enables in-line characterization of materials [46]. The use of wide area illumination (WAI) probes enhances its accuracy in analyzing heterogeneous materials, making it suitable for industrial applications, including mineralogy [47].

LIBS is an advanced analytical technique used for the characterization of raw materials by providing elemental compositions based on plasma emission [48]. Similar to how hyperspectral sensors analyze a wide range of the electromagnetic spectrum for identifying minerals to generate a plasma from the surface of a sample. This plasma emits light at wavelengths unique to the elements within the sample, which are then collected and analyzed. The laser operates typically at 1064 nm, and the spectral range covered by LIBS usually spans the ultraviolet (UV) to visible regions, which includes most elements of the periodic table. LIBS is highly effective for rapid, in situ identification of both major and trace elements in raw materials. For instance, LIBS is capable of detecting elements, such as Si, Al, Fe, Ca, Na, and Mg, as well as trace elements, such as Li, which are difficult to detect using other portable instruments. The emitted light is collected and analyzed to provide real-time data that aids in material characterization. LIBS features several advantages, including the lack of sample preparation, making it particularly useful for field analysis and geological exploration. It also allows for micro-scale spatial analysis, depth profiling, and the detection of elements with low atomic numbers, which are key for understanding mineral compositions and geological histories [49].

In addition, LIBS provides the flexibility of working in challenging field environments due to its ability to operate in different atmospheric conditions, including open-path and stand-off configurations. These features enable it to characterize heterogeneous materials, such as geological samples and archaeological artifacts, by collecting elemental data from multiple spots on the surface and averaging them for comprehensive analysis [50].

Recent advancements in LIBS technology have enhanced its application in geology and mining, particularly for in situ and real-time elemental analysis of raw materials. Innovations, such as improved spectrometer designs and adaptive calibration techniques enable high-resolution characterization of heterogeneous geological samples and mineral ores. These developments address challenges such as matrix effects and enhance LIBS' accuracy and repeatability in harsh field conditions [51,52]. Furthermore, LIBS's high-resolution capabilities facilitate the detection of elements, including rare earth elements and toxic components, in complex samples, such as mining waste, solidifying its role in raw material characterization and monitoring [53].

The spectra and prompt gamma peaks shown in Figure 3 are entirely synthetic and were generated solely for illustrative purposes. In Figure 3A, a 532 nm Nd: YAG laser pulse produces a micro-plasma on the sample; the emitted light is collected through a 600 μm optical fiber and analyzed with an echelle spectrometer, yielding an exemplary single-shot LIBS spectrum with discrete elemental lines. Figure 3B depicts the PGNAA arrangement, where a DT neutron source irradiates ore on a moving conveyor and the resulting prompt γ-rays are detected by a $LaBr_3$ scintillator behind an HDPE shield, producing the illustrative multi-keV energy peaks presented.

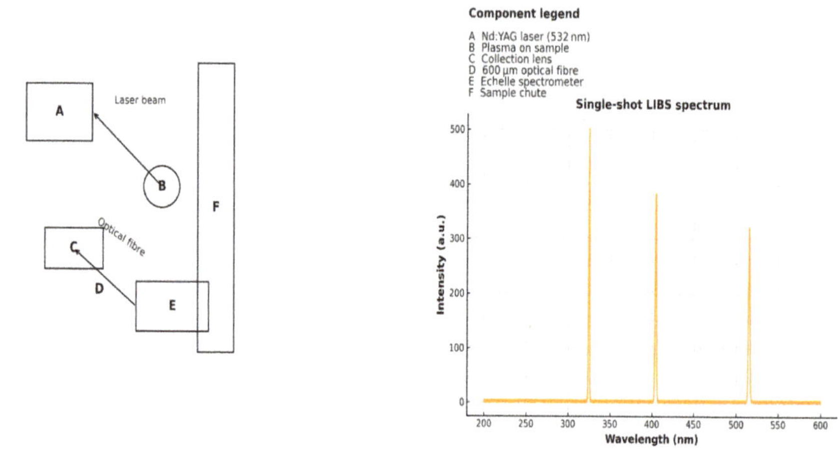

(**A**)

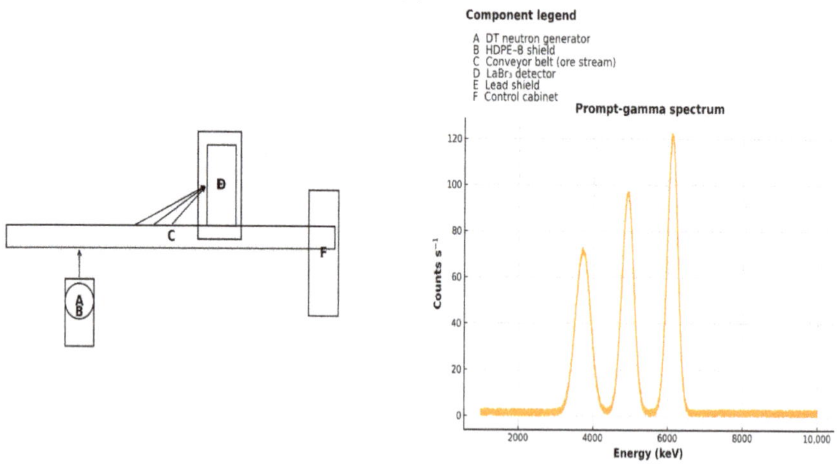

(**B**)

Figure 3. Schematic layouts and illustrative (synthetic) outputs of the two sensing technologies investigated. (**A**) Laser-induced breakdown spectroscopy (LIBS), (**B**) prompt gamma neutron activation analysis (PGNAA).

LiDAR is an active remote sensing technology used to characterize raw materials by providing detailed geometric and radiometric information through the use of laser pulses. LiDAR works by emitting laser pulses at different wavelengths and then detecting the reflected backscatter from target objects to obtain spatial information in the form of point clouds. It is capable of capturing 3D geometric information with high accuracy [54–56]. Multispectral LiDAR (MSL) systems combine both spectral and geometric data, which is particularly beneficial for characterizing complex materials in diverse conditions. MSL operates in multiple spectral channels, typically including visible to short-wave infrared (SWIR) wavelengths. These channels provide a combination of height and intensity data that enables precise classification of minerals and geological features based on their reflectance characteristics [57,58]. As illustrated in recent studies, MSL sensors can collect both spatial and spectral information, allowing for the efficient classification of rock surfaces and the identification of subtle mineral variations [59].

In addition to these advanced sensor technologies, several other sensors are widely used in mining, particularly in material characterization, resource evaluation, and decision making throughout the value chain. RGB sensors are frequently used for color detection and qualitative analysis in mineralogical applications. These sensors capture visible light in three bands (red, green, and blue) and are particularly useful for recognizing minerals with distinct colors and textures, making them valuable for mineral mapping and fragmentation analysis [60,61].

Another important technology is hyperspectral imaging (VNIR/SWIR), which captures data across a wide range of the electromagnetic spectrum. VNIR (visible-near infrared) covers the 0.4–1.0 μm range, and SWIR (short-wave infrared) covers the 1.0–2.5 μm range. These sensors enable precise mineral identification and sorting based on their spectral signatures. Specifically, VNIR is effective in distinguishing sulfide minerals, while SWIR is useful for detecting carbonates, clays, and sulfates. In addition, mid-wave infrared (MWIR) and long-wave infrared (LWIR) sensors, operating in the 2.5–7.0 μm (MWIR) and 7.0–15 μm (LWIR) ranges, respectively, are used for detecting and classifying minerals based on their reflectance or emission characteristics. LWIR is particularly effective for analyzing rock-forming minerals [62,63].

Electromagnetic and X-ray-based technologies, such as dual energy X-ray transmission (DE-XRT) and electromagnetic sensors, are also employed to detect mineral density and other physical properties [64]. These technologies are commonly used in pre-concentration processes to separate ore from waste, improving overall efficiency in mining operations [20,61]. The comparison of sensors can be seen in Table 2.

Table 2. Strengths, weaknesses, and opportunities of material characterization sensors in mining.

Sensor Technology	Application in Mining	Strengths	Weaknesses	Opportunities
Measurement While Drilling (MWD)	Blast optimization, rock mass blastability analysis	Provides real-time geomechanical data, enables M2M optimization	Limited to drilling data, high setup cost	Enhanced bench characterization, integration with M2M systems
Prompt Gamma Neutron Activation Analysis (PGNAA)	Elemental analysis of primary crushed ore	Real-time ore composition, high accuracy	Expensive, radiation safety concerns	In situ metal grade estimation, reduced sampling errors
Hyperspectral Imaging (VNIR/SWIR)	Mineral identification, ore–waste discrimination	High spectral resolution, rapid, non-contact	Sensitive to environmental conditions (dust, moisture)	Potential for in situ sorting, dynamic technology
Mid-Wave Infrared (MWIR)/Long-Wave Infrared (LWIR)	Ore–waste discrimination, mineral classification	Effective for rock analysis, high classification rates	Limited instrument development, weak spectral features	Automation potential, integration with chemometric tools
Raman Spectroscopy	Mineral identification	Detailed fingerprints, in situ handheld instruments	Fluorescence interference, limited elemental correlation	Real-time applications, expanded mineral libraries
Laser-Induced Breakdown Spectroscopy (LIBS)	Raw material characterization, elemental analysis	No sample prep, effective for major/trace elements	Limited detection range, affected by environmental factors	In situ exploration, micro-scale spatial analysis
Light Detection and Ranging (LiDAR)	Raw material geometry, mineral mapping	Provides 3D spatial data, operates in various conditions	Limited spectral data, expensive equipment	Multispectral LiDAR for mineral classification

Table 2. *Cont.*

Sensor Technology	Application in Mining	Strengths	Weaknesses	Opportunities
RGB Imaging	Mineral mapping, fragmentation analysis	Portable, rapid data processing, non-destructive	Limited to surface characteristics, affected by dust	Enhanced imaging, ruggedized systems
Electromagnetic Sensors	Mineral density detection, ore–waste separation	Effective for physical property measurement	Affected by mineral mixture complexity	Improved accuracy, expanded detection range
Dual Energy X-Ray Transmission (DE-XRT)	Pre-concentration, ore–waste separation	Accurate density-based discrimination	Limited to specific mineral densities	Application in automated sorting processes

Despite their potential, continuous advancements in sensor robustness and data fusion are essential to address challenges such as environmental interference and complex mineral compositions [20]. The convergence of advanced sensor technology and data analytics heralds a transformative shift in mining exploration and operational efficiency.

4. Mine Planning and Optimization

Mine planning is a comprehensive process that analyzes the economic conditions of mining operations using geological, structural, and mineralogical data collected during the exploration phase. The primary goal is to determine the optimal sequence for ore block extraction while accounting for blending and geometric constraints, with the objective of maximizing the net present value (NPV) of the operation. Effective mine planning is crucial to the performance of mining operations, as deviations from the planned schedule can significantly affect profitability. Typically, mine planning is divided into distinct phases: short-term and long-term planning. Each phase incorporates elements of production control and logistics, varying in detail and timeframe, but they are interdependent; plans for each phase must align with both immediate operational needs and overarching strategic objectives.

4.1. Long-Term Mine Planning: Strategic Production Scheduling and Ultimate Pit Definition

The aim of long-term mine planning is to establish an optimal production strategy for the entire lifespan of a mining operation. This involves analyzing extraction sequences and destination policies to optimize the operation's economic value. A resource block model characterizes rock types and mineral grades, enabling the identification of ultimate pit limits (UPL) and the development of an open-pit production schedule (OPS) for the entire project duration [65,66].

Traditional deterministic methods, such as the Lerchs–Grossmann algorithm, are widely used for defining ultimate pit limits and establishing life-of-mine schedules [67]. However, these methods do not account for geological uncertainty, which can lead to suboptimal planning outcomes. To address this limitation, stochastic approaches have been developed. For example, Albor Consuegra and Dimitrakopoulos [68] utilized simulated annealing to enhance pit limit determination, resulting in a 17% larger pit and a 10% higher net present value (NPV) compared to conventional methods. Another example includes the hybrid augmented Lagrangian relaxation method applied by Moosavi, et al. [69], which effectively solved large-scale long-term production scheduling problems, improving convergence speed and achieving better solutions compared to traditional linear methods.

4.2. Short-Term Mine Planning: Detailed Scheduling, Production Control, and RTM

Short-term mine planning focuses on developing a detailed production schedule covering a timeframe from several weeks to a few years. This level of planning involves equipment allocation, resource scheduling, and operational decisions to ensure production targets are met. It is operational in nature, optimizing detailed production sequences and allocating equipment such as trucks and shovels [12].

For example, Both and Dimitrakopoulos [70] presented a joint stochastic optimization model for short-term production scheduling and fleet management, simultaneously addressing uncertainties related to geological conditions, equipment performance, and operational logistics. Their method reduced shovel movement costs by 56% and improved truck allocation efficiency, demonstrating the value of integrated short-term planning approaches. Additionally, Mousavi et al. [71] and Mousavi et al. [72] compared metaheuristic algorithms, such as simulated annealing and Tabu search, for short-term open-pit block sequencing, demonstrating that hybrid approaches outperformed individual algorithms.

It is important to mention that the significance of short-term mine planning has increased due to its critical role in RTM operations. In RTM, dynamic adjustments are often required to deal with uncertainties inherent in mining activities, such as equipment breakdowns, geological variations, and market demand changes. The ability to adapt production schedules in real time ensures that the mine operates efficiently, minimizing production delays and optimizing resource utilization. Techniques such as simulation-based optimization and metaheuristic approaches enable planners to adjust schedules dynamically, incorporating new data and real-time feedback from the mine site [6].

4.3. Stochastic and Deterministic Approaches in Mine Planning

Mine planning can generally employ both deterministic and stochastic approaches. Stochastic methods account for geological uncertainties, resulting in more robust solutions. For instance, Goodfellow and Dimitrakopoulos [73] presented a two-stage stochastic optimization framework for mining complexes that integrates material extraction, blending, and transportation, leading to enhanced production target achievement and increased NPV compared to conventional deterministic methods. Similarly, Lamghari et al. [74] introduced a hybrid approach combining linear programming and variable neighborhood descent, which efficiently addressed the complexities of open-pit production scheduling, providing new best-known solutions for benchmark instances.

In contrast, traditional deterministic methods, such as the Lerchs–Grossmann algorithm, are foundational for defining ultimate pit limits and developing long-term schedules. However, they do not consider geological uncertainty, which can lead to suboptimal planning outcomes [67]. Recent approaches have also sought to combine deterministic and stochastic elements, such as the hybrid augmented Lagrangian relaxation combined with genetic algorithms, which has shown improved performance for large-scale long-term scheduling [69].

4.4. Digital Twin-Enabled Mine Planning and Optimization

Digital twins (DTs) introduce a cyber–physical replica of the mining system into the planning workflow. A DT is a multi-physics, multi-scale simulation that mirrors mine behavior while being continuously updated with field data [75]. The physical layer represents drilling, loading, haulage, processing and logistics, while the cyber layer links orebody models, process simulators and economic evaluators through a real-time data thread. Engineers can therefore test schedules in silico and receive immediate feedback on net present value, energy intensity and emissions.

Research and industrial reports published since 2020 show rapidly growing interest in DTs, a trend fueled by cheaper sensors, edge–cloud connectivity and machine-learning surrogates that cut the time needed for each simulation cycle, which place planning twins at the enterprise or process layer where geological, production and energy data converge and can be re-optimized whenever fresh measurements arrive [75]. Common enabling tools include LiDAR and UAV photogrammetry for geometry updates, plant historians for mass–balance reconciliation and data-driven models that replace slower finite-element codes during iterative pit design.

The connection with RTM is direct. RTM already streams grade, equipment and environmental data into a closed-loop optimization framework. Embedding a DT at the center of that loop turns raw measurements into foresight. Sensor feeds recalibrate geological and process sub-models; the up-to-date twin forecasts key performance indicators (KPIs) and uncertainty; optimizers then issue new long-, medium- and short-term mine planning that can be executed immediately. Planning, monitoring and control thus merge into a self-updating system capable of reacting to grade variability or energy-price shocks within minutes rather than weeks.

At the operational scale, integrated mine-to-mill twins have delivered throughput gains of 5 to 20 percent by jointly tuning blasting parameters, crusher settings and grinding-circuit controls [76]. These implementations underscore three key advantages of digital twins in mine planning and optimization: rapid, what-if evaluation under geological and market uncertainty, holistic KPI dashboards that connect strategic and operational layers, and a seamless hand-off between planning horizons as the twin self-updates.

4.5. Optimization Approaches

Beyond traditional deterministic and stochastic methods, advanced optimization algorithms have been increasingly applied to mine planning to tackle complex, large-scale problems. Particle swarm optimization (PSO) and evolutionary strategies (ES) are examples of nature-inspired algorithms that mimic biological processes to explore the solution space efficiently [77,78]. These algorithms are particularly effective in handling multi-objective optimization problems where trade-offs between different objectives, such as cost, NPV, and environmental impact, must be balanced.

Reinforcement learning, as part of AI methods, has recently emerged as a promising approach in mine planning optimization. Reinforcement learning algorithms learn optimal policies through interactions with the environment, enabling adaptive and dynamic decision-making processes. Reinforcement learning has gained significant attention in recent years for mining optimization applications [79–81].

Hybrid approaches, which combine multiple optimization techniques, have also shown significant promise. For instance, the integration of genetic algorithms (GA) with simulated annealing (SA) leverages the global search capability of GA and the local search efficiency of SA, resulting in superior performance in finding optimal or near-optimal solutions [71]. Similarly, combining linear programming (LP) with variable neighborhood descent (VND) [74] has proven effective in addressing the intricacies of open-pit production scheduling, yielding new best-known solutions for benchmarking.

In the following section, we discuss recent advancements in mining optimization techniques, which are summarized in Table 3.

Table 3. Advancements in mining optimization techniques over time.

Author(s)	Method	Case Study	Result	Advantages
Pendharkar and Rodger [82]	Nonlinear programming and genetic search	Coal mines (Illinois, Virginia, Pennsylvania)	Improved production scheduling under cost and geological constraints	Potential to enhance decision support for coal production and blending
Leite and Dimitrakopoulos [83]	Stochastic optimization, simulated annealing	Copper deposit	26% increase in NPV	Efficient scheduling with risk analysis; reduced likelihood of production target deviations compared to conventional methods
Askari-Nasab, Frimpong and Szymanski [67]	Discrete stochastic simulation	Iron ore deposit	$422 million NPV vs. $414 million (conventional)	Better optimization of pit limit using a stochastic production simulator, outperforming Lerchs–Grossmann algorithm
Albor Consuegra and Dimitrakopoulos [68]	Simulated annealing	Copper deposit	17% larger pit limits, 10% higher NPV	Integration of uncertainty to derive optimal pit limits and production schedule improvements
Sayadi, et al. [84]	Artificial neural network	Esfordi phosphate mine (Iran)	Generated pit with higher profit under impurity constraints	Improved pit limit classification and profitability compared to Lerchs–Grossmann algorithm
Moosavi, Gholamnejad, Ataee-Pour and Khorram [69]	Hybrid augmented Lagrangian relaxation and genetic algorithm	Long-term production scheduling	Better feasible solutions than traditional linearization method	Effective for large-scale problems; faster convergence than standard methods
Souza, et al. [85]	Hybrid heuristic (GRASP and GVNS)	Open-pit mining operational planning	Near-optimal solutions (less than 1% gap)	Dynamic truck allocation, minimized number of trucks while meeting production goals; efficient computing time
Mousavi, Kozan and Liu [72]	Hybrid branch-and-bound and simulated annealing	Short-term open-pit block sequencing	Optimized extraction sequences over short intervals	Efficient in solving short-term sequencing under machine capacity and precedence constraints
Mousavi, Kozan and Liu [71]	Comparative analysis of metaheuristics	Short-term open-pit sequencing	Hybrid TS-SA was superior to SA and TS	Enhanced performance and feasibility for real-scale open-pit problems
Goodfellow and Dimitrakopoulos [73]	Two-stage stochastic mixed integer nonlinear programming	Mining complexes	Improved production targets and NPV compared to deterministic methods	Joint optimization of extraction, blending, processing, and transportation under uncertainty

Table 3. *Cont.*

Author(s)	Method	Case Study	Result	Advantages
Both and Dimitrakopoulos [70]	Stochastic mixed integer programming	Real-world mining complex	56% reduction in shovel movement costs and 3.1% reduction in truck costs	Improved short-term production scheduling and fleet management by integrating shovel relocation, truck allocation, and uncertainty management
Shishvan and Benndorf [86]	Simulation-based optimization	Continuous mining system	Minimized idle time, improved material dispatching	Integrated deterministic optimization and stochastic simulation for adaptive material dispatching
Lamghari, Dimitrakopoulos and Ferland [74]	Hybrid method (linear programming and variable neighborhood descent)	Open-pit mine scheduling	New best-known solutions for benchmark instances	Efficient and superior in computational performance compared to recent literature methods
Lamghari and Dimitrakopoulos [87]	Hyper-heuristic with reinforcement learning and tabu search	Generic mining operations	Comparable or better results than state-of-the-art methods	Robust against problem-specific details, effective for complex scheduling involving multiple processing streams
LaRoche-Boisvert and Dimitrakopoulos [88]	Simultaneous stochastic optimization	Open-pit gold mining complex	Maximized NPV; SAG mill identified as bottleneck	Integrated production scheduling across three mines and stockpiles under supply uncertainty
Levinson and Dimitrakopoulos [89]	Stochastic programming and reinforcement learning	Copper mining complex	Jointly optimized short- and long-term schedules	Reduced risk of misalignment across timescales, enhanced operational feasibility
Levinson and Dimitrakopoulos [90]	Reinforcement learning and stochastic optimization	Copper mining complex	Improved production schedule using infill drilling data	Reduced uncertainty in production schedule, optimized drilling location selection

Mining optimization involves a complex interplay of data-driven decision-making processes aimed at maximizing economic returns while managing geological uncertainties and operational constraints. Short-term and long-term planning are interdependent phases that, when effectively synchronized, lead to significant improvements in production efficiency and economic value. By integrating mathematical, statistical, and AI-based optimization techniques, mining operations can address uncertainties, optimize resource allocation, and enhance profitability. The advancements presented demonstrate the critical role that innovative approaches play in achieving greater accuracy, reducing costs, and maximizing value in mining projects.

Building on these foundational concepts, RTM bridges short-term and long-term planning, offering substantial opportunities to revolutionize production control and logistics through continuous data collection and adaptive decision making. By integrating advanced analytics, real-time adjustments, and AI-driven optimization, RTM refines resource allocation, reduces risks, and boosts profitability. Despite its considerable promise in enhancing both production control and logistics, gaps in current research indicate ample scope for further exploration and development of RTM in mining contexts.

5. Data Assimilation in Closed-Loop RTM Management

The closed-loop concept in RTM involves continuously updating resource models by integrating real-time sensor measurements with predictive models, thus enhancing decision making and operational efficiency. Central to achieving this integration is the process of data assimilation, defined within the RTM framework as the recursive merging of new observations into the current resource model state. These observations may encompass additional spatial data types (spatial fusion) or arrive as sequential, time-stamped production updates, enabling the model to dynamically evolve and maintain accurate representations of the operational environment. Figure 4 presents a closed-loop workflow where sensor measurements from production monitoring are compared with model-based predictions to quantify discrepancies, thereby driving ongoing model updates. Data from exploration, geodesy, and grade control inform these models, accounting for ore deposit characteristics and the inherent uncertainties. Through continual iteration, sampling information and real-time sensor feedback align the estimated resource and grade control models with actual operational outcomes. The discrepancy quantification process ensures that any divergence from predictions is captured and integrated into subsequent model refinements. This adaptive feedback loop facilitates timely decision making for mine production and planning, ultimately optimizing operational strategies. By assimilating up-to-date information at each iteration, the system maintains robust, data-driven resource management.

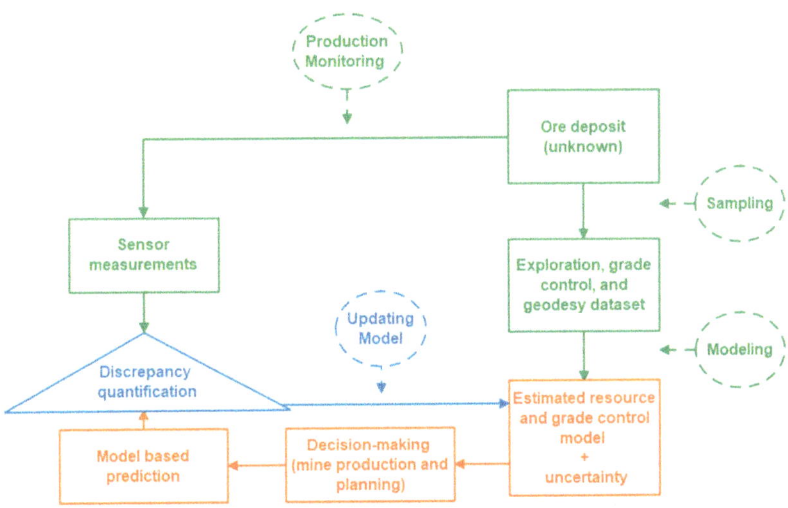

Figure 4. Closed-loop workflow for real-time resource and grade control management in mining operations (adapted from after Benndorf [6]).

5.1. Methods for Updating Data Assimilation in RTM Concept

Over the past decade, model-updating methodologies have advanced considerably, evolving from traditional linear Kalman filters to cutting-edge machine-learning approaches, such as reinforcement learning. This progression responds to the growing complexity and scale of RTM applications.

One of the most widely used techniques, the Kalman filter, offers a simple yet effective way to handle linear systems and parameter estimation. Its key limitation, however, lies in its reliance on linear dynamics and Gaussian assumptions, making it less suitable for capturing highly nonlinear relationships. Consequently, it performs best in scenarios where system updates are relatively moderate and largely linear in nature [16,91,92].

Another prevalent approach is the ensemble Kalman filter (EnKF), which accommodates moderate nonlinearities and sequential data assimilation better than the classic Kalman filter. Nevertheless, EnKF still depends on Gaussian assumptions, limiting its performance when variables deviate significantly from Gaussianity or when discontinuities are present. In complex or highly dynamic environments, hybridizing EnKF with other nonlinear methods can improve its robustness and assimilation accuracy [30,93,94].

Most recently, reinforcement learning has emerged as a particularly promising strategy for RTM, thanks to its ability to adapt to complex, high-dimensional datasets and rapidly changing mining conditions [32,95]. Although it demands extensive training data and robust computational resources, its capacity for real-time, adaptive decision making holds significant potential for optimizing operations in dynamic mining environments.

Moreover, other AI and machine-learning approaches, especially deep-learning models, such convolutional networks (e.g., 1-D, 2-D, 3-D CNNs) and recurrent networks (e.g., LSTM, GRU, Bidirectional LSTM, ConvLSTM (recurrent + convolutional)), as well as lightweight transformer encoders and transfer-learning pipelines, are now being deployed as learned surrogates within the red 'data-assimilation' loop (Figure 2); they ingest raw or minimally processed sensor streams, output probabilistic risk scores, and feed those to the EnKF or reinforcement learning policy, which then updates the dynamic resource state accordingly.

The diversity of data-assimilation approaches for RTM [31,32,96–99] illustrates a tension between simplicity and comprehensiveness. Although linear methods, such as the Kalman filter [16,92], have relatively low computational cost and can be easily implemented, they are typically effective only when the state-updating process remains close to linear. By contrast, EnKF-based approaches [30,93,100–102] and state of the art machine-learning methods, such as reinforcement learning [31,32], are better suited to severe nonlinearity and higher-dimensional problems. However, they demand significant computational resources and expertise. Thus, the all-important question is how much complexity (and, hence, computational overhead) can an RTM application withstand and still reach its project goals. Some practitioners exist on a spectrum of trade-offs: simpler algorithms are often adequate in relatively stable or mildly non-Gaussian conditions (more examples can be found in Table 4) while more complex hybrid and, most especially, reinforcement-learning-based methods are likely to become increasingly important for sharp dynamic, nonlinear, and high-dimensional mining challenges. The essential element for successful RTM is to find the sweet spot between computational tractability and accuracy within sense of the-RTM extent.

Table 4. Strengths, weaknesses, and opportunities of RTM methods.

Methodology	Strengths	Weaknesses	Opportunities
Kalman Filter	Simple and effective for linear systems; well-established methodology for parameter estimation	Assumes linearity and Gaussianity; struggles with nonlinear relationships	Use in systems with predominantly linear relationships and smaller-scale updates
EnKF	Widely used for sequential data assimilation; handles nonlinear problems reasonably well	Limited by Gaussian assumptions; can fail with strongly non-Gaussian or discontinuous variables	Hybridize with other nonlinear techniques for improved assimilation of complex datasets

Table 4. *Cont.*

Methodology	Strengths	Weaknesses	Opportunities
Normal-Score EnKF	Addresses non-Gaussian characteristics by transforming variables into Gaussian distributions	Transformation may not ensure joint multi-Gaussianity; requires multiple transformations	Use in settings with moderate non-Gaussianity where the linear update assumption mostly holds
Indicator-Based Data Assimilation	Suitable for non-Gaussian relationships; effective for transforming model parameters	Limited in handling very complex relationships; covariance-based association may be insufficient	Extend the approach to hybrid models using additional nonlinear statistics
EnKF with Compositional Data	Suitable for geometallurgical variables; uses transformations to maintain compositional consistency	Log-ratio transformations add complexity; requires careful handling of relationships between variables	Apply in compositional geoscience models where maintaining data consistency is crucial
MPS with EnKF	Maintains geological realism in facies models; handles complex spatial patterns effectively	Computationally expensive; requires careful calibration to maintain geological realism	Improve computational algorithms to allow for faster and more efficient assimilation in realistic models
MPS with Soft Data Integration	Enhances facies model calibration by incorporating soft data; maintains geological continuity	Sensitive to initial conditions; computationally intensive	Utilize for integrating additional soft data in highly uncertain geological environments
EnKF with P-Field Simulation	Enhances facies modeling by incorporating probability maps; better integrates geological information	Requires additional assimilation steps; increased computational requirements	Useful for complex geological settings where standard EnKF struggles to reproduce accurate facies
EnKF with TPG Models	Assimilates data while maintaining facies realism; handles categorical variables through truncation maps	Difficult to handle non-monotonic truncation maps; requires derivative adjustments for accurate data matching	Use in channel facies modeling where categorical boundaries are critical
EnKF with DCT	Helps reduce dimensionality and improves history matching for high-dimensional problems	Increases computational complexity; sensitive to non-Gaussian data distributions	Apply for history matching in high-dimensional geological models where data compression is critical
SEOD with EnKF	Provides optimized sampling strategies and enhances model parameter estimation	Time-consuming optimization; computational overhead when dealing with complex geological data	Combine with simpler methods to balance accuracy and computational efficiency
EnKF with DWT	Helps reduce geological boundary uncertainty; enhances real-time data assimilation	High computational requirement; depends on quality of initial realizations	Extend application for sensor-based geological boundary identification
Reinforcement Learning (DDPG)	Learns adaptively in complex environments; able to handle dynamic, high-dimensional datasets	Requires extensive training data; computationally intensive and relies on strong computational resources	Apply in dynamic mining environments where adaptive decision making is beneficial
Actor-Critic Reinforcement Learning	Learns adaptively in dynamic environments; can self-optimize based on incoming data	Computationally heavy; requires large datasets for effective learning	Use in industrial-scale mining for optimizing resource extraction and adaptive decision making

This methodological evolution is illustrated further in Table 4, as it begins with simple linear algorithms (Kalman filter) and works its way up to state-of-the-art machine-learning techniques (reinforcement learning). Simpler approaches perform well for linear or mildly non-Gaussian systems, whereas more complex algorithms are better suited to dynamic, high-dimensional, or nonlinear problems, albeit with increased computational cost. Therefore, selecting the most appropriate method typically involves balancing computational efficiency with the level of accuracy required for reliable decision making in RTM.

5.2. Applications of Updating Data Assimilation in RTM Concept

Several studies have demonstrated the potential of closed-loop data assimilation for RTM. For instance, Benndorf [16] introduced a closed-loop framework employing sequential resource model updating via the Kalman filter. By using synthetic data (representing a fully known environment), this approach showcased significant improvements in prediction accuracy whenever sensor-derived data were assimilated. The mean squared error (MSE) between estimated and actual block values served as the primary evaluation metric, showing substantial reductions in uncertainty, even in scenarios that involved multiple extraction sites and blending.

The updating approach uses the Kalman filter to estimate unknown state parameters recursively based on differences between predictions and observations. Let $Z(x)$ represent a spatial random field with elements $Z(x_i)$, where $i = 1, \ldots, n$ denotes the index of discrete extraction locations. The production matrix A describes the contribution of each mining block to the total production during a specific time interval. The relationship between the predicted resource model and sensor measurements can be expressed as (Equation (1)):

$$z_t^* = Az^*(x_t) \tag{1}$$

where z_t^* is the model-based prediction of extracted material for time intervals. Kalman filter updating can then be represented as (Equation (2)):

$$z^*(x)_{t+1} = z^*(x)_t + K(y_t - Az^*(x)_t) \tag{2}$$

where K is the Kalman gain matrix, determined by Equation (3):

$$K = C_{t,t}A^T\left(AC_{t,t}A^T + C_{v,v}\right)^{-1} \tag{3}$$

where $C_{t,t}$ represents the covariance of the prior model, while $C_{v,v}$ is the covariance of the measurement noise. The Kalman gain controls how much the observed differences (or innovations) between predicted and measured values influence the updated resource model. If the measurement noise is low and the prior model error is high, the Kalman gain increases, resulting in significant updates to the resource model.

Building upon this foundation, Yüksel, Thielemann, Wambeke and Benndorf [94] implemented a closed-loop concept for real-time resource model updating in the Garzweiler lignite mine, focusing on coal quality control. By continuously integrating online ash content measurements from the KOLA system into the model, they achieved up to a 70% reduction in model uncertainty. The EnKF gain determined the weight assigned to new sensor data versus the existing model, guiding the sequential updating process.

Extending this work, Yüksel, Benndorf, Lindig and Lohsträter [93] applied the closed-loop concept to a lignite mining case with multiple production benches. Online ash content measurements were taken using a radiometric sensor on the central conveyor belt, measuring blended material from multiple excavators. The forward simulation technique was used to associate sensor measurements with respective blocks, considering travel time from different excavators. Significant improvements were reported, with an average absolute error (AE) reduction of up to 73% using the updated model, where AE is defined as (Equation (4)):

$$AE = \frac{1}{n}\sum_{i=0}^{n}|l_i - z^*(x)_i| \tag{4}$$

where l_i represents the sensor measurements and $z^*(x)_i$ are the updated model values. Significant improvements were reported, with an average error reduction of up to 73% when using the updated model compared to the prior model.

Furthermore, the study compared two approaches for generating prior models: one based on geostatistical simulation using drill hole data and the other based on a simplified short-term model, which introduced fluctuations around the company's mining plan. The simplified approach offered comparable improvements in model predictions, demonstrating the practicality of using a less computationally intensive method for RTM applications. By reducing the computational burden, the short-term model approach facilitated easier operational implementation without the need for complex geostatistical simulations.

Moreover, Wambeke and Benndorf [29] expanded the closed-loop concept through the development of geostatistical models for grade control reconciliation, utilizing the EnKF to handle the complexities involved in large-scale mining operations. They developed a sequential estimation method for updating the grade control (GC) model in real time based on new sensor observations. The algorithm is designed to integrate blended observations from multiple extraction points, making it particularly useful in mining environments where material streams are combined before measurements are taken.

The forward simulation step is crucial for estimating the contribution of individual blocks to blended material streams. This allows for linking each measurement back to its constituent blocks, despite blending from different extraction locations. Unlike previous methods, their forward simulator does not require an analytical formulation of the observation model, enhancing practicality in complex mining scenarios.

Subsequently, Wambeke and Benndorf [30] analyzed the influence of system parameters, such as measurement volume, blending ratios, and sensor precision on algorithm performance. Conducting 125 experiments, they explored how these factors affect the EnKF-based grade control model updating. The blending ratio defines the proportion of material in the blended measurement volume originating from different extraction zones, while the measurement volume represents the amount of material characterized by the sensor over a given time interval.

The updated GC model significantly improved the alignment of predicted versus actual production data. Specifically, the reduction in root-mean-square deviation ($RMSE$) between predicted and measured values was used to quantify improvements by Equation (5):

$$RMSE_t = \sqrt{\frac{1}{N}\sum_{n=1}^{N}(z^*(n) - E[z_t(n)])^2} \tag{5}$$

where $z^*(n)$ represents the true state, and $E[z_t(n)]$ is the expected value of the estimated block values at time t. By incorporating new sensor observations sequentially, the model progressively reduced uncertainty, resulting in improvements of up to 74% in some scenarios.

Lan, et al. [103] proposed a sequential ensemble-based optimal design (SEOD) method to enhance real-time parameter estimation in groundwater reactive transport models, which has direct parallels to closed-loop concepts in RTM. They employed the SEOD method in both one-dimensional and two-dimensional groundwater models to jointly estimate hydraulic conductivity and geochemical parameters. The results showed significant reductions in RMSE and ensemble spread, indicating the effectiveness of the method in reducing uncertainty and improving model predictions.

The SEOD method utilizes the EnKF to estimate both physical and geochemical parameters, continuously updating the system model based on optimal sensor data collection. This is similar to previous approaches in RTM, where sensor measurements and forward simulations are integrated to update resource models, aiming to reduce uncertainties in model predictions.

The closed-loop concept, as described in this study, involves an iterative process to refine the resource model in real time by assimilating sensor-based observations. The SEOD framework addresses some of the inherent challenges in such systems, including the optimization of sensor placement to maximize information gain, using the Kullback–Leibler divergence (also known as relative entropy, RE) as a metric for determining the value of information obtained from measurements.

Mathematically, the SEOD framework is implemented through the following stages.

1. Forecast Step: The EnKF-based SEOD begins with a forecast step, where the forward model G is used to propagate the current ensemble of state vectors $x^a_{i,j}$ to the next time step (Equation (6)):

$$x^f_{i,j+1} = G\left(x^a_{i,j}\right) \tag{6}$$

where x^f and x^a represent the forecast and analysis states, respectively.

2. Optimal Sampling Design: The SEOD framework aims to determine the optimal locations for collecting new measurements, which is achieved by maximizing the relative entropy (*RE*) between the previous and posterior distributions. For a Gaussian distribution, *RE* is given by Equation (7):

$$RE = J_b + \left[Indet\left(BA^{-1}\right) + Tr\left(AB^{-1}\right) - n\right]/2 \tag{7}$$

where J_b is the signal part of *RE*, and *A* and *B* are the covariance matrices of prior and posterior distributions, respectively. The optimal sampling locations are determined using a genetic algorithm (GA) by solving the following optimization problem (Equation (8)):

$$H_{opt} = argmaxRE(H) \tag{8}$$

where *H* represents the candidate sampling strategy.

3. Analysis Step: After determining the optimal sampling strategy, the analysis step is performed to assimilate the collected data and update the state vector by Equation (9):

$$x^a_{i,j} = x^f_{\daleth,j+1} + C_{YD}(C_{DD} + C_D)^{-1}(d_{obs} - d_i) \tag{9}$$

where C_{YD} is the cross-covariance between forecast states and predicted data, C_{DD} is the covariance of the predicted data, and C_D represents the measurement error covariance.

The SEOD method, thus, represents a dynamic approach to integrating measurement data for the purpose of parameter estimation. This approach shares similarities with RTM practices where data assimilation techniques, such as EnKF, are used to refine geological models and optimize production processes. A critical difference, however, lies in the use of an optimal sampling strategy to enhance the quality of parameter estimation, leading to more efficient computational performance and better model accuracy compared to traditional methods.

Several studies addressed the limitations of the EnKF in handling non-Gaussian distributions and maintaining geological realism. Nejadi, et al. [104] incorporated the EnKF with probability field (P-Field) simulation to improve characterization of facies boundaries in reservoir models. Jafarpour and McLaughlin [101] combined the EnKF with the discrete cosine transform (DCT) parameterization to reduce dimensionality and preserve geological features. Hu, Zhao, Liu, Scheepens and Bouchard [100] applied a closed-loop concept by combining multipoint simulation (MPS) with the EnKF for real-time reservoir history matching, maintaining geological consistency in facies models. Oliver and Chen [105] discussed data assimilation in truncated plurigaussian (TPG) models, addressing non-monotonic relationships between latent variables and facies types.

For preserving non-Gaussian geological features, Kumar and Srinivasan [97] introduced indicator-based data assimilation (InDA), preserving non-Gaussian characteristics of geological models during assimilation. Ma and Jafarpour [98] integrated soft data into MPS simulation for facies model calibration, improving consistency with training images and observed data.

Zhou, et al. [106] transformed the original state vector into a univariate Gaussian using the normal-score EnKF (NS-EnKF), preserving non-Gaussian distributions in the updated models, crucial for maintaining geological realism. This method is particularly useful in RTM, where maintaining the complex spatial features of geological formations, such as channels or facies boundaries, is essential for accurate model updates. The closed-loop approach presented in this study demonstrates how transforming state variables into Gaussian space can improve data assimilation in non-Gaussian contexts, ultimately enhancing the quality of real-time geological models and reducing uncertainty in decision-making processes.

In smaller-scale grade heterogeneity studies, Li, Sepúlveda, Xu and Dowd [92] presented a rapid updating method to predict grade heterogeneity at smaller scales using a Kalman filter framework, integrating production data for near real-time resource model downscaling. The method improved mining selectivity and ore recoverability by accurately predicting grade heterogeneity within larger mining blocks. The advantages of the proposed method include its ability to integrate different types of sensed data without requiring complex co-kriging or simulation techniques, as well as its computational efficiency.

Prior, Benndorf and Mueller [102] proposed a closed-loop framework for updating resource and grade control models in underground mining, integrating sensor-based information using the EnKF. The approach improved prediction accuracy for ore grade and vein thickness, even with sparse initial conditioning information.

Previously, Tolosana-Delgado, van den Boogaart and Benndorf [33] introduced a closed-loop updating framework for compositional geometallurgical variables, employing log-ratio transformations and flow anamorphosis to handle compositional data in the EnKF while preserving their relationships.

Machine-learning approaches have also gained traction. Avalos, et al. [107] applied machine-learning and deep learning techniques in RTM to forecast energy consumption in semi-autogenous grinding mills. By leveraging recurrent neural networks to model temporal dependencies, they achieved significant improvements in prediction accuracy, enabling dynamic energy management and operational efficiency in real-time settings. Ortiz, et al. [108] highlighted the integration of geometallurgical data and process models within a RTM framework to optimize the mining value chain. By addressing uncertainties through stochastic modeling, the study demonstrated enhanced mine designs and operational strategies for improved economic and environmental outcomes.

Kumar and Dimitrakopoulos [32] proposed a novel approach to updating geostatistically simulated models of mineral deposits in real time using a reinforcement learning framework. They developed a self-learning algorithm based on the deep deterministic policy gradient (DDPG) reinforcement learning method, incorporating actor and critic agents. The purpose of this approach is to learn from incoming data collected during mining operations and to update the geostatistical models accordingly. This approach allows the model to account for high-order spatial statistics, enabling more accurate representation of the mineral deposits' geological structure and grade distribution.

The proposed method applies reinforcement learning where the mining grid is sequentially visited in a random path. The environment provides the state for each grid node, which includes the properties of blocks, sensor data, conditioning data events, and other contextual information. The actor agent takes actions that predict updated properties for the grid nodes, and these predictions are evaluated by the critic agent. This setup allows for a dynamic model that learns and adjusts based on new incoming information.

Mathematically, the reward at each step is calculated to ensure that the updated properties align with both high-order spatial statistics and the incoming temporal sensor data. The actor–critic system is designed such that the state at time step is updated by incorporating new sensor data and conditioning events to maximize the reward function, which includes terms for spatial and temporal consistency.

Talesh Hosseini, Asghari, Benndorf and Emery [99] integrated the discrete wavelet transform (DWT) with the EnKF to update and improve geological boundary definitions at the Golgohar Iron Ore Mine, significantly increasing boundary accuracy while maintaining consistency with production data.

Finally, de Carvalho and Dimitrakopoulos [31] presented an actor–critic reinforcement learning approach for short-term production planning and fleet management in mining complexes. By defining shovel and truck allocation policies adaptively, the model addressed operational constraints and uncertainties dynamically, resulting in a 47% improvement in cash flow compared to traditional fixed fleet assignments. The continuous update of orebody models based on sensor-collected data further emphasized the practical advantages of integrating RL for adaptive decision making. The performance of these studies are shown in Table 5.

Table 5. Summary of studies on closed-loop data assimilation.

Author(s)	Method	Case Study	Result	Advantages
Jafarpour and McLaughlin [101]	EnKF with DCT	Synthetic Reservoirs	Reduced dimensionality, preserved geological realism	Reduced computational cost; better preservation of channel connectivity
Zhou, Gomez-Hernandez, Franssen and Li [106]	Normal-Score EnKF	Synthetic Bimodal Aquifer	Preserved non-Gaussian distributions, improved characterization	Preserved complex spatial features; enhanced quality of real-time geological models
Hu, Zhao, Liu, Scheepens and Bouchard [100]	MPS with EnKF	Fluvial Reservoir	Maintained geological consistency, effective history matching	Maintained geological features; effective in real-time history matching
Benndorf [16]	Kalman Filter	Synthetic Data	Reduced MSE, improved prediction accuracy	Improved decision making by incorporating real-time sensor data; reduced uncertainty even with blending

Table 5. *Cont.*

Author(s)	Method	Case Study	Result	Advantages
Nejadi, Trivedi and Leung [104]	EnKF with P-Field Simulation	Reservoir Models	Improved facies boundary characterization	Preserves statistical properties; prevents overfitting; ensures ensemble diversity
Yüksel, Thielemann, Wambeke and Benndorf [94]	EnKF	Garzweiler Lignite Mine	Up to 70% reduction in model uncertainty	Direct incorporation of real-time measurements; better decision making and quality control
Yüksel, Benndorf, Lindig and Lohsträter [93]	EnKF	Lignite Mining with Multiple Benches	Error reduction up to 73%	Practical and less computationally intensive; improved model predictions
Wambeke and Benndorf [29]	EnKF	Grade Control Reconciliation	Enhanced accuracy, practical for complex scenarios	Improved operational efficiency; reduced uncertainties; better resource extraction
Wambeke and Benndorf [30]	EnKF	Analysis of System Parameters	RMSE reduction up to 74%, improved model alignment	Mitigated discrepancies; enhanced production processes
Lan, Shi, Jiang, Sun and Wu [103]	SEOD with EnKF	Groundwater Models	Reduced uncertainty, improved predictions	Optimal sensor data collection; efficient computational performance; better model accuracy
Oliver and Chen [105]	EnKF with TPG Models	Synthetic Reservoir	Improved data match, reduced uncertainty	Handled nonlinearity and non-monotonic relationships; robust in complex geological environments
Kumar and Srinivasan [97]	Indicator-Based Data Assimilation	Synthetic Reservoir	Preserved non-Gaussian distributions, accurate spatial features	Overcomes EnKF limitations; maintains geological features; accurate in non-Gaussian contexts
Ma and Jafarpour [98]	MPS with Soft Data Integration	Facies Model Calibration	Improved consistency with training images and data	Maintains geological consistency; integrates soft data; effective framework
Li, Sepúlveda, Xu and Dowd [92]	Kalman Filter	Synthetic Dataset	Improved model accuracy, better mining selectivity	Integrates sensed data without complex methods; computational efficiency; practical for rapid decision making
Prior, Benndorf and Mueller [102]	EnKF	Underground Mining (Reiche-Zeche)	Improved prediction accuracy for ore grade and vein thickness	Improved mining selectivity; effective even with sparse initial data

Table 5. *Cont.*

Author(s)	Method	Case Study	Result	Advantages
Prior, Tolosana-Delgado, van den Boogaart and Benndorf [33]	EnKF with Compositional Data	Bauxite Deposit	Accurate updates, preserved compositional characteristics	Reduced uncertainty; improved accuracy for decision making
Kumar and Dimitrakopoulos [32]	Reinforcement Learning (DDPG)	Synthetic Dataset	Dynamic model updates, accounted for high-order statistics	Integrates new information dynamically; suitable for complex mining operations
Talesh Hosseini, Asghari, Benndorf and Emery [99]	EnKF with DWT	Golgohar Iron Ore Mine	Improved geological boundary accuracy, high compatibility	Enhances block model quality control; reduces spatial uncertainty; preserves statistical parameters
de Carvalho and Dimitrakopoulos [31]	Actor-Critic Reinforcement Learning	Copper Mining Complex	47% improvement in cash flow	Dynamic fleet allocation and production scheduling based on real-time data

These studies demonstrate significant advancements in data assimilation methods within closed-loop RTM management. The integration of Kalman filters and their variants with innovative techniques, such as DCT, MPS, and reinforcement learning approaches, has led to improved prediction accuracy, reduced uncertainty, and maintained geological realism. The table summarizes these advancements, highlighting the methods employed, case studies undertaken, and the results achieved. Collectively, these works emphasize the practical effectiveness of continuous model updating in modern mining practices, underscoring its value in enhancing operational decision making and efficiency across diverse mining environments.

6. Bridging RTM Implementation Gaps with Advanced Data Analytics

Much progress has been made over the past decade; however, fully integrated RTM concepts remain largely unimplemented in practical, real-world settings. A key tool for assessing the maturity of RTM approaches is the technology readiness level (TRL) framework. Originally developed by NASA, the TRL framework is a metric that ranges from TRL 1, where basic principles are observed, to TRL 9, where systems are fully operational. The framework has been adopted by the European Union and many industries to evaluate innovation across a wide range of sectors, linking technological development to funding and implementation milestones [109–111]. When applied to RTM components, this framework reveals a large gap between research and its practical application, demonstrating the need for strategic pathways to move these technologies toward higher TRLs.

In the context of RTM, the maturity levels of key components are as follows: the real system, comprising physical mining activities, such as drilling, hauling, and processing, stands at TRL 7, reflecting successful prototype demonstration in operational environments but still requiring fully integrated automation and data workflows. Production monitoring, which involves advanced sensor networks that measure ore grades and equipment performance, is at TRL 8, with systems routinely deployed and thoroughly tested, though consistent data standardization remains a challenge. By contrast, system models (e.g., digital twins and resource models) currently reach only TRL 5, as they have been validated in relevant mining environments but are not yet fully operational on a large scale. This is primarily due to the fact that many mines operate with incomplete or inconsistent databases.

Optimization algorithms, providing real-time decision support, stay at TRL 4, with most testing confined to laboratory or pilot-scale settings and lacking robust deployment in live production due to the need for specialized data scientists and IT infrastructure on site. Data assimilation, which updates system models in near real time by blending sensor data and historical records, reaches TRL 6, having been demonstrated in pilot environments but not widely adopted for continuous industrial use. Taken together, these elements form an overall RTM system whose readiness is estimated at around TRL 6, owing primarily to challenges such as a lack of consistent geodata infrastructure, limited availability of geo data scientists in operations, and interoperability complexities that inhibit seamless integration across sensors, models, and algorithms (see Figure 5). Overcoming these hurdles through standardized data protocols, expanded on-site analytics, and interdisciplinary collaboration is essential for reaching TRL 8–9 and achieving a fully operational, field-proven RTM system.

Although key constituents of RTM, ranging from online sensors and data-assimilation schemes to optimization engines and digital-twin platforms, have advanced markedly in laboratory and pilot studies, their collective deployment is still feasible only in narrowly defined scenarios, such as lignite strip mines with continuous conveyor haulage and in highly selective underground operations (e.g., cut-and-fill or sub-level stoping), that already possess ore-tracking infrastructure and relatively homogeneous mineralogy.

RTM adds value when three factors align, namely when the orebody shows meaningful short-range grade variability, the operation uses a continuous shovel–crusher–belt chain in an open pit or a selective, short-cycle underground stoping method, and rugged online sensors feed time-stamped data into haulage and plant databases for rapid model updates [94,102]. If variability is low or digital infrastructure is limited, as in block caving or room-and-pillar coal, RTM offers little economic benefit.

RTM in underground mines is aimed at tactical, short-range optimizations within the existing stoping plan, such as trimming stope boundaries, real-time muck routing and on-belt blend control, rather than wholesale changes to equipment or mining method. These adjustments occur inside the normal drill–blast–haul cycle and therefore do not require costly redesigns or resequencing of developments.

Most conveyor-mounted PGNAA/LIBS grade sensors have demonstrated reliable operation at TRL 8, whereas multimodal face-imaging and LiDAR-fusion units that must survive blasting cycles remain at TRL 5 to 6. Early pilots, most notably the Garzweiler lignite mine (about 70% reduction in block-model uncertainty) on surface [94] and the Reiche-Zeche vein mine (sequential grade-model updating during cut-and-fill) underground [102], confirm the technical soundness of the approach but also highlight four inter-dependent constraints that still block routine adoption: (i) ruggedized, multimodal sensors with verified calibrations across moisture, dust and vibration regimes; (ii) high-fidelity, time-stamped material-tracking systems absent from many legacy mines; (iii) AI-driven optimization engines that require dense labelled data streams and on-site computing rarely available outside flagship demonstrations; and (iv) persistent interoperability gaps among geological databases, process-control historians and enterprise-resource-planning platforms. Until standardized data backbones, edge-to-cloud analytics pipelines and cross-disciplinary human expertise are co-developed at scale, RTM will remain confined to operations whose geological simplicity and digital maturity can tolerate the present limitations, so tackling these systemic hurdles, rather than isolated sensor or algorithm refinements, must be central to any realistic roadmap for closing the RTM implementation gap.

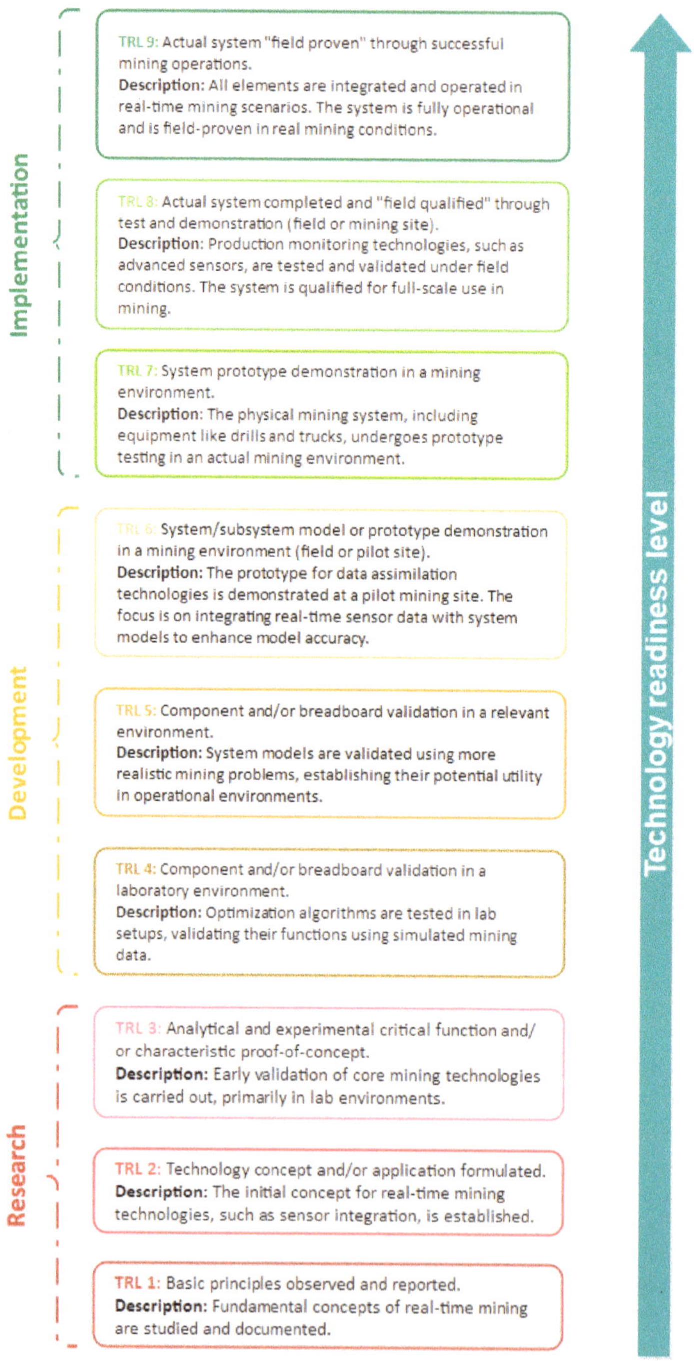

Figure 5. TRLs for RTM from research to full-scale implementation.

To advance RTM components toward higher TRLs, strategic collaboration and targeted innovations are paramount. Advanced data analytics and AI are particularly indispensable in enhancing the adaptability, scalability, and operational efficiency of RTM. One foundational requirement involves the development of robust, field-ready sensors capable of withstanding the challenging conditions inherent to mining environments. Partnering with sensor manufacturers can accelerate this process. Moreover, leveraging machine learning, especially reinforcement learning, can bolster predictive capabilities while enabling real-time operational adjustments. These improvements are further supported by digital twins, which dynamically simulate and optimize mining processes.

Integrating AI into optimization algorithms offers real-time decision support by rapidly analyzing and assimilating large volumes of sensor data. Such integration not only expedites technological development but also ensures operational reliability through proactive equipment failure prevention and efficiency optimization. However, these gains hinge on establishing comprehensive material tracking systems and maintaining consistent geodatabases. The absence of such infrastructure, coupled with a lack of on-site geodata specialists, poses a significant bottleneck to effective RTM implementation.

Moving forward, RTM methodologies stand to benefit from hybrid models that merge AI with geostatistical techniques, thereby increasing both adaptability and precision. Advanced computational platforms, including cloud computing and parallel processing, can alleviate current computational bottlenecks. Additionally, the inclusion of diverse data sources, such as soft data and advanced sensor networks, promises to enhance model resolution and accuracy. By harnessing these advancements, RTM can solidify its role as a cornerstone of intelligent mining practices, driving efficient, adaptive, and data-driven resource management.

Advancing key constituents of the RTM approach requires coordinated efforts among researchers, technology developers, and industry stakeholders. Establishing pilot projects and field demonstrations will showcase practical benefits and return on investment, encouraging adoption. Economic incentives and funding mechanisms can offset initial costs, while standardizing protocols ensures compatibility and ease of integration. By leveraging advanced data analytics and AI, the mining industry can bridge the implementation gaps in RTM, advancing key components to higher TRLs. This progression will transform RTM from a theoretical concept into a practical, efficient, and sustainable operational system, effectively addressing the mineral resource dilemma and meeting global demands.

7. Conclusions

Real-time mining (RTM) now stands at the threshold between proof-of-concept pilots and routine industrial deployment; most subsystems have been validated only to about technology readiness levels (TRLs) 5–6. The remaining practical gaps include ruggedized multimodal sensors, interoperable data backbones, and on-site analytics talent able to convert torrents of geo-environmental data into second-by-second decisions. Closing these gaps demands an explicitly interdisciplinary alliance in which geoscientists, mineral-processing engineers, computer scientists, and social researchers co-design solutions so that grade-control algorithms mesh smoothly with drill-and-blast schedules, labor-safety dashboards, and community air-quality monitors. This holistic integration reframes RTM as more than a productivity tool; it becomes a platform for broad socio-economic benefit, delivering higher resource efficiency, lower energy intensity, transparent supply chains, and an enhanced social license that protects jobs in mining-dependent regions while moderating ecological footprints.

To lift RTM to TRL 8–9, we call for a coordinated policy–industry research agenda. Regulators can accelerate progress by mandating open data standards, embedding near-real-time environmental reporting in permitting schemes, and underwriting demonstration mines that de-risk early adoption. Mining companies should organize pre-competitive consortia with sensor manufacturers, artificial intelligence (AI) providers, recycling firms, and local communities to pilot end-to-end digital value chains and share lessons learned. Immediate research priorities are physics-informed machine learning that fuses sparse geological, geotechnical, and socio-environmental streams; resilient edge–cloud architectures that guarantee sub-second feedback despite harsh connectivity constraints; circular-economy metrics that couple RTM outputs with full life-cycle assessments; and governance models that protect data sovereignty while distributing productivity gains equitably. Progress on these fronts will close current readiness gaps and position RTM as a pivotal lever for meeting the European Critical Raw Materials Act, global net-zero goals, and the United Nations Sustainable Development Goals.

Author Contributions: Conceptualization, K.A. and J.B.; Methodology, K.A. and J.B.; Validation, K.A. and J.B.; Formal Analysis, K.A. and J.B.; Investigation, K.A.; Resources, J.B.; Data Curation, K.A.; Writing—Original Draft Preparation, K.A.; Writing—Review and Editing, K.A. and J.B.; Visualization, K.A.; Supervision, J.B.; Project Administration, J.B. All authors have read and agreed to the published version of the manuscript.

Funding: This research was funded by the Saxonian State Scholarship Program (Sächsische Landesstipendienverordnung, SächsLStipVO), awarded by the Rectorate Commission "Graduiertenförder-ung" at the Technical University Bergakademie Freiberg.

Data Availability Statement: This review article does not involve the generation of new datasets. All data discussed and referenced are derived from previously published sources, which are appropriately cited throughout the paper.

Conflicts of Interest: The authors declare no conflicts of interest.

Abbreviations

The following abbreviations are used in this manuscript:

AE	Absolute Error
AI	Artificial Intelligence
CLRM	Closed-Loop Mineral Resource Management
DCT	Discrete Cosine Transform
DE-XRT	Dual Energy X-Ray Transmission
DTs	Digital twins
DDPG	Deep Deterministic Policy Gradient
EnKF	Ensemble Kalman Filter
GA	Genetic Algorithm
GC	Grade Control
InDA	Indicator-Based Data Assimilation
IoT	Internet of Things
LIBS	Laser-Induced Breakdown Spectroscopy
LiDAR	Light Detection and Ranging
LWIR	Long-Wave Infrared
M2M	Mine-to-Mill
MWD	Measurement While Drilling
MWIR	Mid-Wave Infrared
NS-EnKF	Normal-Score Ensemble Kalman Filter
NPV	Net Present Value
OPS	Open-Pit Production Schedule
P-Field	Probability Field (in EnKF/P-Field simulation)
PGNAA	Prompt Gamma Neutron Activation Analysis
PSO	Particle Swarm Optimization

KPIs	Key Performance Indicators
RL	Reinforcement Learning
RMSE	Root-Mean-Square Error
RTM	Real-Time Mining
SEOD	Sequential Ensemble-Based Optimal Design
SWIR	Short-Wave Infrared
TPG	Truncated Plurigaussian
TRL	Technology Readiness Level
VNIR	Visible-Near Infrared
WAI	Wide Area Illumination

References

1. Watari, T.; Nansai, K.; Nakajima, K. Major metals demand, supply, and environmental impacts to 2100: A critical review. *Resour. Conserv. Recycl.* **2021**, *164*, 105107. [CrossRef]
2. ICA. *Copper—The Pathway to Net Zero*; International Copper Association: Washington, DC, USA, 2023; p. 41.
3. Pickens, N.; Joannides, E.; Laul, B. *Red Metal, Green Demand—Copper's Critical Role in Achieving Net Zero*; Wood Mackenzie: Edinburgh, UK, 2022.
4. Franken, G.; Schütte, P. Current trends in addressing environmental and social risks in mining and mineral supply chains by regulatory and voluntary approaches. *Miner. Econ.* **2022**, *35*, 653–671. [CrossRef]
5. Yu, H.; Zahidi, I.; Fai, C.M.; Liang, D.; Madsen, D.Ø. Elevating community well-being in mining areas: The proposal of the mining area sustainability index (MASI). *Environ. Sci. Eur.* **2024**, *36*, 71. [CrossRef]
6. Benndorf, J. *Closed Loop Management in Mineral Resource Extraction: Turning Online Geo-Data into Mining Intelligence*; Springer Nature: Cham, Switzerland, 2020.
7. Anvari, K.; Mousavi, A.; Sayadi, A.R.; Sellers, E.; Salmi, E.F. Automatic detection of rock boundaries using a hybrid recurrence quantification analysis and machine learning techniques. *Bull. Eng. Geol. Environ.* **2022**, *81*, 398. [CrossRef]
8. Masoumi, I.; Kamali, G.; Asghari, O.; Emery, X. Assessing the impact of geologic contact dilution in ore/waste classification in the Gol-Gohar Iron Ore Mine, Southeastern Iran. *Minerals* **2020**, *10*, 336. [CrossRef]
9. Aghahadi, M.H.; Jozanikohan, G.; Asghari, O.; Hosseini, S.T.; Emery, X.; Rezaei, M. Geochemical anomaly separation based on geology, geostatistics, compositional data and local singularity analyses: A case study from the Kuh Panj copper deposit, Iran. *Appl. Geochem.* **2024**, *173*, 106135. [CrossRef]
10. Matthäus, A.; Graham, P.; Soares, A. *Real-Time Optimization of Extraction and the Logistic Process in Highly Complex Geological and Selective Mining Settings*; European Commission: Brussels, Belgium, 2017.
11. Kumar, A. *Artificial Intelligence Algorithms for Real-Time Production Planning with Incoming New Information in Mining Complexes*; McGill University: Montreal, Canada, 2020.
12. Blom, M.; Pearce, A.R.; Stuckey, P.J. Short-term planning for open pit mines: A review. *Int. J. Min. Reclam. Environ.* **2019**, *33*, 318–339. [CrossRef]
13. Dominy, S.C.; Noppé, M.A.; Annels, A.E. Errors and uncertainty in mineral resource and ore reserve estimation: The importance of getting it right. *Explor. Min. Geol.* **2002**, *11*, 77–98. [CrossRef]
14. Lindi, O.T.; Aladejare, A.E.; Ozoji, T.M.; Ranta, J.-P. Uncertainty Quantification in Mineral Resource Estimation. *Nat. Resour. Res.* **2024**, *33*, 2503–2526. [CrossRef]
15. Ullah, G.W.; Nehring, M.; Kizil, M.; Knights, P. Environmental, Social, and Governance Considerations in Production Scheduling Optimisation for Sublevel Stoping Mining Operations: A Review of Relevant Works and Future Directions. *Min. Metall. Explor.* **2023**, *40*, 2167–2182. [CrossRef]
16. Benndorf, J. Making use of online production data: Sequential updating of mineral resource models. *Math. Geosci.* **2015**, *47*, 547–563. [CrossRef]
17. Sánchez, F.; Hartlieb, P. Innovation in the mining industry: Technological trends and a case study of the challenges of disruptive innovation. *Min. Metall. Explor.* **2020**, *37*, 1385–1399. [CrossRef]
18. Young, A.; Rogers, P. A review of digital transformation in mining. *Min. Metall. Explor.* **2019**, *36*, 683–699. [CrossRef]
19. Aznar-Sánchez, J.A.; Velasco-Muñoz, J.F.; Belmonte-Ureña, L.J.; Manzano-Agugliaro, F. Innovation and technology for sustainable mining activity: A worldwide research assessment. *J. Clean. Prod.* **2019**, *221*, 38–54. [CrossRef]
20. Buxton, M.; Benndorf, J. The use of sensor derived data in optimization along the Mine-Value-Chain. In Proceedings of the 15th International ISM Congress, Aachen, Germany, 16–20 September 2013; pp. 324–336.
21. Desta, F. Sensing and Data Fusion Opportunities for Raw Material Characterisation in Mining: Technology and Data-Driven Approach. Ph.D. Thesis, Delft University of Technology, Delft, The Netherlands, 2021.

22. Lessard, J.; de Bakker, J.; McHugh, L. Development of ore sorting and its impact on mineral processing economics. *Miner. Eng.* **2014**, *65*, 88–97. [CrossRef]
23. Corrigan, C.C.; Ikonnikova, S.A. A review of the use of AI in the mining industry: Insights and ethical considerations for multi-objective optimization. *Extr. Ind. Soc.* **2024**, *17*, 101440. [CrossRef]
24. Jackson, J. Enabling Short Interval Control Using Spatial Dense Geoscience Info Beyond Grade. 2020. Available online: https://smi.uq.edu.au/event/session/6543 (accessed on 18 November 2024).
25. Anvari, K.; Mousavi, A.; Sayadi, A.R.; Afzal, P.; Sellers, E.; Salmi, E.F. A Hybrid Recurrence Analysis and Wavelet Transformation for Rock Boundary Identification. Preprint 2024. *Preprint* **2024**. [CrossRef]
26. Aghahadi, M.H.; Jozanikohan, G.; Asghari, O.; Anvari, K.; Hosseini, S.T. Spatial Clustering of Primary Geochemical Halos Using Unsupervised Machine Learning in Sari Gunay Gold Deposit, Iran. *Min. Metall. Explor.* **2024**, *41*, 2741–2760. [CrossRef]
27. Dalm, M. Sensor-Based Sorting Opportunities for Hydrothermal Ore Deposits: Raw Material Beneficiation in Mining. Ph.D. Thesis, Delft University of Technology, Delft, The Netherlands, 2018.
28. Guatame-Garcia, A.; Buxton, M. Framework for monitoring and control of the production of calcined kaolin. *Minerals* **2020**, *10*, 403. [CrossRef]
29. Wambeke, T.; Benndorf, J. A simulation-based geostatistical approach to real-time reconciliation of the grade control model. *Math. Geosci.* **2017**, *49*, 1–37. [CrossRef]
30. Wambeke, T.; Benndorf, J. A study of the influence of measurement volume, blending ratios and sensor precision on real-time reconciliation of grade control models. *Math. Geosci.* **2018**, *50*, 801–826. [CrossRef] [PubMed]
31. de Carvalho, J.P.; Dimitrakopoulos, R. Integrating short-term stochastic production planning updating with mining fleet management in industrial mining complexes: An actor-critic reinforcement learning approach. *Appl. Intell.* **2023**, *53*, 23179–23202. [CrossRef]
32. Kumar, A.; Dimitrakopoulos, R. Updating geostatistically simulated models of mineral deposits in real-time with incoming new information using actor-critic reinforcement learning. *Comput. Geosci.* **2022**, *158*, 104962. [CrossRef]
33. Prior, Á.; Tolosana-Delgado, R.; van den Boogaart, K.G.; Benndorf, J. Resource model updating for compositional geometallurgical variables. *Math. Geosci.* **2021**, *53*, 945–968. [CrossRef]
34. Bisschoff, R.; Grobbelaar, S. Evaluation of data-driven decision-making implementation in the mining industry. *S. Afr. J. Ind. Eng.* **2022**, *33*, 218–232. [CrossRef]
35. Jansen, J.-D.; Douma, S.; Brouwer, D.R.; Van den Hof, P.; Bosgra, O.; Heemink, A. Closed-loop reservoir management. In Proceedings of the SPE Reservoir Simulation Conference, The Woodlands, TX, USA, 2–4 February 2009. SPE-119098-MS.
36. Babaei Khorzoughi, M.; Hall, R.; Apel, D. Rock fracture density characterization using measurement while drilling (MWD) techniques. *Int. J. Min. Sci. Technol.* **2018**, *28*, 859–864. [CrossRef]
37. Babaei Khorzoughi, M.; Hall, R. Processing of measurement while drilling data for rock mass characterization. *Int. J. Min. Sci. Technol.* **2016**, *26*, 989–994. [CrossRef]
38. Sellers, E.J.; Salmi, E.F.; Usami, K.; Greyvensteyn, I.; Mousavi, A. Detailed Rock Mass Characterization—A Prerequisite for Successful Differential Blast Design. In Proceedings of the YSRM2019—The 5th ISRM Young Scholars' Symposium on Rock Mechanics and REIF2019—International Symposium on Rock Engineering for Innovative Future—An ISRM Specialized Conference, Okinawa, Japan, 1–4 December 2019.
39. Usami, K.; Salmi, E.F.; Sellers, E.J. Genetic Algorithm to Optimise Rock Fragmentation in Selective Blasting for Grade Engineering. In Proceedings of the 26th International Mining Congress and Exhibition of Turkey, Belek, Antalya, Turkey, 16–19 April 2019; pp. 359–372.
40. McKee, D. *Understanding Mine to Mill*; The Cooperative Research Centre for Optimising Resource Extraction: Brisbane, Australia, 2013; Volume 96, p. 2013.
41. Salmi, E.; Mousavi, A.; Usami, K.; Sellers, E. Value chain optimization for grade engineering by mine-wide frag-mentation optimisation. In *Rock Fragmentation by Blasting*; Wang, X.G., Ed.; CRC Press: London, UK, 2022.
42. Salmi, E.F.; Sellers, E.J. A review of the methods to incorporate the geological and geotechnical characteristics of rock masses in blastability assessments for selective blast design. *Eng. Geol.* **2021**, *281*, 105970. [CrossRef]
43. Kurth, H. Measuring ore quality using real-time on-belt elemental analysis—More than grade control. In Proceedings of the International Mining Geology Conference, Hobart, Australia, 20–22 September 2017; pp. 20–22.
44. Charbucinski, J.; Malos, J.; Rojc, A.; Smith, C. Prompt gamma neutron activation analysis method and instrumentation for copper grade estimation in large diameter blast holes. *Appl. Radiat. Isot.* **2003**, *59*, 197–203. [CrossRef]
45. Clarkson, L.; Williams, D. Critical review of tailings dam monitoring best practice. *Int. J. Min. Reclam. Environ.* **2020**, *34*, 119–148. [CrossRef]
46. Jawhari, T.; Roid, A.; Casado, J. Raman spectroscopic characterization of some commercially available carbon black materials. *Carbon* **1995**, *33*, 1561–1565. [CrossRef]

47. Lintvedt, T.A.; Andersen, P.V.; Afseth, N.K.; Wold, J.P. In-line Raman spectroscopy for characterization of an industrial poultry raw material stream. *Talanta* **2024**, *266*, 125079. [CrossRef] [PubMed]
48. Dalm, M.; Sandtke, M. Geochemical mapping of drill core samples using a combined LIBS and XRF core scanning system. In Proceedings of the Real-Time Mining Conference, Freiberg, Germany, 26–27 March 2019; pp. 26–27.
49. Lawley, C.J.; Somers, A.M.; Kjarsgaard, B.A. Rapid geochemical imaging of rocks and minerals with handheld laser induced breakdown spectroscopy (LIBS). *J. Geochem. Explor.* **2021**, *222*, 106694. [CrossRef]
50. Díaz Pace, D.M.; Gabriele, N.A.; Garcimuño, M.; D'Angelo, C.A.; Bertuccelli, G.; Bertuccelli, D. Analysis of minerals and rocks by laser-induced breakdown spectroscopy. *Spectrosc. Lett.* **2011**, *44*, 399–411. [CrossRef]
51. Guo, L.-B.; Zhang, D.; Sun, L.-X.; Yao, S.-C.; Zhang, L.; Wang, Z.-Z.; Wang, Q.-Q.; Ding, H.-B.; Lu, Y.; Hou, Z.-Y. Development in the application of laser-induced breakdown spectroscopy in recent years: A review. *Front. Phys.* **2021**, *16*, 22500. [CrossRef]
52. Palleschi, V. Laser-induced breakdown spectroscopy: Principles of the technique and future trends. *ChemTexts* **2020**, *6*, 18. [CrossRef]
53. Benndorf, J.; Restrepo, D.A.; Merkel, N.; John, A.; Buxton, M.; Guatame-Garcia, A.; Dalm, M.; de Waard, B.; Flores, H.; Möllerherm, S. TRIM4Post-Mining: Transition Information Modelling for Attractive Post-Mining Landscapes—A Conceptual Framework. *Mining* **2022**, *2*, 248–277. [CrossRef]
54. Andreas Gunke, S.J. Derivation of Mine from correlates ALS, TLS and RGB Data. In *19. Geokinematischer TU Bergakademie Freiberg*; Department of Mine Surveying and Geodesy: Freiberg, Germany, 2018; pp. 79–86.
55. Jiang, C.; Chen, Y.; Tian, W.; Feng, Z.; Li, W.; Zhou, C.; Shao, H.; Puttonen, E.; Hyyppä, J. A practical method utilizing multi-spectral LiDAR to aid points cloud matching in SLAM. *Satell. Navig.* **2020**, *1*, 29. [CrossRef]
56. Martienßen, T.; Geier, A.; Wand, R. *Untersuchungen Zur Entfernungsmessung Terrestrischer Long-Range Laserscanner*; Technische Universitaet Bergakademie Freiberg Universitaetsbibliothek "Georgius Agricola": Freiberg, Germany, 2016.
57. Donner, R.; Rabel, M.; Scholl, I.; Ferrein, A.; Donner, M.; Geier, A.; John, A.; Köhler, C.; Varga, S. *Die Extraktion Bergbaulich Relevanter Merkmale Aus 3d-Punktwolken Eines Untertagetauglichen Mobilen Multisensorsystems*; Institutionelles Repositorium der Leibniz Universität Hannover: Hannover, Germany, 2019.
58. Köhler, C.; Donner, M.; Donner, R. *Semantische Klassifizierung von 3D-Punktwolken*; TU Bergakademie Freiberg: Freiberg, Germany, 2018.
59. Chen, Y.; Jiang, C.; Hyyppä, J.; Qiu, S.; Wang, Z.; Tian, M.; Li, W.; Puttonen, E.; Zhou, H.; Feng, Z. Feasibility study of ore classification using active hyperspectral LiDAR. *IEEE Geosci. Remote Sens. Lett.* **2018**, *15*, 1785–1789. [CrossRef]
60. Naseri, A.; Rezaei Nasab, A. Automatic identification of minerals in thin sections using image processing. *J. Ambient Intell. Humaniz. Comput.* **2023**, *14*, 3369–3381. [CrossRef]
61. Desta, F.; Buxton, M. Evaluation of sensor technologies for on-line raw material characterization in "Reiche Zeche" underground mine-outcomes of RTM implementation. In Proceedings of the 2019 REAL TIME MINING-Conference on Innovation on Raw Material Extraction, Freiberg, Germany, 26–27 March 2019; pp. 32–47.
62. Riley, D.N.; Hecker, C.A. Mineral mapping with airborne hyperspectral thermal infrared remote sensing at Cuprite, Nevada, USA. In *Thermal Infrared Remote Sensing: Sensors, Methods, Applications*; Springer: Dordrecht, The Netherlands, 2013; pp. 495–514.
63. Prabhavathy, P.; Tripathy, B.; Venkatesan, M. Unsupervised learning method for mineral identification from hyperspectral data. In *Innovations in Bio-Inspired Computing and Applications, Proceedings of the 10th International Conference on Innovations in Bio-Inspired Computing and Applications (IBICA 2019), Gunupur, Odisha, India, 16–18 December 2019*; Springer: Cham, Switzerland, 2021; Volume 10, pp. 148–160.
64. Neubert, K.; Wotruba, H. Investigations on the detectability of rare-earth minerals using dual-energy X-ray transmission sorting. *J. Sustain. Metall.* **2017**, *3*, 3–12. [CrossRef]
65. Noriega, R.; Pourrahimian, Y. A systematic review of artificial intelligence and data-driven approaches in strategic open-pit mine planning. *Resour. Policy* **2022**, *77*, 102727. [CrossRef]
66. Whittle, D. Open-pit planning and design. In *SME Mining Engineering Handbook*; Society for Mining, Metallurgy, and Exploration: Englewood, NJ, USA, 2011; Volume 1, pp. 877–901.
67. Askari-Nasab, H.; Frimpong, S.; Szymanski, J. Modelling open pit dynamics using discrete simulation. *Int. J. Min. Reclam. Environ.* **2007**, *21*, 35–49. [CrossRef]
68. Albor Consuegra, F.; Dimitrakopoulos, R. Stochastic mine design optimisation based on simulated annealing: Pit limits, production schedules, multiple orebody scenarios and sensitivity analysis. *Min. Technol.* **2009**, *118*, 79–90. [CrossRef]
69. Moosavi, E.; Gholamnejad, J.; Ataee-Pour, M.; Khorram, E. A hybrid augmented Lagrangian multiplier method for the open pit mines long-term production scheduling problem optimization. *J. Min. Sci.* **2014**, *50*, 1047–1060. [CrossRef]
70. Both, C.; Dimitrakopoulos, R. Joint stochastic short-term production scheduling and fleet management optimization for mining complexes. *Optim. Eng.* **2020**, *21*, 1717–1743. [CrossRef]
71. Mousavi, A.; Kozan, E.; Liu, S.Q. Comparative analysis of three metaheuristics for short-term open pit block sequencing. *J. Heuristics* **2016**, *22*, 301–329. [CrossRef]

72. Mousavi, A.; Kozan, E.; Liu, S.Q. Open-pit block sequencing optimization: A mathematical model and solution technique. *Eng. Optim.* **2016**, *48*, 1932–1950. [CrossRef]
73. Goodfellow, R.; Dimitrakopoulos, R. Simultaneous stochastic optimization of mining complexes and mineral value chains. *Math. Geosci.* **2017**, *49*, 341–360. [CrossRef]
74. Lamghari, A.; Dimitrakopoulos, R.; Ferland, J.A. A hybrid method based on linear programming and variable neighborhood descent for scheduling production in open-pit mines. *J. Glob. Optim.* **2015**, *63*, 555–582. [CrossRef]
75. Nobahar, P.; Xu, C.; Dowd, P.; Faradonbeh, R.S. Exploring digital twin systems in mining operations: A review. *Green Smart Min. Eng.* **2024**, *1*, 474–492. [CrossRef]
76. Saldana, M.; Gallegos, S.; Arias, D.; Salazar, I.; Castillo, J.; Salinas-Rodríguez, E.; Navarra, A.; Toro, N.; Cisternas, L.A. Applications of Kuz–Ram Models in Mine-to-Mill Integration and Optimization—A Review. *Minerals* **2024**, *14*, 1162. [CrossRef]
77. Guyaguler, B.; Horne, R.N.; Rogers, L.; Rosenzweig, J.J. Optimization of well placement in a Gulf of Mexico waterflooding project. *SPE Reserv. Eval. Eng.* **2002**, *5*, 229–236. [CrossRef]
78. Yeten, B.; Durlofsky, L.J.; Aziz, K. Optimization of nonconventional well type, location, and trajectory. *SPE J.* **2003**, *8*, 200–210. [CrossRef]
79. Levinson, Z.; Dimitrakopoulos, R.; Keutchayan, J. Simultaneous stochastic optimization of an open-pit mining complex with preconcentration using reinforcement learning. *Appl. Soft Comput.* **2023**, *138*, 110180. [CrossRef]
80. Noriega, R.; Boisvert, J. Optimization of underground mining production layouts considering geological uncertainty using deep reinforcement learning. *Eng. Appl. Artif. Intell.* **2025**, *139*, 109493. [CrossRef]
81. Noriega, R.; Pourrahimian, Y.; Askari-Nasab, H. Deep Reinforcement Learning based real-time open-pit mining truck dispatching system. *Comput. Oper. Res.* **2025**, *173*, 106815. [CrossRef]
82. Pendharkar, P.C.; Rodger, J.A. Nonlinear programming and genetic search application for production scheduling in coal mines. *Ann. Oper. Res.* **2000**, *95*, 251–267. [CrossRef]
83. Leite, A.; Dimitrakopoulos, R. Stochastic optimisation model for open pit mine planning: Application and risk analysis at copper deposit. *Min. Technol.* **2007**, *116*, 109–118. [CrossRef]
84. Sayadi, A.R.; Fathianpour, N.; Mousavi, A.A. Open pit optimization in 3D using a new artificial neural network. *Arch. Min. Sci.* **2011**, *56*, 389–403.
85. Souza, M.J.; Coelho, I.M.; Ribas, S.; Santos, H.G.; Merschmann, L.H.d.C. A hybrid heuristic algorithm for the open-pit-mining operational planning problem. *Eur. J. Oper. Res.* **2010**, *207*, 1041–1051. [CrossRef]
86. Shishvan, M.S.; Benndorf, J. Simulation-based optimization approach for material dispatching in continuous mining systems. *Eur. J. Oper. Res.* **2019**, *275*, 1108–1125. [CrossRef]
87. Lamghari, A.; Dimitrakopoulos, R. Hyper-heuristic approaches for strategic mine planning under uncertainty. *Comput. Oper. Res.* **2020**, *115*, 104590. [CrossRef]
88. LaRoche-Boisvert, M.; Dimitrakopoulos, R. An application of simultaneous stochastic optimization at a large open-pit gold mining complex under supply uncertainty. *Minerals* **2021**, *11*, 172. [CrossRef]
89. Levinson, Z.; Dimitrakopoulos, R. Connecting planning horizons in mining complexes with reinforcement learning and stochastic programming. *Resour. Policy* **2023**, *86*, 104136. [CrossRef]
90. Levinson, Z.; Dimitrakopoulos, R. A reinforcement learning approach for selecting infill drilling locations considering long-term production planning in mining complexes with supply uncertainty. *Min. Technol.* **2024**, *133*, 176–187. [CrossRef]
91. Kalman, R.E. A new approach to linear filtering and prediction problems. In *Control Theory: Twenty-Five Seminal Papers (1932–1981)*; Wiley-IEEE Press: Hoboken, NJ, USA, 1960.
92. Li, Y.; Sepúlveda, E.; Xu, C.; Dowd, P. A rapid updating method to predict grade heterogeneity at smaller scales. *Math. Geosci.* **2021**, *53*, 1237–1260. [CrossRef]
93. Yüksel, C.; Benndorf, J.; Lindig, M.; Lohsträter, O. Updating the coal quality parameters in multiple production benches based on combined material measurement: A full case study. *Int. J. Coal Sci. Technol.* **2017**, *4*, 159–171. [CrossRef]
94. Yüksel, C.; Thielemann, T.; Wambeke, T.; Benndorf, J. Real-time resource model updating for improved coal quality control using online data. *Int. J. Coal Geol.* **2016**, *162*, 61–73. [CrossRef]
95. Avalos, S.; Ortiz, J.M. Multivariate Geostatistical Simulation and Deep Q-Learning to optimize mining decisions. *Math. Geosci.* **2023**, *55*, 673–692. [CrossRef]
96. Benndorf, J.; Jansen, J.D. Recent developments in closed-loop approaches for real-time mining and petroleum extraction. *Math. Geosci.* **2017**, *49*, 277–306. [CrossRef]
97. Kumar, D.; Srinivasan, S. Ensemble-based assimilation of nonlinearly related dynamic data in reservoir models exhibiting non-Gaussian characteristics. *Math. Geosci.* **2019**, *51*, 75–107. [CrossRef]
98. Ma, W.; Jafarpour, B. Integration of soft data into multiple-point statistical simulation: Re-assessing the probability conditioning method for facies model calibration. *Comput. Geosci.* **2019**, *23*, 683–703. [CrossRef]

99. Talesh Hosseini, S.; Asghari, O.; Benndorf, J.; Emery, X. Real-time uncertain geological boundaries updating for improved block model quality control based on blast hole data: A case study for Golgohar iron ore mine in Southeastern Iran. *Math. Geosci.* **2023**, *55*, 541–562. [CrossRef]
100. Hu, L.; Zhao, Y.; Liu, Y.; Scheepens, C.; Bouchard, A. Updating multipoint simulations using the ensemble Kalman filter. *Comput. Geosci.* **2013**, *51*, 7–15. [CrossRef]
101. Jafarpour, B.; McLaughlin, D.B. History matching with an ensemble Kalman filter and discrete cosine parameterization. *Comput. Geosci.* **2008**, *12*, 227–244. [CrossRef]
102. Prior, Á.; Benndorf, J.; Mueller, U. Resource and grade control model updating for underground mining production settings. *Math. Geosci.* **2021**, *53*, 757–779. [CrossRef]
103. Lan, T.; Shi, X.; Jiang, B.; Sun, Y.; Wu, J. Joint inversion of physical and geochemical parameters in groundwater models by sequential ensemble-based optimal design. *Stoch. Environ. Res. Risk Assess.* **2018**, *32*, 1919–1937. [CrossRef]
104. Nejadi, S.; Trivedi, J.; Leung, J.Y. Estimation of facies boundaries using categorical indicators with P-Field simulation and ensemble Kalman filter (EnKF). *Nat. Resour. Res.* **2015**, *24*, 121–138. [CrossRef]
105. Oliver, D.S.; Chen, Y. Data assimilation in truncated plurigaussian models: Impact of the truncation map. *Math. Geosci.* **2018**, *50*, 867–893. [CrossRef]
106. Zhou, H.; Gomez-Hernandez, J.J.; Franssen, H.-J.H.; Li, L. An approach to handling non-Gaussianity of parameters and state variables in ensemble Kalman filtering. *Adv. Water Resour.* **2011**, *34*, 844–864. [CrossRef]
107. Avalos, S.; Kracht, W.; Ortiz, J.M. Machine learning and deep learning methods in mining operations: A data-driven SAG mill energy consumption prediction application. *Min. Metall. Explor.* **2020**, *37*, 1197–1212. [CrossRef]
108. Ortiz, J.M.; Avalos, S.; Riquelme, A.I.; Leuangthong, O.; Madani, N.; Frenzel, M. Uncertainty and value: Optimising geometallurgical performance along the mining value chain. *Elements* **2023**, *19*, 377–383. [CrossRef]
109. Clay, R.L.; Shneider, M.S.; Marburger, S.; Trucano, T.G. *Modeling and Simulation Technology Readiness Levels*; Sandia National Laboratories (SNL): Albuquerque, NM, USA; Livermore, CA, USA, 2006.
110. Héder, M. From NASA to EU: The evolution of the TRL scale in Public Sector Innovation. *Innov. J.* **2017**, *22*, 1.
111. Manning, C.G. Technology Readiness Levels. Available online: https://www.nasa.gov/directorates/somd/space-communications-navigation-program/technology-readiness-levels/ (accessed on 21 November 2024).

mining

MDPI

Review

Harnessing Emerging Technologies in the Global Mining Sector from a Bibliometric Standpoint

Chijioke Emere [1], Olusegun Oguntona [1,*], Ifije Ohiomah [2] and Emmanuel Ayorinde [2]

[1] Department of Built Environment, Faculty of Engineering, Built Environment and Information Technology, Walter Sisulu University, Butterworth 4960, South Africa

[2] Faculty of Engineering, Built Environment and Information Technology, Walter Sisulu University, Butterworth 4960, South Africa

* Correspondence: architectoguntona12@gmail.com or ooguntona@wsu.ac.za

Abstract: The numerous challenges facing the global mining industry and the adverse impacts on the natural and human environment call for urgent action. In the present industry 4.0, the signature influx of emerging technologies (ETs) has seen various sectors of the economy embracing their application. To improve the safety, operational efficiency, and sustainability of the mining value chain, there has been a significant increase in the adoption, incorporation and application of ETs such as digital twins, artificial intelligence, the internet of things, and blockchain. Through a bibliometric analysis of scholarly publication outputs on ETs in the mining industry, this study visualises and ascertains the development and trends of these technologies from 1986 until now. Bibliometric datasets made up of 135 articles drawn from the popular Scopus database were employed. Dataset analysis revealed influential scholarly outputs, authors, and research clusters. The study provides relevant stakeholders in the sector with firsthand insight into the state of ET integration and use in the mining sector. Further studies are recommended to explore innovative technological interventions in other industries that can be adapted to enhance and optimise the activities and processes of the mining sector.

Keywords: artificial intelligence; digital technologies; health and safety; industry 4.0; mining industry; sustainable development goals

1. Introduction

The mining sector globally is one of the core industries responsible for gross domestic product (GDP) growth, economic development, and employment generation avenues. Despite its linkage to large social impacts, environmental disturbances, and inequalities, the mining sector provides the majority of the minerals that we are dependent on to provide energy, infrastructure, and other core basic amenities [1]. With the widespread clamour for sustainable development in all sectors of the economy, the mining sector remains behind owing to the continued pressure it experiences. Such pressure arises from the fact that replacing fossil fuel-based technologies requires mined metals such as cobalt, lithium, indium, and neodymium to enable sustainable technologies like distributed transmission grids, solar panels, electric vehicles, and wind turbines, among others [2]. Also, most African countries depend on mining as a source of income and GDP growth. Sub-Saharan Africa is an important region for the global supply of precious and significant minerals. For example, the region accounts for over 10% of global mine production, a 37% share of world diamond productions, and more than a 10% market share in six minerals (such as uranium, rutile, manganese, cobalt, and bauxite) with mostly all African countries

https://doi.org/10.3390/mining5010013

known for one mineral or other [3]. Also, countries such as China, India, and Indonesia are regarded as top mining economies, with China widely known to be the leading mining country for most commodities. According to Statista [4], as shown in Figure 1, the top ten leading gold mining countries in the world based on production in metric tons are China, Australia, Russia, Canada, the United States, Kazakhstan, Mexico, Indonesia, South Africa, and Uzbekistan. Hence, the economic derivatives of the mining sector seem to overshadow the need to consider the various adverse impacts of the industry on the environment. To overcome these challenges, among numerous others, the mining sector will have to reinvent itself, embrace innovative approaches, and ensure minerals can be efficiently and sustainably exploited to support global development that leads to the people's welfare and quality of life [5]. As a result, voluntary initiatives are undertaken by major mining entities to meet their social and environmental responsibilities [6]. A major game changer and intervention is the influx and application of emerging technologies (ETs) to solve the various challenges experienced in the mining sector.

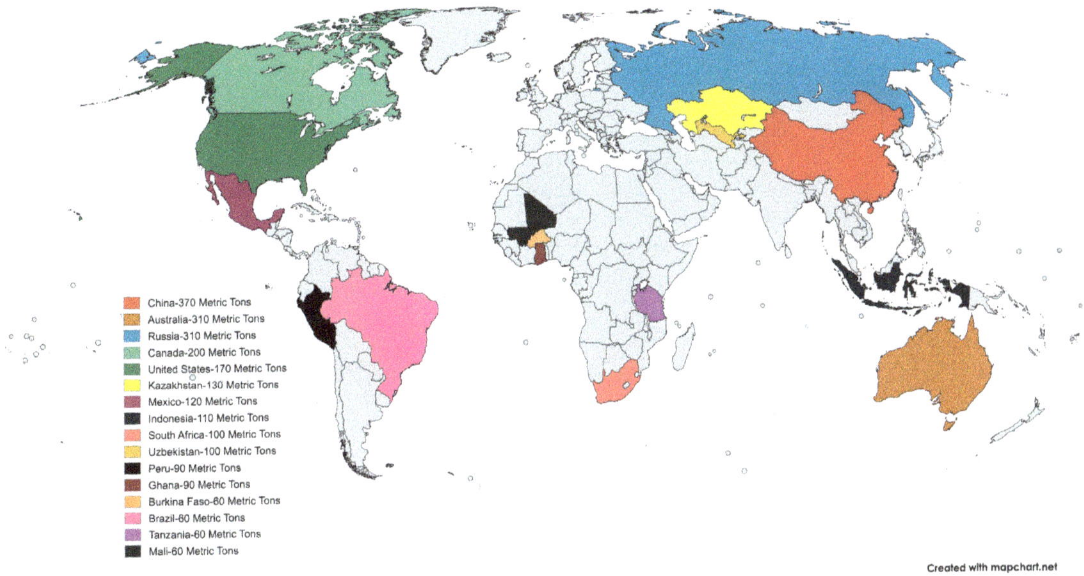

Figure 1. World-leading gold mining countries in 2023 [4].

2. Emerging Technologies for Mining Solutions

Despite the mining industry's socioeconomic impact, the sector faces numerous challenges, such as accidents, safety concerns, and environmental degradation. A study by Nicholaus Chusi et al. [7] identified poor capital, poor infrastructure system, lack of skilled labour, decline of mineral deposits due to over-exploitation, lack of water and energy supply, poor condition of workers, political barriers, and local market issues, among others. The numerous problems facing the mining sector were grouped into environmental, operational, and social challenges with a lack of automation and technological intervention [8]. Hence, some of the opportunities to address these challenges are to develop and incorporate clean technologies and utilise intelligent industry 4.0 solutions to improve mining efficiency and occupational safety [9]. The essence of technological interventions lends credence to the general understanding that ETs can revolutionise various sectors of the economy when adopted and embraced holistically. For an industry such as mining, it is, therefore, imperative that ETs play a significant role in reinventing the operations, processes, and activities of the industry to minimise the challenges that have plagued the sector drastically.

Digitalisation in the mining industry entails using ETs or innovative digital systems to transform mining practices, improve productivity, and reduce costs [10]. Technological systems and innovations greatly benefit mining firms in improving mineral recovery, increasing production, lessening environmental impacts, and reducing costs associated with mining processes and activities [11,12]. Some notable ETs in the mining sector are 3D printing, machine learning, the internet of things, drones, remote operating centres, autonomous guided vehicles, quantum computing, power generation and storage, cybersecurity and encryption, text, image, and voice processing, digital trade and e-commerce, augmented reality, advanced process control, digital twins, cloud-based solutions, real-time data, real-time mine planning, artificial intelligence, enterprise resource planning, predictive maintenance, virtual reality mine training, real-time location sensing, cloud and edge computing, supply chain planning, robotics, automation and process integration, and blockchain, to mention a few [10,12–14]. Against the different mining value chains (exploration, establishment, mining, processing, waste management, smelting and refining, sales, and closure), these technologies have the potential to significantly affect and transform the mining sector [15]. As further alluded to by a study by Lazarenko et al. [12], ETs will improve resource management, enhance performance management, and advance marketing management for the core stages of the mining value chain that consist of exploration, planning and mine development, engineering and construction, operations and logistics, and sales and marketing. Hence, this study explores the exploitation of ETs in the global mining sphere, utilising a bibliometric analysis to visualise and ascertain their development and trends.

3. Research Methodology

This study is intended to determine the key subjects of interest in publications about ETs in the global mining industry. Consequently, a bibliometric methodology was employed to identify research trends and map out areas of knowledge. Relevant publications for the investigation were found using the Scopus search engine. This is because Scopus has wider coverage and indexes most papers in other databases, including Web of Science, ScienceDirect, and Google Scholar [16,17]. Similarly, Scopus is adjudged as the largest citation and abstract database of peer-reviewed scholarly publications and is highly subscribed to by most higher education institutions [18]. Some studies have acknowledged the significance of Scopus as a more recent and rapidly expanding database and a favoured option for literature search [17,19]. The search keywords were "Emerging Technologies" OR "Emerging Technology" OR "Digital Technologies" OR "Digital Technology" AND "Mining Industry" OR "Mining Sector". All mining-related fields were considered, including engineering, environmental science, computer science, energy, earth and planetary science, etc. The period that was considered was 1986–2025. The search was conducted on 17 November 2024. Journals, conference articles, and book chapters were all included. The search, which was limited to English, produced 136 publications. VOSviewer, a program appropriate for bibliometric literature evaluations, was used to graphically analyse the data [20]. Figure 2 depicts the research process and outputs, namely "Publication per year", "publications per country", "publications by document source", "most cited publications", and "co-occurring analysis".

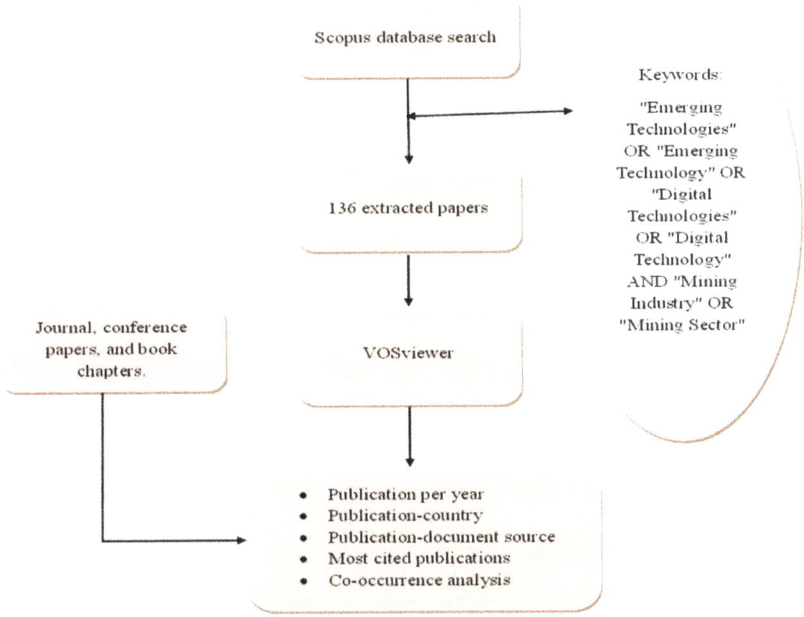

Figure 2. Research methodology flowchart.

4. Results and Discussions

This section presents the results and discusses the analysis of the 136 articles that constitute the dataset employed for the bibliometric analysis in this study.

4.1. Publications per Year

Figure 3 indicates yearly publications in the Scopus database relating to emerging technologies in the mining sector from 1986 to 2025. Results showed that one publication was made yearly from 1986 to 2003 except for the years 1989, 1990, 1993, 1995, 1996, and 1997 with no publication. Similarly, one publication was made in 2006 and 2007. Also, other years with no publications included 2004, 2013, 2014, and 2016. Conversely, two publications were made in 2008, 2011, and 2015, while three were made in 2009, 2010, and 2012. From 2017, a major improvement and upward growth was seen with four publications, followed by six publications in 2018, and twelve and thirteen publications in 2019 and 2020, respectively. However, there was an increase to nineteen publications in 2021 and another decrease to sixteen and thirteen publications in 2022 and 2023, respectively. Notably, the highest number of publications (twenty-four) was recorded in 2024. Generally, the results indicate a rollercoaster ride and very scanty publications, especially from 1986 to 2016. However, findings revealed a significant improvement in published works from 2017 to 2024. Overall, results indicate a need for more scholarly output in the mining sector relating to the adoption, influence, and utilisation of emerging technologies.

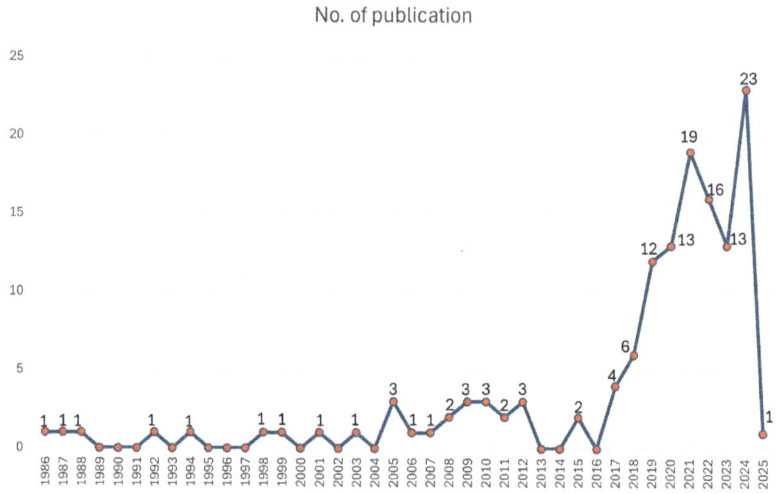

Figure 3. Number of publications per year.

4.2. Publications per Country

The 136 extracted documents were published in 48 countries. However, ten countries dominated the field at a threshold of five documents. The revealed countries are shown in Figure 4. Russia recorded the highest number of publications, with 22 documents and 184 citations. This was followed by Australia (21 documents; 239 citations), the United States (16 documents; 131 citations), China (12 documents; 137 citations), and Canada (11 documents; 42 citations). These were the top 5 countries among the listed 10 countries in Figure 3. India and South Africa were bracketed with 10 publications and ranked sixth position. Germany (nine documents; 161 citations) was ranked 8th, while Kazakhstan (six documents; 17 citations) and Chile (five documents; 27 citations) were ranked 9th and 10th, respectively. The results indicate a very wide range of significant publications on the subject in Asian, North American, and European countries. Notably, these countries are among the top mining countries in the world, known for their significant contributions to global mineral production [21,22]. Similarly, South Africa has contributed significantly to the subject compared to other African countries and is one of the leading African mining countries. Regrettably, Chile was the only South American country represented in the top category. Consequently, there is a great opportunity for African and South American countries, especially Congo (DRC), Zambia, and Brazil, which are among the top mining countries in their continent, to make significant research contributions to using emerging technologies in the mining sector [21–23]. Lumadi and Nyasha [24] highlighted frequently mentioned difficulties in adopting emerging mining technologies. These challenges include the sector's cyclical nature, high cost, uncertainty, poor involvement with external stakeholders, and significant risk associated with the implementation of unproven technologies and performance systems centred on volumetric production [24,25]. Consequently, proper learning culture and knowledge sharing were recommended as critical to adopting mining technology [24,25].

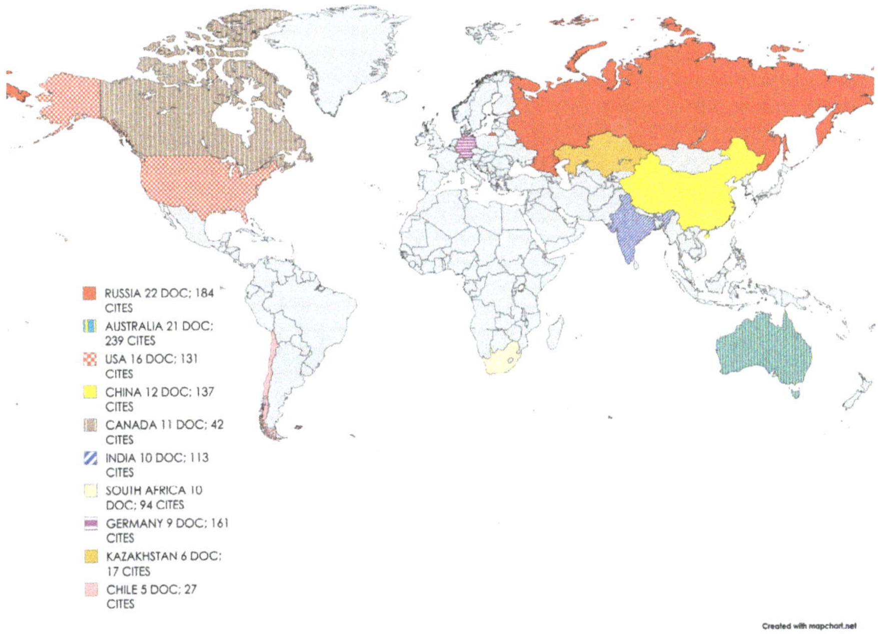

Figure 4. Number of publications per country.

4.3. Publications per Document Source

The 136 downloaded documents had 105 sources. However, with the VOSviewer default threshold of three documents, nine sources were unearthed and displayed in Table 1. According to Table 1, the ranked source included *E3S Web of Conferences* (six documents; 39 citations). This was seconded by *Resources Policy* (five documents; 85 citations). The rest of the sources had three documents. However, results showed that articles published in *Eurasian Mining*, MDPI's *Sustainability* (Switzerland), and the *Journal of the Southern African Institute of Mining and Metallurgy* had more citations compared to the rest with three documents. Overall, the results indicate scantiness in published works. Hence, there is a need for more publications of documents that Scopus indexes in reputable publication sources that are not listed in Table 1.

Table 1. Number of documents per source.

Source	Number of Documents	Citations
E3S Web of Conferences	6	39
Resources Policy	5	85
Eurasian Mining	3	36
Sustainability (Switzerland)	3	14
Journal of the Southern African Institute of Mining and Metallurgy	3	14
IOP Conference Series: Earth and Environmental Science	3	8
News of the National Academy of Sciences of the Republic of Kazakhstan, Series of Geology and Technical Sciences	3	8
Journal Of Mines, Metals and Fuels	3	5
Canadian Mining Journal	3	0

4.4. Most Cited Publications

The 136 retrieved articles were examined to identify the most often cited papers and their primary topics for understanding emerging technologies research in the mining sector. Twelve relevant publications emerged when a criterion of 25 citations was applied. The significance and ambition to further the topic are reflected in the study with significant citation counts as seen in Table 2. Table 2 shows that reviews and experiments were the most employed approaches. Most of the focus was on blockchain technology, productivity, and supply chain improvement in the mining sector, especially in oil, gas, and energy. The results suggest that the concept is at the developing stage. It is critical to comprehend how emerging technologies can be adopted in the mining industry. Finding the ideal combination for the mining industry can only be accomplished through case studies and experimentation. Consequently, more studies on experiments and case studies are recommended.

Table 2. Most cited publications.

Author(s)	Title	Citations	Method
Johnson and Hallberg [26]	"Acid mine drainage remediation options: a review".	1715	Review
Barnewold and Lottermoser [10]	"Identification of digital technologies and digitalisation trends in the mining industry".	121	Review
Samylovskaya et al. [27]	"Digital Technologies in Arctic Oil and Gas Resources Extraction: Global Trends and Russian Experience".	62	Review
Mann et al. [28]	"Blockchain technology for supply chain traceability, transparency and data provenance".	42	Review
Divyasudha et al. [29]	"Analysis of Smart Helmets and Designing an IoT based Smart Helmet: A cost-effective solution for Riders".	32	Review
Quayson et al. [30]	"Building blockchain-driven dynamic capabilities for developing circular supply chain: Rethinking the role of sensing, seizing, and reconfiguring".	30	Case Study
Onifade et al. [31]	"Challenges and applications of digital technology in the mineral industry".	29	Review
Smith et al. [32]	"Spatial distribution and management of total actual acidity in an acid sulfate soil environment, McLeods Creek, northeastern NSW, Australia".	29	Experimental
Lacey et al. [33]	"Public perceptions of established and emerging mining technologies in Australia".	28	Survey
Bissell [34]	"Encountering automation: Redefining bodies through stories of technological change".	28	Interview
More et al. [35]	"Automated measurement systems in mine water management and mine workings–A review of potential methods".	26	Review
Carr et al. [36]	"Development of a method to determine operator location using electromagnetic proximity detection".	25	Experimental

4.5. Publications Based on Co-Occurring Keywords

A keywords co-occurrence map was made to properly characterise the focus subjects from the prior investigation. Among the keywords were author and indexed keywords. Consequently, the 136 papers that were downloaded contained 1241 keywords. Using the VOSviewer's default threshold of five co-occurrences, 24 keywords were left. Three clusters were produced as a result. The mapped keywords are displayed in Figure 5 and tabulated in Table 3.

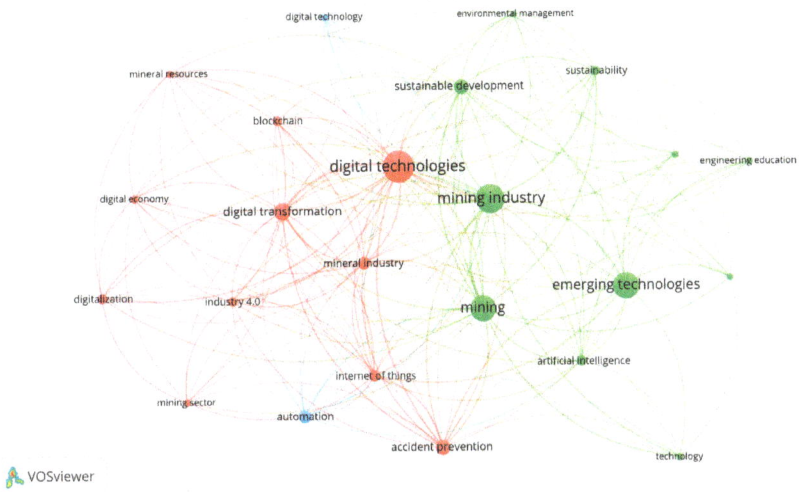

Figure 5. Network visualisation map for co-occurring keywords.

Table 3. Clusters of co-occurring keywords.

Clusters	Keywords	Occurrence	Link Strength
Cluster 1: Digital Transformation and Safety Management	Digital technologies	36	94
	Digital transformation	17	62
	Accident prevention	14	50
	Mineral industry	11	54
	Internet of things	10	41
	Digitalisation	9	26
	Blockchain	8	23
	Industry 4.0	8	34
	Digital economy	7	26
	Mineral resources	6	26
	Mining sector	6	15
	Total	132	451
Cluster 2: Sustainability Growth	Mining industry	32	93
	Emerging technologies	28	41
	Mining	28	87
	Sustainable development	14	54
	Artificial intelligence	9	35
	Sustainability	8	32
	Engineering education	6	14
	Environmental technology	6	18
	Technology	6	15
	Environmental management	5	21
	Mining operations	5	15
	Total	147	425
Cluster 3: Automation	Automation	12	43
	Digital technology	6	26
	Total	18	69

4.5.1. Cluster 1: Digital Transformation and Safety Management

This cluster comprised eleven keywords mapped with red with a total link strength of 451. Based on priorities/number of occurrences, the keywords included digital technologies, digital transformation, accident prevention, mineral industry, internet of things, digitalisation, blockchain, industry 4.0, digital economy, mineral resources, and mining sector. Consequently, the keywords in this cluster relate to Digital Revolution in the Mineral Industry: Leveraging IoT and Blockchain and Industry 4.0 for Transformation and Safety Management. For instance, routine tasks can be automated using IoT, improving workflows and lowering human error [37,38]. Costs are reduced, and productivity is raised as a result [38]. When utilised in an industrial setting, the insights gathered can be used right at once to improve asset management, worker safety, machine health, and process optimisation [37,39]. Similarly, with little human assistance, IoT can help real-time platforms remotely monitor and manage a complicated manufacturing system [37,39].

Conversely, blockchain technology is transforming the mining industry by enhancing transparency, security, and efficiency [40,41]. It makes adhering to rules and standards simpler by offering a safe and unchangeable record of transactions [40,42]. It can help with smart contracts (self-executing contracts) by automating procedures like payments and compliance, reducing the need for intermediaries [40,43].

4.5.2. Cluster 2—Sustainability Growth

This cluster, represented in green, is characterised by 15 items with a total link strength of 425. The keywords/items included mining industry, emerging technologies, mining, sustainable development, artificial intelligence, sustainability, engineering education, environmental technology, technology, environmental management, and mining operations. These keywords, therefore, relate to emerging technologies for sustainable growth in the mining industry. A typical example is the use of artificial intelligence (AI). AI is used for exploration, predictive maintenance, and environmental monitoring [44,45]. It lowers maintenance costs and downtime by improving the accuracy of mineral deposit identification and anticipating equipment faults before they happen [44,45]. Technological progress is fueled by the desire for sustainability, which results in cleaner and more effective mining techniques [46]. According to the International Institute for Sustainable Development (IISD) [47], sustainable mining methods limit emissions, waste, and land disturbance. Sustainable mining will improve the well-being of employees and local communities through enhanced safety protocols and cleaner operations [48]. Particularly through community engagement, it can offer communities long-term economic benefits, including infrastructure development and job generation [48].

4.5.3. Cluster 3—Automation

This cluster is represented in blue and has two keywords (automation and digital technology) with a total link strength of 69. The use of automation in the mining industry has been acknowledged to enhance efficiency, safety, and productivity [14,49]. The applications of automation in mining include autonomous haulage systems (AHSs), remote operation centres, automated drilling systems, and process optimisation [49,50]. AHSs' mining trucks and vehicles are equipped with autonomous technology, which enables them to function without direct human supervision [50]. AHSs lower operating costs, increase safety, and boost efficiency [49,50]. Similarly, remote operation centres enable mines to be operated remotely using sophisticated control systems and monitoring technologies [50]. Operational efficiency is increased through quick responses to possible problems made possible by real-time data collecting and analysis [50]. Conversely, automated drilling systems increase the speed and accuracy of drilling operations, reducing the need for manual labour and improving safety [50]. Generally, the primary advantages of using digital technologies in the mining industry include "increased productivity, cost reduction, occupational safety, better working conditions, improved operational efficiency and reliability of information" [14].

4.6. Research Focus by Year of Publication

Figure 6 displays the "network map overlay visualisation" for the co-occurring keywords. It was observed that, in at least five occurring keywords, the predominant focus of the studies on emerging technologies in the mining industry included digital transformation and safety management, sustainability growth, and automation. Focus on keywords like automation, environmental management, and engineering education became evident in 2018. From 2019, studies majored in IoT, accident prevention, and sustainable development. From 2020 to 2021, keywords like digital transformation, digital technologies, digital economy, industry 4.0, and AI were viral. Also, from 2022 until now, studies on digitisation and digital technologies have further grown with blockchain technology and sustainability becoming more evident. However, keywords like 3D printing and additive manufacturing, augmented reality (AR), virtual reality (VR), bio-mining, phytomining, renewable energy integration, water recycling and treatment technologies, and clean tech solutions were missing. This indicates a major gap in studies on emerging technologies in the mining industry.

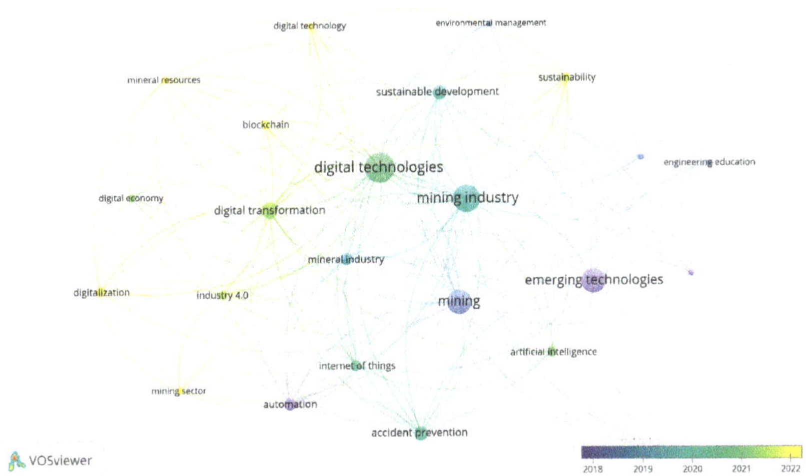

Figure 6. Overlay visualisation map for co-occurring keywords.

Nevertheless, the gaps provide opportunities for future researchers to contribute majorly to the highlighted areas. For instance, 3D printing can produce on-site mining equipment and spare parts, enabling less storage and transportation [45]. It also makes it possible to create equipment specifically suited to certain mining operations [45]. AR and VR technologies are used for training, simulation, and remote mining equipment operation, providing immersive experiences that improve worker safety and operational efficiency [45,51]. Renewable energy integration will reduce reliance on fossil fuels and lower greenhouse gas emissions [52]. Similarly, mines can reuse water through sophisticated water recycling and treatment systems, which minimise wastewater discharge and the demand for freshwater [52]. Reverse osmosis and membrane filtration are two technologies that guarantee more ecologically conscious mining operations [52]. Furthermore, clean tech solutions like carbon capture, storage, and utilisation, as well as waste management technologies, can reduce the environmental impact of mining operations [45]. These technologies, therefore, should be carefully considered in newer research.

5. Conclusions and Recommendations

This study used a bibliometric technique to ascertain the focus on emerging technology research in the mining industry. The extracted papers span more than 20 years and are listed in the Scopus database. The results indicate a rollercoaster ride and very scanty publications, especially from 1986 to 2016. However, countries like Russia, Australia, the US, China, Canada, India, Germany, Kazakhstan, and Chile have made significant contributions. The highest number of publications on the subject in Scopus-indexed research was in 2024. Nevertheless, countries, especially in developing nations, are encouraged to make a substantial contribution towards the subject matter. The advantages of emerging technologies cannot be overemphasised.

Emerging/digital technologies have been found to improve operational efficiency, save costs, improve occupational safety, improve working conditions, raise production, and increase information dependability. Considering the clusters revealed, research on emerging technologies in the mining industry focused on digital transformation and safety management, sustainability growth, and automation. Furthermore, the study discovered that recent research in this field indicates enhancing transparency, security, and efficiency using blockchain technology.

However, there was a dearth of research on emerging technologies such as AR and VR, 3D printing and additive manufacturing, clean tech solutions, and water recycling and treatment technologies. These areas need to be more researched. Consequently, more studies on these technologies are required. For instance, the potential of AR and VR to enhance operational efficiency, training, and safety needs to be researched. Similarly, the efficacy of 3D printing and additive manufacturing in reducing inventory needs, minimisation of downtime, and cost reductions should be thoroughly studied. Similarly, energy raw material consumption is expected to climb by 2035, and the switch to renewable energy sources like solar and wind, and electric vehicles (EVs) necessitates a significant rise in vital minerals like lithium, cobalt, nickel, and rare earth elements. To guarantee the availability and affordability of raw materials, it is imperative to make investments in mining, refining, and supporting infrastructure.

This research has contributed to the body of knowledge by highlighting areas where studies on emerging technology in the mining industry have focused on. Its conclusions have also indicated possible directions for future research. However, this study only looked at papers from the Scopus database. For a more thorough grasp of the research topic, it is recommended that further studies be conducted to explore other databases or merge Scopus with other databases.

Author Contributions: Conceptualisation, O.O. and C.E.; methodology, O.O. and C.E.; software, C.E.; validation, O.O. and C.E.; formal analysis, C.E.; investigation, O.O. and C.E.; resources, I.O., E.A., O.O. and C.E.; data curation, O.O. and C.E.; writing—original draft preparation, I.O., E.A., O.O. and C.E.; writing—review and editing, I.O., E.A., O.O. and C.E.; visualisation, O.O. and C.E.; supervision, I.O., E.A., O.O. and C.E. All authors have read and agreed to the published version of the manuscript.

Funding: This research received no external funding.

Data Availability Statement: The datasets employed in this study are available on request from the corresponding author.

Acknowledgments: The authors acknowledge the National Research Foundation (NRF), and the Directorate of Research and Innovation, Faculty of Engineering, Built Environment and Information Technology, Walter Sisulu University, South Africa.

Conflicts of Interest: The authors declare no conflicts of interest.

References

1. Carvalho, F.P. Mining industry and sustainable development: Time for change. *Food Energy Secur.* **2017**, *6*, 61–77. [CrossRef]
2. Upadhyay, A.; Laing, T.; Kumar, V.; Dora, M. Exploring barriers and drivers to the implementation of circular economy practices in the mining industry. *Resour. Policy* **2021**, *72*, 102037. [CrossRef]
3. Akabzaa, T.; Darimani, A. Impact of mining sector investment in Ghana: A study of the Tarkwa mining region. *Third World Netw.* **2001**, *11*, 47–61.
4. Statista. Leading Gold Mining Countries Worldwide in 2023. Available online: https://www.statista.com/statistics/264628/world-mine-production-of-gold/ (accessed on 1 October 2024).
5. Mateus, A.; Martins, L. Challenges and opportunities for a successful mining industry in the future. *Boletín Geológico Y Min.* **2019**, *130*, 99–121. [CrossRef]
6. Dashwood, H.S. Sustainable development and industry self-regulation: Developments in the global mining sector. *Bus. Soc.* **2014**, *53*, 551–582. [CrossRef]
7. Nicholaus Chusi, T.; Bouraima, M.B.; Yazdani, M.; Jovčić, S.; Hernández, V.D. Addressing the challenges facing developing countries in the mining sector: Moving towards sustainability. *J. Appl. Res. Ind. Eng.* **2024**, *11*, 333–349.
8. Marimuthu, R.; Sankaranarayanan, B.; Ali, S.M.; de Sousa Jabbour, A.B.L.; Karuppiah, K. Assessment of key socioeconomic and environmental challenges in the mining industry: Implications for resource policies in emerging economies. *Sustain. Prod. Consum.* **2021**, *27*, 814–830. [CrossRef]
9. Brodny, J.; Tutak, M. Challenges of the Polish coal mining industry on its way to innovative and sustainable development. *J. Clean. Prod.* **2022**, *375*, 134061. [CrossRef]
10. Barnewold, L.; Lottermoser, B.G. Identification of digital technologies and digitalisation trends in the mining industry. *Int. J. Min. Sci. Technol.* **2020**, *30*, 747–757. [CrossRef]
11. Jang, H.; Topal, E. Transformation of the Australian mining industry and future prospects. *Min. Technol.* **2020**, *129*, 120–134. [CrossRef]
12. Lazarenko, Y.; Garafonova, O.; Marhasova, V.; Tkalenko, N. Digital transformation in the mining sector: Exploring global technology trends and managerial issues. In *E3S Web Conferences*; EDP Sciences: Les Ulis, France, 2021; Volume 315, p. 04006.
13. Bhattacharyya, S.S.; Shah, Y. Emerging technologies in Indian mining industry: An exploratory empirical investigation regarding the adoption challenges. *J. Sci. Technol. Policy Manag.* **2022**, *13*, 358–381. [CrossRef]
14. Flores-Castañeda, R.O.; Olaya-Cotera, S.; López-Porras, M.; Tarmeño-Juscamaita, E.; Iparraguirre-Villanueva, O. Technological advances and trends in the mining industry: A systematic review. *Miner. Econ.* **2024**, *in press*. [CrossRef]
15. Sánchez, F.; Hartlieb, P. Innovation in the mining industry: Technological trends and a case study of the challenges of disruptive innovation. *Min. Metall. Explor.* **2020**, *37*, 1385–1399. [CrossRef]
16. Okafor, C.C.; Aigbavboa, C.; Thwala, W.D. A bibliometric evaluation and critical review of the smart city concept–making a case for social equity. *J. Sci. Technol. Policy Manag.* **2023**, *14*, 487–510. [CrossRef]
17. Aliu, J.; Emere, C.; Oguntona, O. Mapping Smart City and Industry 4.0 Research in Construction-Related Studies. *Balt. J. Real Estate Econ. Constr. Manag.* **2024**, *12*, 258–275. [CrossRef]
18. Oguntona, O.A.; Aigbavboa, C.O.; Thwala, W.D. A scientometric analysis and visualisation of green building research in Africa. *J. Green Build.* **2021**, *16*, 83–86. [CrossRef]
19. Aghimien, D.O.; Aigbavboa, C.O.; Oke, A.E.; Thwala, W.D. Mapping out research focus for robotics and automation research in construction-related studies: A bibliometric approach. *J. Eng. Des. Technol.* **2020**, *18*, 1063–1079. [CrossRef]
20. van Eck, N.J.; Waltman, L. Accuracy of citation data in Web of Science and Scopus. *arXiv* **2019**, arXiv:1906.07011.
21. Buchholz, K. The Countries That Are the Biggest Miners in the World. Available online: https://www.statista.com/chart/19839/biggest-miners-among-countries/ (accessed on 21 December 2024).
22. Leighton, M. Top Mining Countries in the World 2024. Available online: https://www.cruxinvestor.com/posts/top-mining-countries-in-the-world-2024 (accessed on 21 December 2024).
23. AFRIKTA. Top 10 Mining Companies in Africa. 2023. Available online: https://afrikta.com/top-10-mining-companies-in-africa/ (accessed on 21 December 2024).
24. Lumadi, V.W.; Nyasha, S. Technology and growth in the South African mining industry: An assessment of critical success factors and challenges. *J. South. Afr. Inst. Min. Metall.* **2024**, *124*, 163–171. [CrossRef]

25. Ediriweera, A.; Wiewiora, A. Barriers and enablers of technology adoption in the mining industry. *Resour. Policy* **2021**, *73*, 102188. [CrossRef]
26. Johnson, D.B.; Hallberg, K.B. Acid mine drainage remediation options: A review. *Sci. Total Environ.* **2005**, *338*, 3–14. [CrossRef] [PubMed]
27. Samylovskaya, E.; Makhovikov, A.; Lutonin, A.; Medvedev, D.; Kudryavtseva, R.E. Digital technologies in arctic oil and gas resources extraction: Global trends and Russian experience. *Resources* **2022**, *11*, 29. [CrossRef]
28. Mann, S.; Potdar, V.; Gajavilli, R.S.; Chandan, A. Blockchain technology for supply chain traceability, transparency and data provenance. In Proceedings of the 2018 International Conference on Blockchain Technology and Application, Xi'an, China, 10–12 December 2018; pp. 22–26.
29. Divyasudha, N.; Arulmozhivarman, P.; Rajkumar, E.R. Analysis of Smart helmets and Designing an IoT based smart helmet: A cost effective solution for Riders. In Proceedings of the 2019 1st International Conference on Innovations in Information and Communication Technology (ICIICT), Chennai, India, 25–26 April 2019; pp. 1–4.
30. Quayson, M.; Bai, C.; Sun, L.; Sarkis, J. Building blockchain-driven dynamic capabilities for developing circular supply chain: Rethinking the role of sensing, seizing, and reconfiguring. *Bus. Strategy Environ.* **2023**, *32*, 4821–4840. [CrossRef]
31. Onifade, M.; Adebisi, J.A.; Shivute, A.P.; Genc, B. Challenges and applications of digital technology in the mineral industry. *Resour. Policy* **2023**, *85*, 103978. [CrossRef]
32. Smith, J.; Van Oploo, P.; Marston, H.; Melville, M.D.; Macdonald, B.C.T. Spatial distribution and management of total actual acidity in an acid sulfate soil environment, McLeods Creek, northeastern NSW, Australia. *Catena* **2003**, *51*, 61–79. [CrossRef]
33. Lacey, J.; Malakar, Y.; McCrea, R.; Moffat, K. Public perceptions of established and emerging mining technologies in Australia. *Resour. Policy* **2019**, *62*, 125–135. [CrossRef]
34. Bissell, D. Encountering automation: Redefining bodies through stories of technological change. *Environ. Plan. D Soc. Space* **2021**, *39*, 366–384. [CrossRef]
35. More, K.S.; Wolkersdorfer, C.; Kang, N.; Elmaghraby, A.S. Automated measurement systems in mine water management and mine workings–A review of potential methods. *Water Resour. Ind.* **2020**, *24*, 100136. [CrossRef]
36. Carr, J.L.; Jobes, C.C.; Li, J. Development of a method to determine operator location using electromagnetic proximity detection. In Proceedings of the 2010 IEEE International Workshop on Robotic and Sensors Environments, Phoenix, AZ, USA, 15–16 October 2010; pp. 1–6.
37. Molaei, F.; Rahimi, E.; Siavoshi, H.; Afrouz, S.G.; Tenorio, V. A comprehensive review on Internet of Things (IoT) and its implications in the mining industry. *Am. J. Eng. Appl. Sci.* **2020**, *13*, 499–515. [CrossRef]
38. Zealous Systems. IoT In Mining—Benefits, Examples, Challenges. Available online: https://www.zealousys.com/blog/iot-in-mining-industry/ (accessed on 26 December 2024).
39. Zvarivadza, T.; Onifade, M.; Dayo-Olupona, O.; Said, K.O.; Githiria, J.M.; Genc, B.; Celik, T. On the impact of Industrial Internet of Things (IIoT)-mining sector perspectives. *Int. J. Min. Reclam. Environ.* **2024**, *38*, 771–809. [CrossRef]
40. Jha, A.; Young, A.; Sattarvand, J. Blockchain technology and mining industry: A review. *Min. Metall. Explor.* **2023**, *40*, 2269–2280. [CrossRef]
41. Onifade, M.; Adebisi, J.A.; Zvarivadza, T. Recent advances in blockchain technology: Prospects, applications and constraints in the minerals industry. *Int. J. Min. Reclam. Environ.* **2024**, *38*, 497–533. [CrossRef]
42. Wipro. How Will Blockchain Technology Change the Mining Industry. Available online: https://www.wipro.com/content/dam/nexus/en/industries/natural-resources/latest-thinking/how-will-blockchain-technology-change-the-mining-industry.pdf (accessed on 26 December 2024).
43. World Economic Forum. 4 Ways Blockchain Will Transform the Mining and Metals Industry. 2018. Available online: https://tinyurl.com/3hrz5ruj (accessed on 26 December 2024).
44. Onifade, M.; Zvarivadza, T.; Adebisi, J.A.; Said, K.O.; Dayo-Olupona, O.; Lawal, A.I.; Khandelwal, M. Advancing toward sustainability: The emergence of green mining technologies and practices. *Green Smart Min. Eng.* **2024**, *1*, 157–174. [CrossRef]
45. StartUs Insights. Mining Technology Insights (2025 & Beyond). Available online: https://www.startus-insights.com/innovators-guide/mining-technology/ (accessed on 26 December 2024).
46. Columbia University. Mining and the Sustainable Development Goals. Available online: https://ccsi.columbia.edu/content/mining-and-sustainable-development-goals (accessed on 26 December 2024).
47. International Institute for Sustainable Development (IISD). How to Advance Sustainable Mining. Available online: https://www.iisd.org/articles/deep-dive/how-advance-sustainable-mining (accessed on 26 December 2024).
48. Phakathi, S.; van der Woude, S. Responsible Mining and Sustainable Development Goals: A Shared Value Perspective. Engineering News. Available online: https://tinyurl.com/s4dacfzr (accessed on 26 December 2024).
49. Duff, J. How Automation Is Transforming the Mining Industry. 2022. Available online: https://tinyurl.com/27wmve4b (accessed on 26 December 2024).

50. Kappa. How Automation Is Revolutionising the Mining Industry. Available online: https://kappaeng.co.za/automation-is-revolutionising-the-mining-industry/ (accessed on 26 December 2024).
51. Strzałkowski, P.; Bęś, P.; Szóstak, M.; Napiórkowski, M. Application of Virtual Reality (VR) Technology in Mining and Civil Engineering. *Sustainability* **2024**, *16*, 2239. [CrossRef]
52. Emmerick Group. Sustainable Mining: Green Technologies Transforming the Industry. Available online: https://emmerick-group.com/sustainable-mining-green-technologies-transforming-the-industry/ (accessed on 27 December 2024).

MDPI AG
Grosspeteranlage 5
4052 Basel
Switzerland
Tel.: +41 61 683 77 34

Mining Editorial Office
E-mail: mining@mdpi.com
www.mdpi.com/journal/mining

www.ingramcontent.com/pod-product-compliance
Lightning Source LLC
Chambersburg PA
CBHW040811250926
51850CB00007BA/1135